Phase Change Heat Transfer

Phase Change Heat Transfer covers theory and experiments in heat transfer associated with boiling and condensation. It presents the basics of solid–liquid–vapor phase interactions and methods to measure surface wettability and interfacial tension between liquid and vapor.

Exploring the mechanistic details of pool boiling heat transfer, this book gives correlations for the prediction of nucleate boiling, critical heat flux, and post-critical heat flux. It describes and analyzes homogeneous nucleation and heterogeneous nucleation. This book also discusses both film and dropwise condensation including integral and differential solutions for laminar film condensation. The book also includes flow boiling and numerical simulations of pool boiling.

This book is intended for upper-level undergraduate mechanical, aerospace, nuclear, and chemical engineering students taking courses in Phase Change Heat Transfer, Boiling and Condensation, Nuclear Thermal Hydraulics, and Advanced Heat and Mass Transfer.

Instructors will be able to utilize a Solutions Manual and Figure Slides for their course.

Vijay K. Dhir, Distinguished Professor of Mechanical and Aerospace Engineering, served as Dean of UCLA's Henry Samueli School of Engineering & Applied Science from 2003 to 2016. Born in India, Dhir received his Bachelor of Science degree from Punjab Engineering College in Chandigarh, India, and his Master of Technology degree from the Indian Institute of Technology in Kanpur, India. He received his PhD from the University of Kentucky, USA. He joined the faculty at UCLA in 1974 and served in several leadership roles before his appointment as Dean. He leads the Boiling Heat Transfer Lab, which has conducted pioneering work in fundamental and applied sciences involving boiling, an efficient process of heat removal. More than 46 PhD students and 40 MS students have graduated under his supervision.

In 2006, he was elected to the National Academy of Engineering—among the highest honors awarded to engineers—for his work in boiling heat transfer and nuclear reactor thermal hydraulics and safety. He received the 2004 Max Jakob Memorial Award of ASME and AIChE. He is a fellow of ASME and the American Nuclear Society. He is also a member of European Academy of Sciences and Arts. He is the author or co-author of over 350 papers published in archival journals and proceedings of conferences.

[illegible]

[illegible]

[illegible]

Phase Change Heat Transfer

Boiling and Condensation

Vijay K. Dhir

With Chapter 13 by Professor Gihun Son

CRC Press
Taylor & Francis Group
Boca Raton London New York

CRC Press is an imprint of the
Taylor & Francis Group, an **informa** business

Designed cover image: © 2025 Shutterstock

First edition published 2025
by CRC Press
2385 NW Executive Center Drive, Suite 320, Boca Raton FL 33431

and by CRC Press
4 Park Square, Milton Park, Abingdon, Oxon, OX14 4RN

CRC Press is an imprint of Taylor & Francis Group, LLC

ISBN: 978-1-032-66837-6 (hbk)
ISBN: 978-1-032-66842-0 (pbk)
ISBN: 978-1-032-66843-7 (ebk)

DOI: 10.1201/9781032668437

Typeset in Times
by codeMantra

Access the Support Material: https://www.routledge.com/9781032668376

Contents

Preface

This book is the result of intermittent effort spanning almost three decades. Initially, it started as a joint effort between Professor John H. Lienhard of the University of Houston and myself. However, we quickly got absorbed in other activities. I got involved in academic and administrative duties, and his interests shifted to the history of technology.

Finally, I have tried to complete the task with intermittent effort, but the book lacks John's wisdom and command of the English language. The book, in its present form, is based on the graduate course that I teach at UCLA, which lasts a quarter. The book covers the basics of physical phenomena, important correlations that have been proposed in the past to describe the phenomena and limited numerical simulations of the process. In a field that has continued to evolve over the last nine decades, the material contained in this book is not comprehensive.

I would like to express my gratitude to the students who used partially completed notes over the years and have worked on the book. Dr. Tung Vu reviewed the manuscript, and Nicholas Veracruz and Atindra Krishnan helped with word processing and the drawing of figures, respectively. Ultimately, it is my family that deserves the most credit—my wife Komal; my daughters Vinita and Vashita; and my grandchildren Keshiv and Shaila. They showed considerable patience and understanding during the development of this book. Keshiv also helped in the preparation of the index of the book and editing.

1 Boiling and Condensation

INTRODUCTION

Boiling is a phase change process in which vapor bubbles are formed either on a heater surface or in a superheated liquid layer adjacent to the heated surface. It differs from evaporation at predetermined vapor/gas–liquid interfaces, as it also involves creation of these interfaces at discrete sites on the heater surface. Boiling (nucleate) is a very efficient mode of heat transfer, and it is utilized in various energy conversion and heat exchange systems and in cooling of high-energy-density electronic components. Figure 1.1 shows a photograph of nucleate boiling on a horizontal plate placed in a pool. Pool boiling refers to boiling under natural convection conditions, whereas in forced flow boiling, liquid flow over the heater surface is imposed by external means. Forced flow boiling is subdivided into external and internal flow boiling. In external boiling, liquid flow occurs over unconfined heated surfaces, whereas internal flow boiling refers to flow inside tubes

Condensation is also a phase change process in which vapor is converted into a liquid as it contacts a surface held at a temperature below the saturation temperature of vapor at system pressure. Condensation is divided into two broad categories: film condensation and dropwise condensation. In film condensation, a liquid film forms on the cooled surface, which is the case for surfaces that are well wetted. However,

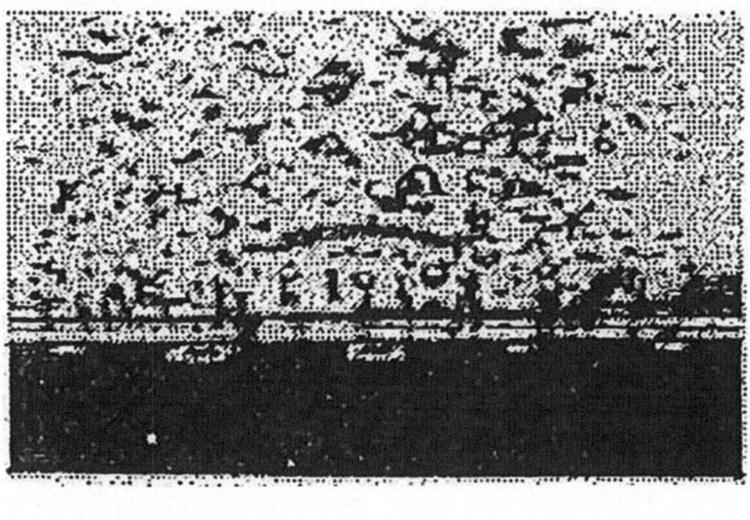

FIGURE 1.1 Nucleate boiling on a horizontal surface at heat flux 4×10^4 W/m^2 [1].

DOI: 10.1201/9781032668437-1

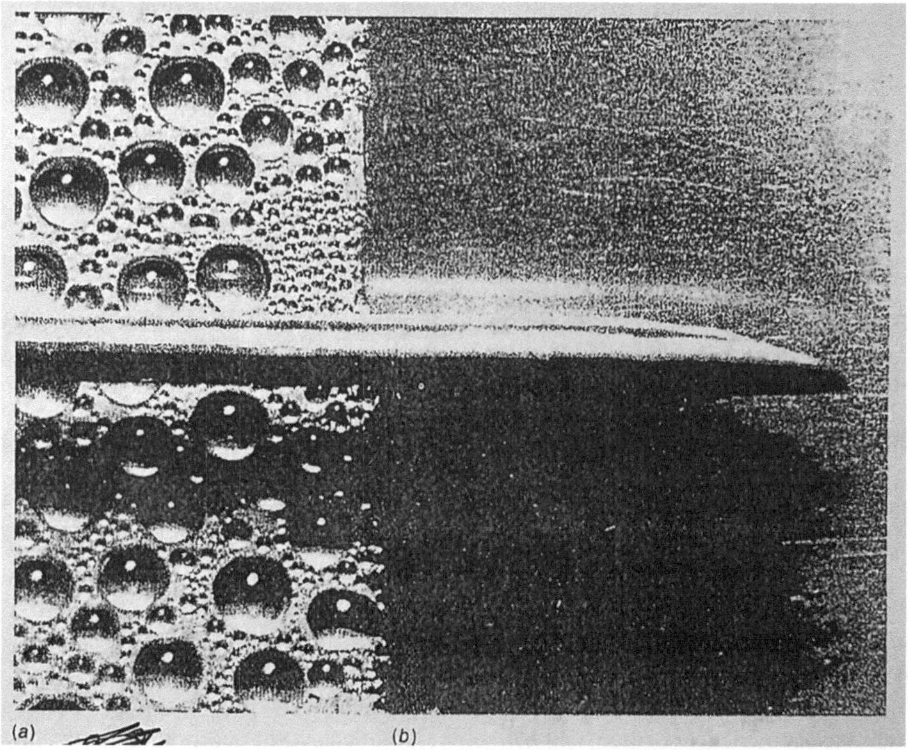

FIGURE 1.2 Condensation on a vertical surface. (a) Dropwise. (b) Film. Photograph courtesy of Professor J. W. Westwater, University of Illinois at Champaign-Urbana [2].

dropwise condensation occurs on surfaces that are not wetted or partially wetted by the condensate. In this case, drops are of micron or submicron size. Heat transfer coefficients with dropwise are much higher than those for film condensation. Figure 1.2 shows a photograph of dropwise and film condensation on a vertical surface. This figure is from the work of Prof. Westwater and reproduced from reference [2]. It should be noted that direct condensation can also occur at the liquid–vapor interface when subcooled liquid contacts vapor at such an interface.

1.1 POOL BOILING

In 1934, Shiro Nukiyama [3], a Japanese scientist, was the first to experimentally identify different modes of pool boiling, A schematic diagram of the experimental apparatus Nukiyama used in his experiments is shown in Figure 1.3 along with resistance temperature relation and the boiling curve obtained through the experiments. In his experiments, Nukiyama held a silver or nickel-plated copper wire between two electrodes. The 3 mm diameter wire was submerged in a pool of water maintained at its saturation temperature ($\approx$ 100°C at one atmosphere pressure). Water pool was formed in a pyrex glass vessel. The vessel in turn was placed in a sand

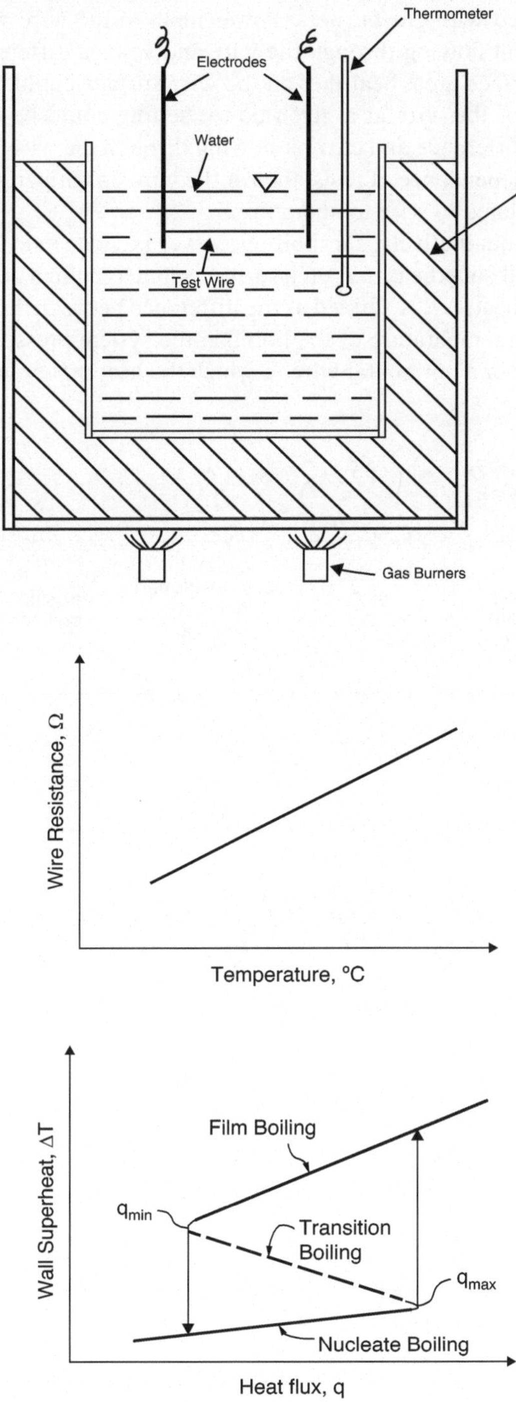

FIGURE 1.3 Nukiyama's pool boiling experiment [3].

bath, which was heated with gas burners. Power input to the wire was controlled. By knowing the current flowing through the wire and voltage difference across the wire, and the wire surface area, heat flux on the wire surface could be determined. Similarly, resistance of the wire at a given power setting could be determined by knowing the voltage difference and current flowing through the wire. Having a calibration curve for the dependence of resistance of the wire on temperature, measured resistance could be related to wire temperature.

Figure 1.4 shows, qualitatively, the boiling curve, i.e., dependence of the wall heat flux, q, on the wall superheat, ΔT, on a surface submerged in a pool of saturated liquid. The wall superheat, ΔT, is defined as the difference between the wall temperature and the saturation temperature of the liquid at the system pressure. The plotted curve is for a flat plate or a horizontal wire to which the heat input rate is controlled.

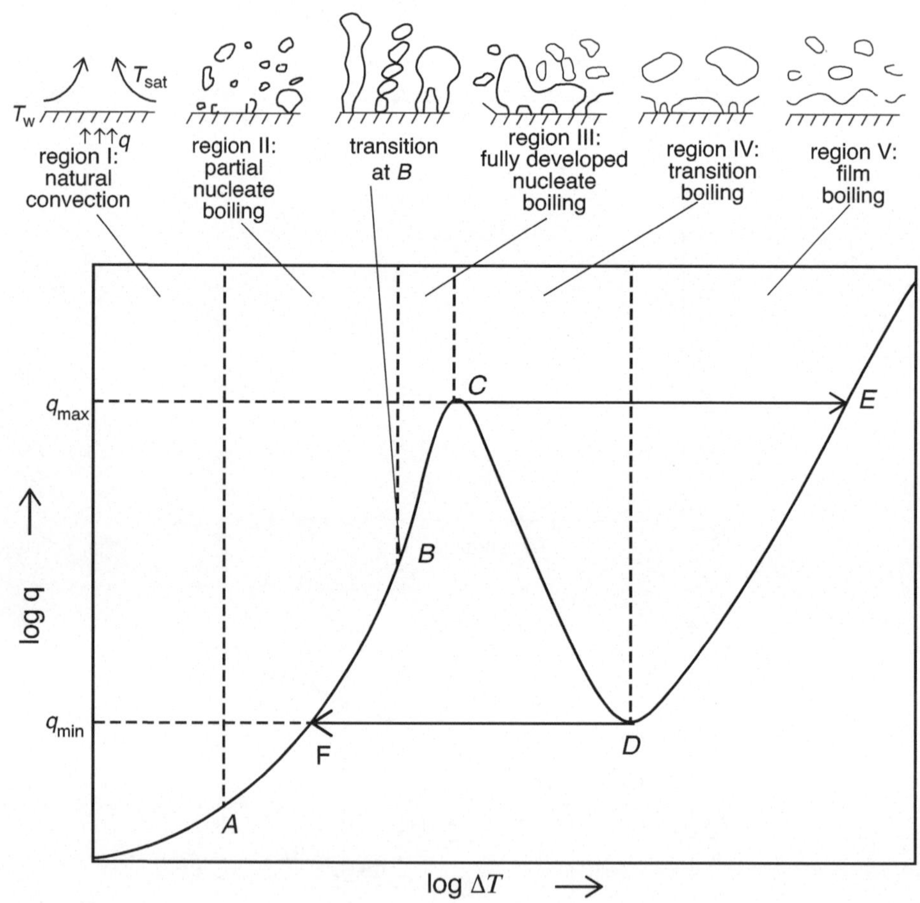

FIGURE 1.4 Typical boiling curve, showing qualitatively the dependence of the wall flux q on the wall superheat ΔT, defined as the difference between the wall temperature T_w and the saturation temperature T_{sat} of the liquid. Schematic drawings show the boiling process in regions I–V [4].

As the heating rate to the surface is increased, starting with the surface and liquid in thermal equilibrium at the saturation temperature of the liquid, the first mode of heat transfer to appear in a gravitational field is natural convection. At a certain value of the wall superheat (Point A), vapor bubbles appear on the heater surface. This is the onset of nucleate boiling. The bubbles form on defects, cavities, or scratches on the surface that contain preexisting gas/vapor nuclei. In liquids that wet the surface well, most cavities are filled with liquid and the onset of nucleation may be delayed. For these liquids, a sudden activation of many cavities accompanied by improved rate of heat transfer at an increased wall superheat causes a reduction in the surface temperature while the imposed heat flux is held constant. This behavior is not observed when the boiling curve is obtained by reducing the heat flux; thus, hysteresis results. After inception, a dramatic increase in the slope of the boiling curve is observed. In partial nucleate boiling, corresponding to region II (curve *AB*) in Figure 1.4, discrete bubbles are released from randomly located active sites on the heater surface. The number density of active sites and the frequency of bubble release increase with increase in superheat. Fully developed nucleate boiling (region III) occurs when bubbles at a given site merge to form vapor jets. The condition of the formation of jets also coincides (approximately) with the merger of vapor bubbles at the neighboring sites. After lateral merger, vapor structures appear like mushroom-type bubbles with several stems (Gaertner [4]).

Figure 1.5 shows a photograph of a large vapor structure supported by several smaller bubbles (stems) [5] during fully developed nucleate boiling on a vertical

FIGURE 1.5 Photograph of fully developed nucleate boiling of saturated water on a vertical plate at 1 atm. $q = 25\ \text{W/cm}^2$, $\Delta T = 12.5$ K [5].

surface. A small change in the slope of the boiling curve can occur upon transition from partial to fully developed nucleate boiling. The heat flux on polished surfaces varies with wall superheat roughly as

$$q \sim \Delta T^m \tag{1.1}$$

Here m has a value between 3 and 4. This functional dependence shows that the heat transfer coefficient increases as ΔT^2 or ΔT^3 as wall superheat increases, thus a strong decrease in resistance to transfer of heat from the wall to fluid.

The maximum, peak, burnout, or critical heat flux, q_{max}, sets the upper limit of fully developed nucleate boiling or safe operation of equipment. After the occurrence of the maximum heat flux condition, most of the surface is very rapidly covered with vapor. As such, the surface becomes nearly insulated, and the surface temperature rises quickly as the rate of heat transfer from the surface degrades. When the heat input rate to the heater is controlled, the heater surface will very rapidly pass through regions IV and V in Figure 1.4 and stabilize at point E. If the temperature at E exceeds the melting temperature of the heater material, the heater will fail (Nukiyama called this state as the "maximum" or "burnout" heat flux). The curve *ED* (region V) represents stable film boiling, and the system can be made to follow this curve by reducing the heat flux. In stable film boiling, the surface is totally covered with a vapor film, and liquid does not contact the solid. On a horizontal surface, the vapor release pattern is governed by the Taylor instability of the vapor–liquid interface. Figure 1.6 shows film boiling pattern on a horizontal disc. With a reduction of heat flux in film

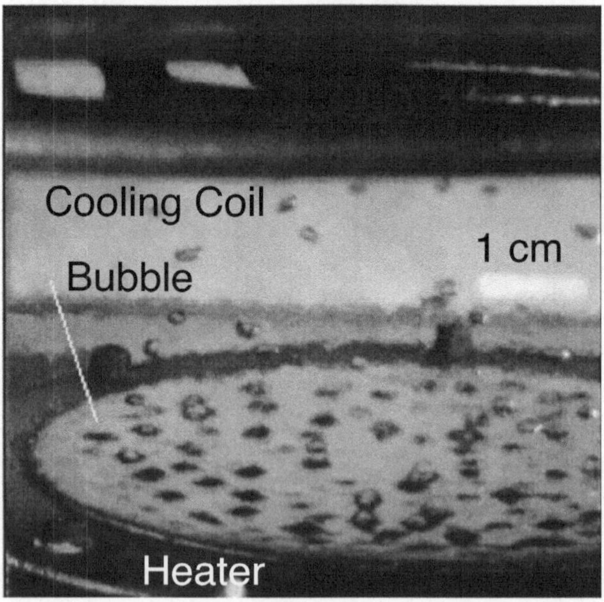

FIGURE 1.6 Digital image of non-departing bubbles arranged on concentric rings on a horizontal circular plate during subcooled film boiling [6].

boiling, the film eventually collapses, heat removal rate increases, and surface for a fixed heat flux cools down very rapidly and stabilizes at point F, in nucleate boiling. This heat flux is called "minimum" heat flux. The mode of boiling between points C and D is called *transition boiling.* It is a mixed mode of boiling that has features of both nucleate and film boiling. Transition boiling is inherently very unstable since it is accompanied by a reduction in the heat flux (current) with an increase in the wall superheat or driving potential. As a result, it is very difficult to obtain steady-state data in transition boiling, except when the heater surface temperature is controlled. Nukiyama, in his experiments, could not access the region between C and D, but rightly speculated that such a region existed. Transient transition boiling data can be obtained either by quenching or by accessing from the nucleate boiling side when heat input to the heater is controlled with a feedback loop.

Condensation of vapor on a solid surface is of considerable practical interest in the process and power industry. Nusselt [7] in 1916 was the first one to analyze laminar film condensation on a vertical wall and on a horizontal tube. He carried out his analysis under the assumption of pure vapor, inertia-less film, plane interface, and no drag of the quiescent vapor. However, after a short distance from the leading edge, ripples may appear on the interface, and subsequently, the film may become turbulent. Also, vapor flowing parallel to the surface can introduce shear on the interface. The presence of non-condensables such as air in steam can significantly reduce the condensation heat transfer rate.

Film condensation occurs on surfaces that are well wetted by the condensate. On non-wetting or partially wetting surfaces, dropwise condensation can occur. Dropwise condensation is associated with much higher heat transfer coefficients than film condensation. We will discuss both processes in a later chapter in the book.

1.2 FLOW BOILING

A boiling curve similar to that for pool boiling is obtained when liquid flow occurs over the heater surface (external boiling) or inside a heated tube (internal boiling). Both of these processes are of interest in power and chemical industries. In internal flow boiling, the vapor/liquid flow configurations change due to the addition of vapor along the flow direction. Figure 1.7 shows a typical forced flow boiling curve for a tube in which subcooled liquid enters the tube. The curve is drawn for different flow rates and fixed system pressure. The first mode of heat transfer as subcooled liquid enters the tube is forced convection. The magnitude of the heat transfer coefficient increases with flow velocity. Since the heat transfer coefficient depends very weakly on the wall superheat in single-phase forced convection, the heat flux varies nearly linearly with wall superheat. After inception, a marked improvement in heat transfer coefficient occurs as wall superheat is increased. The region $A'B'$, spanning wall superheats from inception to fully developed nucleate boiling, is termed partial nucleate boiling. In this region, bubbles are formed discretely on the heated surface, and both single-phase convection and phase change heat transfer contribute to the total heat transfer rate. As such, the functional dependence of heat flux on wall superheat is stronger than that for single-phase forced convection, but it is weaker than that for fully developed nucleate boiling.

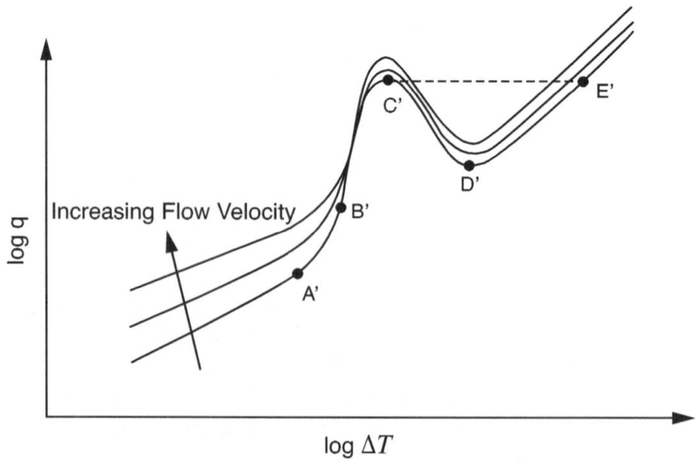

FIGURE 1.7 Flow boiling curve.

Fully developed nucleate boiling curves ($B'C'$) at different velocities nearly overlap each other, suggesting the dominance of the heat removal by phase change. In fully developed nucleate boiling, bubble merger occurs at the heated surface. However, the phasic structure is influenced by the flow regime that exists in the bulk. Possible flow regimes are bubbly, slug, and annular. In annular flows, when liquid film on the wall becomes very thin, nucleate boiling may be suppressed. Now, the heat removal is through evaporation at the liquid–vapor interface of the thin film. The heat transfer coefficient in very thin film annular flow may exceed that given by extension of the nucleate boiling curve. As a result, for a fixed heat flux, the wall superheat in very thin film flow may become smaller than that for fully developed nucleate boiling. If the liquid film dries (dryout heat flux), the surface will be cooled by forced flow of vapor. Now the temperature of the surface will increase significantly. Under low flow/heat flux condition, the critical heat flux occurs when the thin liquid film in annular flow dries out. Such a heat flux is called *dryout heat flux*, and this condition generally occurs in steam generators. Near the dryout front surface will alternately wet and dry out. As such, large variation in temperature will occur as the surface transitions from wet to dry conditions. Some liquid droplets may be entrained in the vapor, and single-phase forced convection correlations modified to account for the presence of droplets can be applied to determine the post dry-out wall heat flux/temperature. However, under high flow/high heat flux conditions, the critical heat flux occurs under conditions similar to pool boiling. The critical heat flux condition results when vapor removal rate from the heater surface falls short of the vapor generation rate or liquid at the heater surface is not replenished fast enough to sustain the evaporation rate. In the literature, critical heat flux condition has also been referred to as boiling crisis or burnout heat flux. The magnitude of the critical heat flux at a given system pressure depends on the fluid properties, local quality, and mass velocity. For a surface subjected to axially uniform heat flux, the critical heat flux condition will first

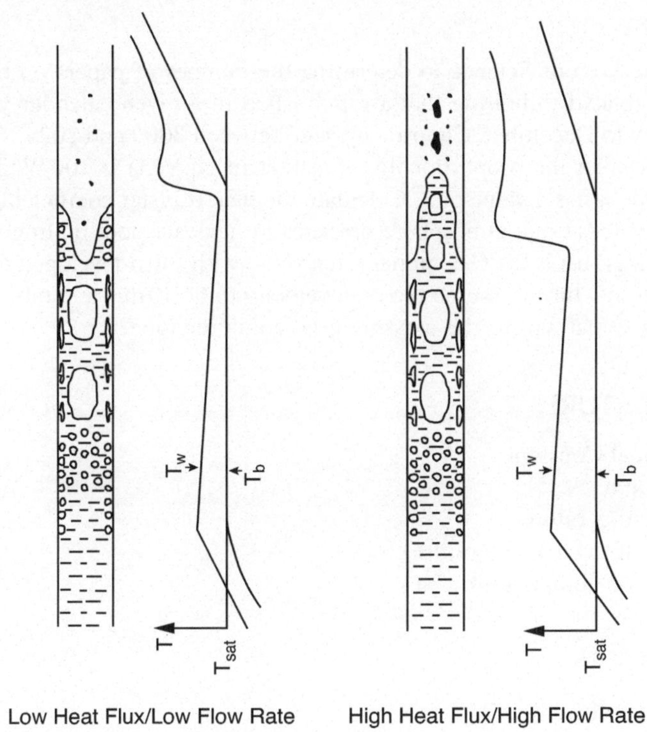

FIGURE 1.8 Flow regimes during flow boiling in a tube.

occur at the exit of the tube where the local quality is the highest. After occurrence of the critical heat flux, the temperature (for a heat-flux-controlled surface) rises very rapidly, and after passing through regions $C'D'$ and $D'E'$, the surface settles down in film boiling at E'. As for pool boiling, in forced flow, a vapor film separates the heater surface from the liquid core. This flow configuration is called *inverse annular flow* boiling. In the region $D'E'$, film boiling prevails.

In the transition boiling region $C'D'$, partial contacts of the liquid with the solid occur. As in pool boiling, this mode of boiling is unstable, and it is difficult to obtain steady-state transition boiling data on a heat flux-controlled surface. Figure 1.8 shows, qualitatively, the flow regimes and variations in tube wall temperature for low flow/low heat flux and high flow/high heat flux conditions.

1.3 SUMMARY

In this chapter, we have qualitatively described the phase change processes of pool and flow boiling and condensation. Condensation can be film or dropwise. Boiling curve (q vs. ΔT) has been described including different regimes of boiling. Maximum or critical heat flux condition has been delineated. This represents the "most efficient" cooling condition in nucleate boiling.

EXERCISES

1. Use the Web of Science to determine the number of papers on boiling or topics related to boiling that are published in a given calendar year from January to December. Choose any year between 2009 and 2024.
2. Assume that the proportionality constant in eq. (1.1) is $100\,\mathrm{W/m^2\,K^3}$ and constant m has a value of 3. Calculate the heat transfer coefficient for saturated water at one atmosphere pressure in nucleate pool boiling when the wall superheat is 20°C. Compare this value with fully developed force convection in a tube. Assume the flow velocity to be 10 m/s and tube diameter is 1 cm. What will be the pressure gradient in the tube?

NOMENCLATURE

m	Empirical Constant
q	Heat Flux
T_b	Bulk Temperature
T_w	Heater Surface Temperature
T_{sat}	Saturation Temperature

GREEK LETTERS

ΔT	Wall Superheat $T_w - T_{sat}$

REFERENCES

1. Nishikawa, K., Fujita, Y., Uchida, S., and Ohta, H. (1984), "Effect of Surface Configuration on Nucleate Boiling Heat Transfer." *Int. J. Heat Mass Transfer*, Vol. 29 (9), pp. 1559–1571.
2. Incropera, F. P. and Dewitt, D.P. (1985), *Fundamentals of Heat and Mass Transfer*, John Wiley and Sons, New York.
3. Nukiyama, S. (1934), "Maximum and Minimum Values for Heat Transmitted from a Metal to Boiling Water under Atmospheric Pressure." *J. Soc. Mech Engr. Japan*, Vol. 37, p. 206.
4. Gaertner, R.F. (1965), "Photographic Study of Nucleate Pool Boiling on a Horizontal Surface." *J. Heat Transfer*, Vol. 97, pp. 17–29.
5. Liaw, S.P. and Dhir, V.K. (1989), "Void Fraction Measurement during Saturated Pool Boiling of Water on Partially Wetted Vertical Surfaces." *ASME Journal of Heat Transfer*, Vol. 111, pp. 731–738.
6. Banerjee, D. and Dhir, V.K. (2005), "Study of Subcooled Film Boiling on a Horizontal Disc: Parts 1 & 2." *J. Heat Transfer*, Vol. 47 (4), pp. 271–293.
7. Nusselt, W. (1918), "Die Oberflächenkondensation des Wasserdamfes." *Z. Yer. Deutsch. Ing.*, Vol. 60 (27 and 28), pp. 541–546, 569–575.

2 Thermodynamic Considerations

INTRODUCTION

In this chapter, we will review thermodynamic aspects of the phase change processes. We limit ourselves to pure compressible substances. For these substances, for example, pressure can be considered as a function of temperature and specific volume. We also introduce the relationship between saturation temperature and pressure; the maximum superheat liquids can sustain before spontaneously breaking into the vapor phase and introduce concepts of Helmholtz and Gibbs functions.

2.1 RELATION BETWEEN PRESSURE, VOLUME, AND TEMPERATURE AND PHASE CHANGE

For a pure compressible substance, pressure can be considered as a function of temperature and specific volume, $p = f(T, v)$. This functional dependence, when graphed, would lead to a surface. Here we begin the discussion by projecting the *p-v-T* surface on temperature-specific volume plane with constant pressure lines (isobars). Figure 2.1 shows qualitatively *T-v* diagram for water with constant pressure lines. In the figure, we see the liquid alone region on the left-hand side and the vapor alone region on the right-hand side of the two-phase dome. Following the line for 1 atm pressure in the liquid region, 1–2, as the liquid is heated and its temperature increases, the liquid expands and specific volume increases while pressure is held constant. The same is true in the vapor region, 3–4. In region 1–2, the liquid is in a subcooled state until it reaches the temperature corresponding to point 2. The temperature at point 2 is called the saturation temperature. As more energy is added, a portion of the liquid is converted into vapor with a significant increase in volume. While the temperature and pressure remain constant, the two-phase mixture is in thermal and thermodynamic equilibrium with liquid as saturated liquid and vapor as saturated vapor. As more thermal energy is added, the process moves along line 2–3 until all the liquid is converted into vapor at state 3. Thereafter, additional heat input will cause vapor temperature to increase accompanied by an increase in specific volume. Vapor is now in the superheated state as the temperature is above the saturation temperature. Similar behavior is seen for a higher pressure (5 MPa) except that now the two-phase region is smaller or increase in specific volume as liquid is converted into vapor is less and saturation temperatures of liquid and vapor are higher. This behavior continues for other pressures up to the critical pressure of 22.09 MPa. For this pressure, the two-phase region disappears, or liquid converts into vapor without any addition of heat (zero latent heat of vaporization) or no distinction exists between liquid and vapor. This is the apex of the dome. The corresponding temperature is

DOI: 10.1201/9781032668437-2

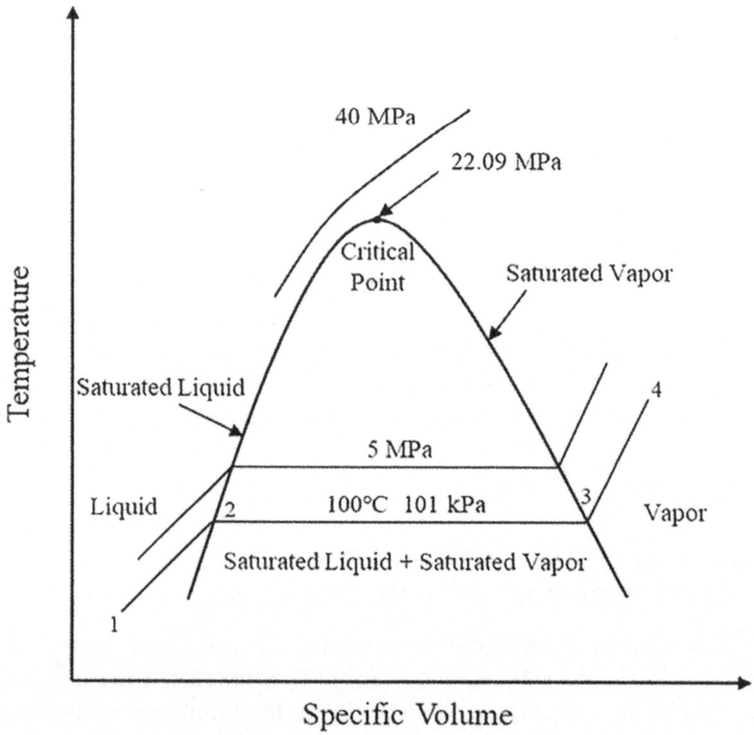

FIGURE 2.1 Temperature-specific volume relationship for water.

called critical temperature, and for water it is 374.14°C. At pressure above the critical pressure, the fluid temperature continues to increase accompanied by an increase in specific volume, but no phase change occurs. The distinction between liquid and vapor no longer exists, and the concept of saturation temperature is not applicable.

On the temperature-specific volume diagram shown in Figure 2.1, left-hand side of the dome represents all liquid at the saturation temperature and the right-hand side all vapor at the saturation temperature. Saturation temperature increases with increase in pressure until critical condition is reached. Specific volume of liquid increases with an increase in saturation temperature. However, specific volume of vapor decreases as saturation temperature (pressure) increases. This is reflected in the shape of the two-phase region or the region under the dome. Later, we will see how change in saturation temperature with change in pressure is related through latent heat of vaporization, increase in specific volume from liquid to vapor, and the saturation temperature itself.

In Figure 2.2, we show qualitatively a phase diagram for water. In this figure, we not only include phase change from liquid to vapor but also from solid to liquid (melting) and from solid to vapor (sublimation). Triple point represents the coexistence of all three phases. We also see that melting temperature increases with a decrease in pressure, whereas sublimation and vaporization temperatures increase with pressure. One should also note that for ice, specific volume decreases upon melting, whereas for other solids, specific volume increases upon melting.

The changes in states during melting, evaporation, and sublimation are shown in Figure 2.2 as well. Starting with a solid at a temperature, corresponding to point *s*, as the solid is heated, its temperature increases with fixed pressure until it reaches the melting temperature indicated by point *sl*. At this temperature, the solid begins to melt. With the addition of heat, melting continues without change in temperature until all the solid is melted. Thereafter, the addition of heat leads to sensible heating of the liquid with an increase in specific volume as was discussed earlier. This process is depicted by the line segment *sl-l*. After liquid reaches saturation temperatures, further addition of heat will lead to the conversion of a certain fraction of liquid into vapor. This will continue until all of the liquid is converted into vapor at the saturation temperature, further addition of heat will cause the vapor to be superheated at a fixed pressure. Sublimation is a process in which solid directly converts into vapor/gas upon heating without going first through liquid phase. In Figure 2.2, sublimation process is shown by the line *s-sv-v*. Starting with a temperature corresponding to point *s*, as solid is heated at a fixed pressure, its temperature increases until it attains a temperature corresponding to point *sv*. This is the sublimation temperature. With further heating, the temperature of the solid will remain constant until all of it becomes vapor/gas at

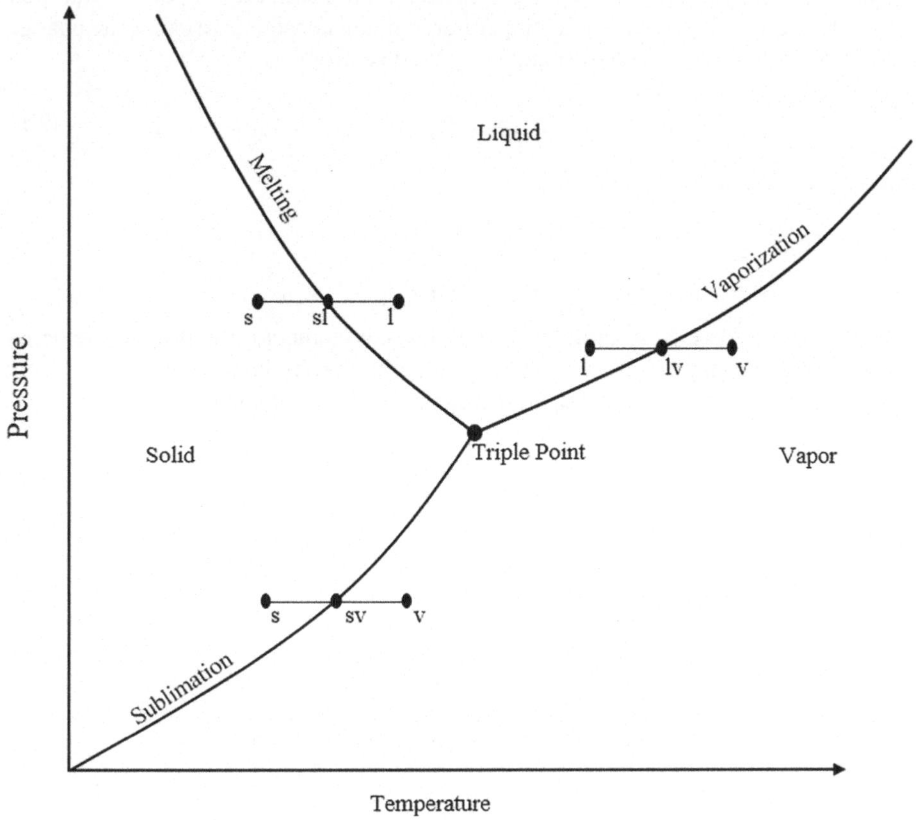

FIGURE 2.2 Phase diagram for water.

the same temperature. Thereafter, the process will move along segment *sv-v* as gas/vapor is superheated with further addition of heat. Similarly, the line *s-sl-l* represents the transitions from solid to solid–liquid and liquid. Starting with a temperature corresponding to point *s*, as the solid is heated at a fixed pressure, its temperature will increase until melting temperature is reached denoted by the point *sl*. Thereafter, with further heating, the temperature of solid–liquid mixture will remain at the melting temperature until all the solid is melted. Subsequently, with further energy input, liquid temperature will increase until it reaches its saturation temperature.

In this text, our focus is on the development of a mechanistic understanding of the phase change process related to the transfer of heat as liquid is converted into vapor (boiling) or vapor is converted into liquid (condensation). We will not be going into details on melting or sublimation processes except provide a few examples of those processes in the context of explanation of modeling of various subphenomena.

Next let us develop a relationship for change in saturation temperature with pressure. From Figure 2.2, we see that for melting of ice into water, the melting temperature increases as pressure is decreased, whereas the opposite is true for vaporization and sublimation. The key reason for the difference is the reduction in specific volume with melting, whereas the specific volume increases during sublimation and evaporation. Focusing just on phase change processes from saturated liquid to saturated vapor, during which pressure and temperature remain constant, we can write change in specific enthalpy in terms of change in specific entropy as

$$dh = Tds \tag{2.1}$$

From Maxwell relation, we know

$$\left(\frac{\partial s}{\partial v}\right)_T = \left(\frac{\partial p}{\partial T}\right)_v \tag{2.2}$$

During phase change from liquid to vapor at a fixed temperature, pressure remains constant and is independent of specific volume, thus right-hand side of eq. (2.2) is determined by the saturation temperature alone, and we can write

$$\left(\frac{\partial p}{\partial T}\right)_v = \left(\frac{dp}{dT}\right)_{\text{sat}} \tag{2.3}$$

or eq. (2.2) can be rewritten as

$$\left(\frac{\partial s}{\partial v}\right)_T = \left(\frac{dp}{dT}\right)_{\text{sat}} \tag{2.4}$$

Integration of eq. (2.4) between liquid and vapor phases yields

$$s_v - s_l = \left(\frac{dp}{dT}\right)_{\text{sat}} (v_v - v_l) \tag{2.5}$$

Similarly, integration of eq. (2.1) at saturation temperature gives [1]

$$h_v - h_l = T_{\text{sat}}(s_v - s_l) \tag{2.6}$$

Eliminating $(s_v - s_l)$ between eqs. (2.5) and (2.6), we obtain

$$\left(\frac{dp}{dT}\right)_{\text{sat}} = \frac{h_v - h_l}{T_{\text{sat}}(v_v - v_l)} \tag{2.7}$$

The above equation relates the slope of the pressure and saturation temperature curve at a particular saturation temperature to the ratio of difference between enthalpies of vapor and liquid and difference of specific volumes of vapor and liquid. This equation is known as Clapeyron equation. If one further assumes that specific volume of liquid is small in comparison to the specific volume of vapor, which is true only at low pressures where the ideal gas law applies, eq. (2.7) can be rewritten as

$$\frac{1}{p}\left(\frac{dp}{dT}\right)_{\text{sat}} = \frac{h_v - h_l}{RT_{\text{sat}}^2}$$

or

$$\left(\frac{d(\ln p)}{dT}\right)_{\text{sat}} = \frac{h_v - h_l}{RT_{\text{sat}}^2} \tag{2.8}$$

Equation (2.8) is known as Clausius Clapeyron equation, and we will use either eq. (2.7) or eq. (2.8) in later chapters.

2.2 METASTABLE STATES OF LIQUID AND VAPOR

In the previous section, we assumed that phase transitions occurred at melting, boiling, or sublimation temperatures. At these temperatures for a fixed pressure, two phases: solid–liquid, liquid–vapor, and solid–vapor/gas exist in equilibrium. Limiting ourselves to a liquid–vapor system, as energy is added to the liquid at saturation temperature, equilibrium temperature corresponding to the imposed pressure (left side of the dome), a portion of the liquid converts to vapor at the same temperature. The process continues until all of the liquid is converted into vapor (right side of the dome). During this process, two phases are in equilibrium, while the chemical potential of the system remains constant. However, in reality, liquid can stay in a metastable state at a temperature higher than the saturation temperature at a particular pressure. Similarly, vapor can stay in metastable vapor state at a temperature well below the saturation temperature at a given pressure. To investigate the metastable state, we begin with the van der Waals equation for pure compressible fluids. This equation represents a relationship between pressure, specific volume, and temperature. The relationship is also referred to as equation of state and is written as

$$p = \frac{RT}{(v-b)} - \frac{a}{v^2} \tag{2.9}$$

where p is the pressure, R is the gas constant, T is the absolute temperature, and v is specific volume. Constant, b, in eq. (2.9) accounts for the finite volume occupied by the fluid molecules, whereas the second term in eq. (2.9) accounts for intermolecular forces, which tend to reduce the system pressure. In Figure 2.3, we plot the saturation curve on p-v graph along with several isotherms—at critical temperature, below critical temperature, and above critical temperature. Constants a and b in van der Waals equation can be determined by recognizing that the isotherm corresponding to critical temperature, T_c, has a zero slope and passes through a point of inflection. This condition corresponds to

$$\left(\frac{\partial p}{\partial v}\right)_{T_c} = 0 \tag{2.10}$$

and

$$\left(\frac{\partial^2 p}{\partial v^2}\right)_{T_c} = 0 \tag{2.11}$$

With van der Waals equation evaluated at the critical point, we have the relations as given below

$$p_c = \frac{RT_c}{v_c - b} - \frac{a}{v_c^2} \tag{2.9a}$$

$$\left(\frac{\partial p}{\partial v}\right)_{T_c} = 0 = -\frac{RT_c}{(v_c - b)^2} + \frac{2a}{v_c^3} \tag{2.10a}$$

$$\left(\frac{\partial^2 p}{\partial v^2}\right)_{T_c} = 0 = \frac{2RT_c}{(v_c - b)^3} - \frac{6a}{v_c^4} \tag{2.11a}$$

With the relations given by eqs. (2.9a), (2.10a), and (2.11a), we can solve for a, b, and v_c in terms of p_c and T_c as

$$a = \frac{27}{64}\frac{R^2T_c^2}{p_c} \tag{2.12}$$

$$b = \frac{RT_c}{8p_c} \tag{2.13}$$

$$v_c = \frac{3}{8}\frac{RT_c}{p_c} \tag{2.14}$$

Constants a and b in van der Waals' equation obtained this way are approximate. However, more accurate values can be obtained by using experimental data for pressure in terms of specific volume and temperature for a particular fluid. Using a, b, and v_c from eqs. (2.12), (2.13), and (2.14), respectively in (2.9), we find that

$$p_r = \frac{8T_r}{3v_r - 1} - \frac{3}{v_r^2} \tag{2.15}$$

where p_r, T_r, and v_r are reduced pressure, temperature, and reduced volume and are obtained by dividing p, T, and v by their values at the critical point or

$$p_r = \frac{p}{p_c} \tag{2.16}$$

$$T_r = \frac{T}{T_c} \tag{2.17}$$

$$v_r = \frac{v}{v_c} \tag{2.18}$$

Next, let us pay our attention to an isotherm at temperature $T_1 < T_c$, pressure at which the isotherm crosses liquid saturation, and vapor saturation curves represent the saturation temperature. This is the equilibrium temperature between vapor and liquid as we discussed in the previous section. We also noted that in the dome as heat is added at a given pressure, more and more of the liquid is converted into vapor along the constant pressure line. However, as we see in Figure 2.3, van der Waals equation gives that the isotherm LM continues as long as the stability condition $\frac{\partial p}{\partial v} < 0$ is satisfied. Liquid beyond LM is in a metastable state. In fact, the isotherm shows a minima and a maxima. Minimum condition corresponds to the upper limit of metastable state for liquid. Similarly, region QS represents a metastable state for vapor. Extreme values of metastable states in both liquid and vapor are called the spinodal limits. Any disturbance or perturbation will cause liquid to spontaneously break into vapor and vapor into liquid state. The curves connecting the spinodal limit for both liquid and vapor shown by a dotted line are called liquid and vapor spinodals, respectively. The region NOQ is physically not possible as in this region $\frac{\partial p}{\partial v} > 0$ or the system is unstable. Since chemical potential is same at points M and S in saturated liquid and saturated vapor, respectively, it can be shown that area of the region MNO should be equal to the area of region OQS. The conditions corresponding to metastable or spinodal limit can be determined by invoking the condition that at these limits $\left(\frac{\partial p}{\partial v}\right)_T = 0$.

Let us take an example for water at 1 atm pressure. With $p_c = 220.9$ atm

$$p_r = \frac{1}{220.9} = 0.0045$$

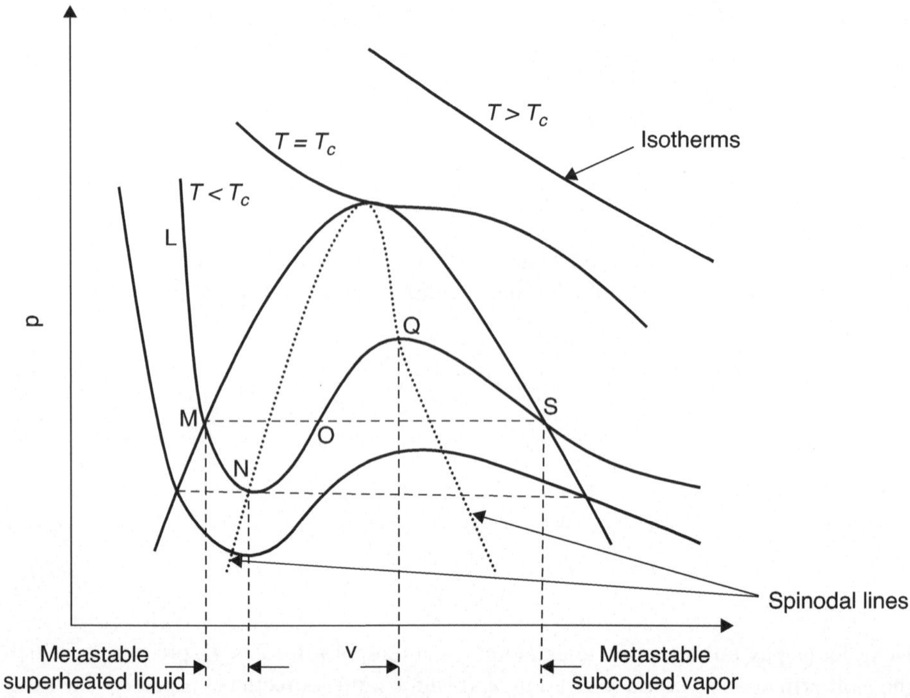

FIGURE 2.3 van der Waals equation and metastable limit.

For determination of the spinodal condition or the upper limit to which water can be superheated before it breaks up into vapor, we set

$$\left(\frac{\partial p_r}{\partial v_r}\right)_{T_r} = 0$$

or from eq. (2.15)

$$\frac{-24T_r}{(3v_r - 1)^2} + \frac{6}{v_r^{\,3}} = 0$$

or

$$T_r = \frac{(3v_r - 1)^2}{4v_r^{\,3}} \tag{2.19}$$

Substituting for T_r from eq. (2.19) into eq. (2.15) and knowing that $p_r = 0.0045$, we obtain

$$0.0045 = \frac{3v_r - z}{v_r^{\,3}} \tag{2.20}$$

Three possible solutions for eq. (2.20) are

$$v_r = 0.667; \quad v_r = 25.4 \quad \text{and} \quad v_r = -26$$

Using the first positive solution for v_r and substituting it in eq. (2.16), we obtain

$$T_r = 0.844$$

or

$$\frac{T}{T_c} = 0.844$$

With $T_c = 647.14$ K, we find the spinodal temperature to be 546 K or 273°C. We have obtained this spinodal temperature for water at 1 atm pressure by following the generalized equation of state or van der Waals equation. Later on in Chapter 4, we will determine this temperature considering molecular dynamics.

2.3 HELMHOLTZ AND GIBBS FUNCTIONS

Helmholtz and Gibbs functions are basic thermodynamic properties of a substance such as pressure, temperature, system volume, internal energy, enthalpy, and entropy. Use of these functions eliminates [2,3] entropy as an independent parameter, which cannot be measured. As an extensive property, Helmholtz function, F, for one-component compressible substance is defined as

$$F = U - TS, \text{ J} \tag{2.21}$$

where U is the internal energy, T is the absolute temperature, and S is the entropy. As such F is a form of energy, the units of which are joules. In an intensive form (per unit mass), Helmholtz free energy is written as

$$f = u - Ts, \text{ J/gm} \tag{2.22}$$

In differential form, Helmholtz function is written as

$$df = du - T\ ds - s\ dT \tag{2.23}$$

however,

$$du = T\ ds - p\ dv \tag{2.24}$$

where v is specific volume.

Substituting eq. (2.24) in (2.23) we obtain

$$df = -s\ dT - p\ dv \tag{2.25}$$

The above relation now utilizes T and v as independent variables.

Similarly Gibbs function, G, as an extensive property is defined as

$$G = H - TS \text{ (J)} \tag{2.26}$$

where H is the enthalpy. As such G is a form of energy and its units are joules. In an intensive form, Gibbs function is written as

$$g = h - Ts, \text{ J/gm} \tag{2.27}$$

In a differential form eq. (2.27) can be written as

$$dg = dh - T\ ds - s\ dT \tag{2.28}$$

however

$$dh = T\ ds + v\ dp \tag{2.29}$$

Using eq. (2.29) in (2.24), we obtain

$$dg = -s\ dT + v\ dp \tag{2.30}$$

Again the above relation now utilizes T and p as independent variables.

2.4 SUMMARY

In this chapter, we have described the relationship between pressure, temperature, and specific volume for pure fluids. We have obtained metastable limits for superheated liquids and subcooled vapor and have defined Gibbs and Helmholtz functions.

EXERCISES

1. Plot saturation temperature normalized with critical temperature as a function of pressure normalized with critical pressure for Water, FC-72, and Sodium. Discuss your results.
2. What are the values of pressure and temperature at triple point for water?
3. Obtain meta-stable temperature for water at pressures of 10 and 100 atm.

NOMENCLATURE

a, b Constants
F Helmholtz function
F Specific Helmholtz function
G Gibbs function
g Specific Gibbs function

H Enthalpy
h Specific enthalpy
p Pressure
R Gas constant
S Entropy
s Specific entropy
T Temperature
v Specific volume

SUBSCRIPTS

c Critical
l Liquid
r Reduced
v Vapor

REFERENCES

1. Moran, M. and Shapiro, H.N. (1988), *Fundamentals of Engineering Thermodynamics*, Vol 1, John Wiley and Sons, Inc., New York.
2. Cengel, Y.A. and Boles, M.A. (1998), *Thermodynamics: An Engineering Approach*, Third Edition, McGraw Hill, New York.
3. Callen, H.B. (1960), *Thermodynamics*, John Wiley and Sons, Inc., New York.

3 Basics of Solid–Liquid–Vapor Phase Interactions

INTRODUCTION

The behavior of the interfaces and rate of heat transfer in boiling and condensation is affected by the interfacial tension that exists between the phases. In this chapter, we develop basic relations that represent interactions between solid–liquid–vapor phases.

3.1 INTERFACIAL TENSION (SURFACE TENSION) BETWEEN A LIQUID AND VAPOR/GAS INTERFACE

The tension that exists between a liquid–vapor/gas interface plays a foremost role in several of the phenomena observed in boiling and condensation. Some examples are vapor/gas bubbles in a liquid and liquid drops in vapor or air. The interface between two immiscible fluids spans a thickness of a few molecules. It is in this region that the density changes from one fluid to another. From molecular considerations, the interfacial tension results from unbalanced intermolecular forces in the few molecules thick layer that exists between the two phases. In the bulk (interior of a phase), the cohesive force between molecules is balanced. However, for the molecules that are at or near the interface, the balance breaks down. For example, at the liquid–gas interface, liquid molecules that are at the boundary share very few neighbors on the gas side and experience unbalanced cohesive force that is directed away from the interface. This in turn leads to pulling of the liquid molecules inward and the tendency of the interface to contract. The interfacial tension, σ, force per unit length, i.e., $\frac{\text{N}}{\text{m}}$ that develops between two immiscible fluids is generally positive and is tangent to the interface. From a thermodynamic point of view, the interfacial tension can also be considered as energy per unit area, $\frac{\text{Nm}}{\text{m}^2}$, i.e., the work done in creating the interface.

Magnitude of surface tension depends on the temperature, and it decreases as fluid temperature increases. Figure 3.1 shows dependence of surface tension of pure water on temperature. It is seen that surface tension decreases nearly linearly with temperature as long as temperature of the fluid is well below the critical temperature. The surface tension data for pure water plotted in Figure 3.1 can be correlated as [1]

$$\sigma = 235.8\times10^{-3}\left(1-\frac{T}{T_C}\right)^{1.256}\left[1-0.625\left(1-\frac{T}{T_C}\right)\right]\ \frac{\text{N}}{\text{m}} \tag{3.1}$$

where T and T_C are the absolute liquid temperature and critical temperatures, respectively, and are measured in degrees Kelvin. For water, $T_C = 647$ K. Presence of a

DOI: 10.1201/9781032668437-3

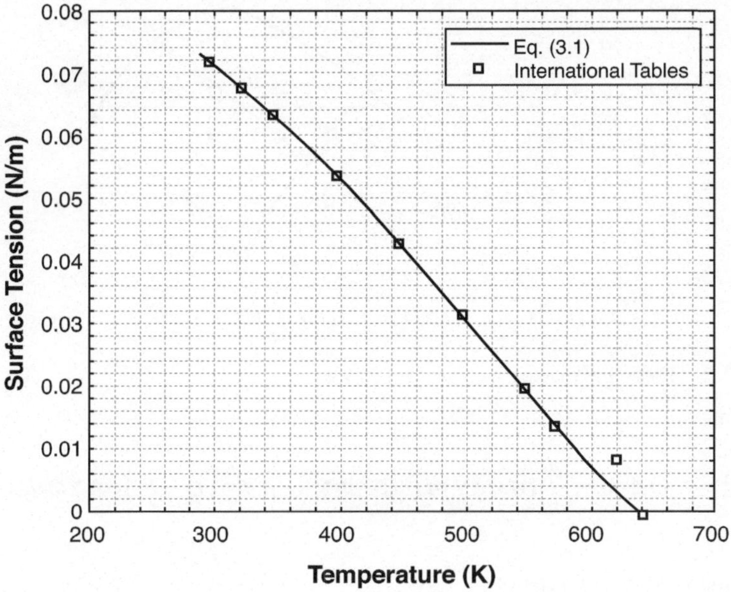

FIGURE 3.1 Surface tension for water as a function of temperature.

foreign material or solute in a pure liquid can also alter the interfacial tension. The solutes that concentrate more at the interface as compared to the bulk affect surface tension more strongly. As described by Miller and Neogi [2], accumulation of solute at the interface is related to the variation of surface tension, σ, with concentration as

$$\Gamma_s = -\left(\frac{x_s}{RT}\right)\left(\frac{\partial \sigma}{\partial x_s}\right)_T \tag{3.2}$$

where x_s is the mole fraction of the solute, R is the gas constant, and T is the absolute temperature. According to eq. (3.2), if surface tension decreases with increase in concentration of the solute, the accumulation of solute at the interface will increase. Similarly, if a solute or parts of it tend to concentrate at the interface, surface tension, σ, will decrease with increase in concentration.

Water being a polar liquid, polar groups of a solute will easily become part of its structure; however, nonpolar components will tend to concentrate at the interface. For example, when a surfactant (e.g., soap) is added to water, polar groups are absorbed in water, whereas nonpolar groups (hydrocarbons) accumulate at the interface, leading to a reduction in surface tension as can be deduced from eq. (3.2).

Presence of interfacial tension leads to a pressure difference between the concave and convex sides of the interface. Figure 3.2 shows an interface in two-dimensional space.

Making a force balance in the y-direction,

$$\left[-\sigma \sin\gamma \,|_{S+dS} + \sigma \sin\gamma \,|_S\right] ds + \left(-P_l \cos\gamma + -P_v \cos\gamma\right) ds \cdot ds = 0 \tag{3.3}$$

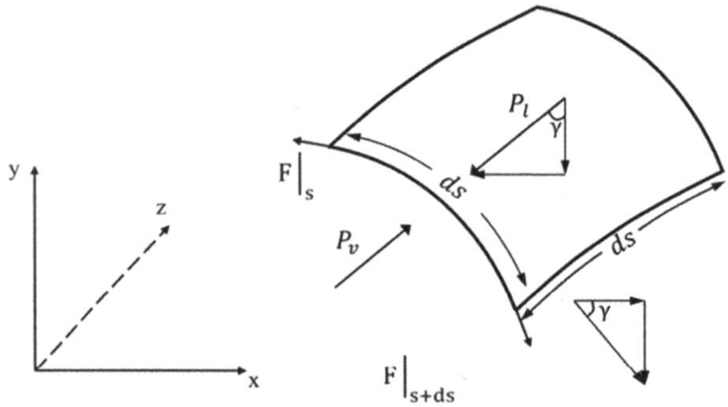

FIGURE 3.2 Force balance on a curved interface.

or

$$\left[-\sigma \cancel{\sin\gamma} \mid_S - \frac{d}{ds}(\sigma \sin\gamma)ds + \sigma \cancel{\sin\gamma} \mid_S\right] ds = (P_l - P_v)\cos\gamma\, ds ds \tag{3.4}$$

or

For constant surface tension,

$$-\sigma \cancel{\cos\gamma} \frac{d\gamma}{ds} \cdot \cancel{ds} = (P_l - P_v)\cancel{\cos\gamma\, ds}$$

$$\sigma \frac{d\gamma}{ds} = P_v - P_l$$

or

$$\sigma\left(\frac{1}{R_1}\right) = P_v - P_l \tag{3.5}$$

where R_1 is the radius of curvature. For a three dimension case, if R_2 is the radius of curvature in the $x-z$ plane, by deduction, we can write

$$\sigma\left(\frac{1}{R_1} + \frac{1}{R_2}\right) = P_v - P_l$$

For $R_1 = R_2 = R$ as for a spherical interface,

$$P_v - P_l = \frac{2\sigma}{R} \tag{3.6}$$

Example:

An air bubble 0.1 mm in diameter exists in water at a temperature of 20°C. (a) What is the pressure in the bubble; (b) If the same size bubble was of water vapor and surrounded by saturated water at 1 atm pressure, what will be the pressure in the bubble and what will be the required temperature of vapor?

Solution:

(a)

$$\Delta p = \frac{2\sigma}{R}$$

$$= \frac{2 \times 73.2 \times 10^{-3}}{0.05 \times 10^{-3}}$$

$$= 2{,}928 \text{ N/m}^2$$

or

Pressure in the bubble will be

$$= 1.01 \times 10^5 + 2{,}928 \text{ N/m}^2$$

$$= 1.04 \times 10^5 \text{ N/m}^2$$

(b)

$$\Delta p = \frac{2\sigma}{R}$$

$$= \frac{2 \times 58.9 \times 10^{-3}}{0.05 \times 10^{-3}}$$

$$= 2{,}356 \text{ N/m}^2$$

or

Pressure in the bubble

$$= 1.01 \times 10^5 + 2356$$

$$= 1.034 \times 10^5 \text{ N/m}^2$$

At this pressure, the saturation temperature of vapor will be 374 K.

3.2 SURFACE WETTABILITY

Wettability is indicative of the ability of a liquid to spread over a solid surface. Examples in everyday life are smearing of rain drops on some surfaces, whereas on others, rain drops continue to persist as beads discretely attached to the surface. Figure 3.3 shows a liquid micro-droplet sitting on a solid surface. The droplet is small in size such that gravity has little effect on its spherical shape. Interfacial forces on the contact line experienced by the droplet surrounded by vapor/gas are also shown. Force balance in the horizontal direction yields

$$\sigma_{lv} \cos\varphi = \sigma_{sv} - \sigma_{sl}$$

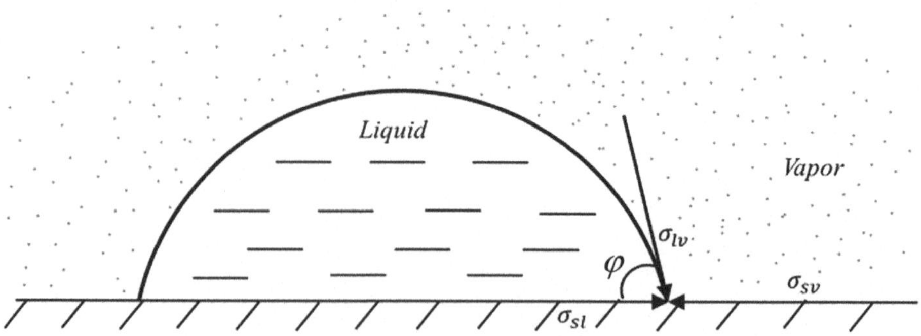

FIGURE 3.3 Forces at the contact lines of a microdroplet sitting on a solid surface.

$$\cos\varphi = \frac{\sigma_{sv} - \sigma_{sl}}{\sigma_{lv}} \tag{3.7}$$

In the above, σ_{sl} is the interfacial tension between solid and liquid, σ_{sv} is that between sold and vapor/gas, and φ is the contact angle measured from the liquid side. Equation (3.7) is known as the Young's [3] equation. For $\varphi = 0$, liquid will spread over the surface while the surface is fully wetted. Such a surface is called hydrophilic. For $\varphi \geq 90°$, surface is non-wetting or hydrophobic. In this case, liquid will not spread as a film on the surface but will break up into droplets. For $0 \leq \varphi \leq 90°$, surface is partially wetting. Most of the ordinary surfaces fall in this latter category. The above description holds good for a static interface. However, φ when the droplet is growing or shrinking, the observed contact angles can be very different from the static contact angle. For an advancing liquid–vapor/gas interface (i.e., a growing droplet), the contact angle measured from the liquid side is called the advancing contact angle, and it is greater than the static contact angle. For a receding interface (i.e., a shrinking droplet), the contact is called the receding contact angle, and it is less than the static contact angle. The magnitude of the advancing and receding contact angles depend on the interface velocity and roughness of the surface, and they can show significant hysteresis. Figure 3.4 shows qualitatively the effect of interfacial velocity on advancing and receding contact angles. We will discuss it further later.

Physiochemical state of the surface and purity of the liquid play an important role in the observed magnitude of the contact angles. For example, presence of liquid molecules on the perceived dry surface can increase σ_{sv} and thereby decrease the measured contact angle. Similarly, roughness of the surface can affect the apparent value of σ_{sv}. Presence of a surfactant such as soap or other contaminants could alter the interfacial tension, σ_{lv}, and thus affect the contact angle or wettability of the surface. With reduced σ_{lv} upon addition of surfactants, contact angle decreases, as the surface becomes more wettable.

A number or approaches have been described in the literature to measure the contact angles. One commonly used approach is the sessile drop method. In this method, a micro-drop of liquid is placed on the "clean," "dry" surface. A photograph of the drop is taken and tangent to the liquid–vapor interface at the solid–liquid–gas

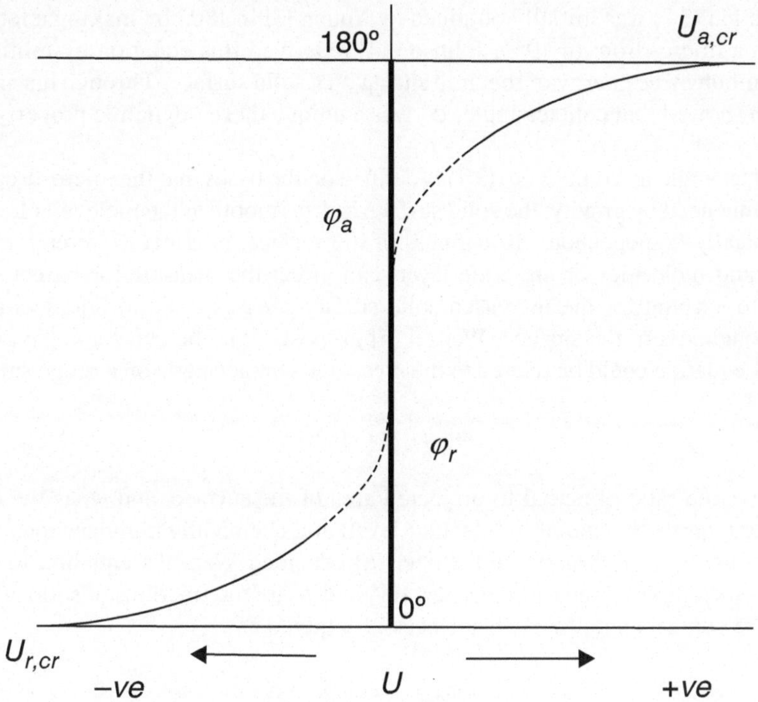

FIGURE 3.4 Advancing and receding contact angle dependence on interface velocity.

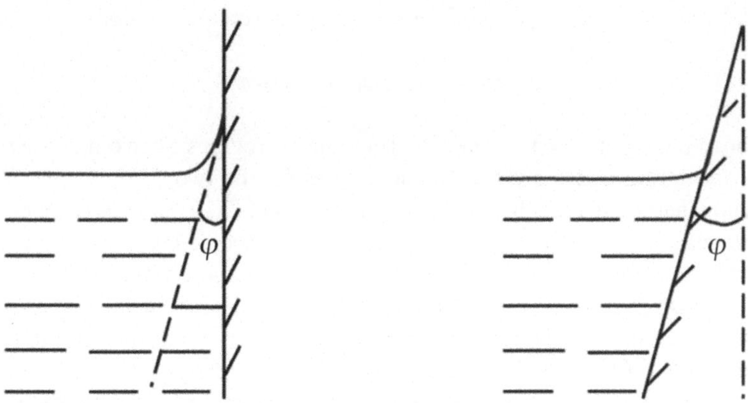

FIGURE 3.5 Tilting plate method for measurement of contact angle.

contact line measured from the solid–liquid side of the interface. Another method is the "tilting plate method." In this method, a vertical plate is submerged in a pool of test liquid. The plate is tiled until the free surface of the liquid intersects the surface of the plate without bending. The tilt angle of the plate with the vertical yields the contact angle. Figure 3.5 shows the method.

Equation (3.7) was initially obtained by Young [3] in 1805 by making a force balance on a micro-drop. In 1959, Johnson [4] rederived this equation by minimizing the Helmholtz free energy of the drop sitting on a solid surface. Through his analysis, Johnson showed that contact angle, φ, was a unique thermodynamic property of the system.

Contact angle in Young's eq. (3.7) is defined on the bases that the micro-drop shape is not influenced by gravity, the solid surface is dry, smooth at the molecular level, and is chemically homogeneous. Roughness of the surface, presence of foreign material (e.g., liquid molecules or an oxide layer) can affect the measured apparent contact angle. To account for the increased solid surface area exposed to liquid as a result of the roughness of the surface, Wenzel [5] proposed that the contact angle given by Young's equation could be related to the measured contact angle on a rough surface as

$$\cos\varphi_w = r\cos\varphi_Y \tag{3.8}$$

where r is the ratio of actual to projected area of the surface, and φ_Y is the contact angle for a perfectly smooth (molecular level) and chemically homogeneous surface as given by eq. (3.7). Cassie and Baxter [6] extended Wenzel's equation to porous surfaces including screens and meshes. If f_1 and f_2 are the fractions of solid and open areas, the contact angle for such surfaces is obtained as

$$\cos\varphi_p = f_1\cos\varphi_Y - f_2 \tag{3.9}$$

where φ_Y is the contact angle as given by eq. (3.7). Later, Cassie [7] extended this approach to a composite solid surface having two components of different surface energies. For such a composite surface, Cassie proposed the relation

$$\cos\varphi_C = f_1\cos\varphi_{Y1} + f_2\cos\varphi_{Y2} \tag{3.10}$$

In the above equation, f_1 and f_2 are the fractions of regions of two different surface energies and φ_{Y1} and φ_{Y2} are the contact angles for the two components of chemically homogeneous surface. It should be noted that eq. (3.10) assumes that composite solid surfaces are distributed in same fashion at microscopic level. Also, for example, for $\varphi_{Y2} = 180°$, eq. (3.10) reduces to eq. (3.9) of Cassie and Baxter [6].

Liaw and Dhir systematically oxidized a polished copper surface by following a particular procedure, thereby presumably creating a nearly uniform and very thin layer of copper oxide on the metal surface. This process in turn increased the difference in interfacial tension, σ_{sv}, between solid and vapor/gas and σ_{sl}, the interfacial tension between solid and liquid. This caused the contact angle for water to decrease from that observed on a clean polished copper surface ($\varphi_Y \approx 90°$). By increasing the heating period, they were able to increase the thickness and uniformity of the oxidized surface and thus decrease the contact angle or make the surface more wetting. Figure 3.6 shows the shapes of the micro-drops on surfaces with different degrees of oxidation.

For a rough or heterogeneous surface in which protrusions or ridges are wetted by the liquid whereas valleys or gas pockets remain dry, Johnson and Dettre [8] modified eq. (3.9) by multiplying the first term in eq. (3.9) by parameter r_w, which is

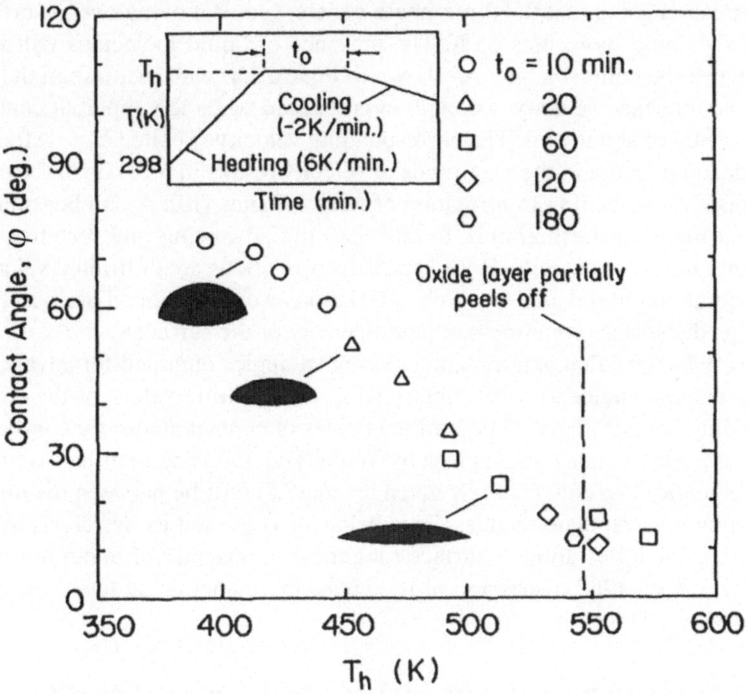

FIGURE 3.6 Dependence of the contact angle measured at room temperature on the highest temperature attained during oxidation (the dark symbols are photographs of droplets) [9]

the ratio of actual wetted area to the projected area of the ridges or protrusions on the surface. Again, one has to be careful in application of eq. (3.9), with respect to feature size of protrusions in relation to the size of micro-drops used to determine the contact angle.

3.3 HYSTERESIS IN CONTACT ANGLE

Contact angle as obtained from Young's eq. (3.7) is for a static solid–liquid–vapor interface. However, if the interface is nonstationary, the apparent contact angle will differ from that given by eq. (3.7). If the liquid–vapor/gas interface is advancing, i.e., liquid is added to the droplet, the contact angle associated with an advancing interface as shown in Figure 3.4 will be higher than the static contact angle. The magnitude of advancing contact angle will increase with the velocity of the interface and eventually approach 180° or non-wetting condition. The interface velocity at which this condition is achieved is called $U_{a,cr}$. After $U_{a,cr}$, any further increase in interface velocity would not change the contact angle but result in the breakup of the interface into micro bubbles at the three-phase contact line. Similarly, if the liquid is removed from the droplet, the three-phase contact line or the liquid vapor interface will shrink. Now the apparent contact angle or receding contact angle will be less than the static contact angle. One reason receding contact angle is smaller than the static contact angle is that the three-phase contact line is moving over a surface that has already liquid molecules on it. The presence of liquid molecules will increase σ_{sv} or reduce the contact angle. As shown in Figure 3.4, with increase in the magnitude of the interface velocity, a condition is reached when the apparent contact line attains a value of about zero. The corresponding velocity is called $U_{r,cr}$. After $U_{r,cr}$ is achieved, any increase in the magnitude of velocity results in breakup of the interface at the contact line leading to formation of micro-islands [10]. A number of attempts have been made in the literature to correlate the advancing and receding contact angles with interface velocity. However, such correlations are of limited value unless experimental conditions under which correlations were developed are nearly duplicated (e.g., the surface roughness or heterogeneity of the surface).

Figure 3.4 shows that extrapolation of contact angles obtained for advancing and receding contact angles to zero velocity will yield limiting values of the advancing and receding contact angles. The limiting values of contact angles for $U = 0$ can differ from the static contact angle given by Young's eq. (3.7) and manifest as hysteresis in contact angle. The contact angle given by eq. (3.7) will lie between the maximum and minimum extrapolated values. This behavior is termed as hysteresis in contact angle and is a manifestation of surface roughness or presence of nonuniformities on the surface. A detailed discussion of hysteresis in contact angle has been given by Huh and Mason [11].

3.4 MOLECULAR DYNAMIC SIMULATION OF INTERFACIAL TENSION AND CONTACT ANGLE

During the past several decades, significant strides have been made to determine the interfacial tensions and surface wettability by modeling molecular/atomic interactions and making a force balance between molecules. The results of such calculations yield density gradients near liquid–vapor, solid–liquid, and solid–vapor interfaces, interfacial tension, and contact angle. Equation of motion for a set of molecules can be written as

$$m_i \frac{d^2\vec{r}_i}{dt^2} = \vec{F}_i = \frac{\partial \Phi}{\partial r_i} \tag{3.11}$$

where m_i, is the mass of molecules, i, r_i is the position vector, and $\vec{F}_i$ is the force vector, which is equal to the negative spatial gradient of the potential function Φ. The potential function Φ represents the sum of pair potentials $\Phi(r_{ij})$ and is defined as

$$\Phi = \sum_i \sum_{j>i} \Phi r_{ij} \tag{3.12}$$

A well-known pair potential is Lennard-Jones that is applicable to monatomic molecules such as Ar or Xe. The potential is expressed as

$$\Phi(r) = 4\varepsilon_o \left[\left(\frac{l_o}{r} \right)^{12} - \left(\frac{l_o}{r} \right)^{6} \right] \tag{3.13}$$

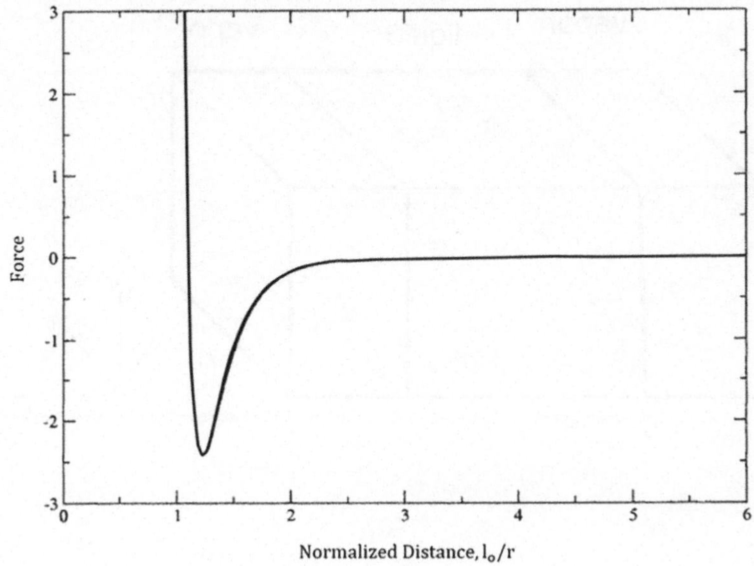

FIGURE 3.7 Lennard-Jones potential as a function of normalized distance.

where r is the instantaneous distance between molecules, and ε_o and l_o are the characteristic energy and length scales. Figure 3.7 shows a plot of the Lennard-Jones potential as a function of normalized spatial distance between molecules. For normalized distances less than 1, intermolecular force is positive (repulsive) and increases very rapidly as distance decreases below 1. However, for distances greater than 1, the force is negative, or attractive, and dies at larger distances. Lennard-Jones potential must be corrected for molecules with point charges such as water molecules. For example, [12], interaction between two such molecules is given by

$$\Phi(r_{ab}) = 4\varepsilon_{oo}\left[\left(\frac{l_{oo}}{r_{oo}}\right)^{12} - \left(\frac{l_{oo}}{r_{oo}}\right)^{6}\right] + \frac{1}{4\pi\varepsilon_o}\sum_{a=1}^{3}\sum_{b=1}^{3}\frac{q_a q_b}{r_{ab}} \tag{3.14}$$

where q_a and q_b are charges on sites a and b respectively and r_{ab} is the distance between sites a and b om two different molecules. The point charge q_o for oxygen is taken to be $-0.8476|e|$ and for hydrogen $q_H = \frac{q_o}{2}$. The parameters l_{oo} and ε_{oo} in the Lennard-Jones potential are taken to be 3.166 Å and 0.6502 $\frac{\text{kJ}}{\text{mole}}$ respectively. Particle–Particle, Particle–Mesh (P^3M) method has been used for the Lennard-Jones term in eq. (3.14), whereas for the charge or Coulomb term, Ewald Sum is used. Details of this approach are given by Shi [12]. The equation of motion was solved by using the Verlet algorithm with a time step of 1×10^{-15} s. After solving the equation of motion for each molecule, the results of molecular motion were averaged to obtain properties such as densities and surface tension.

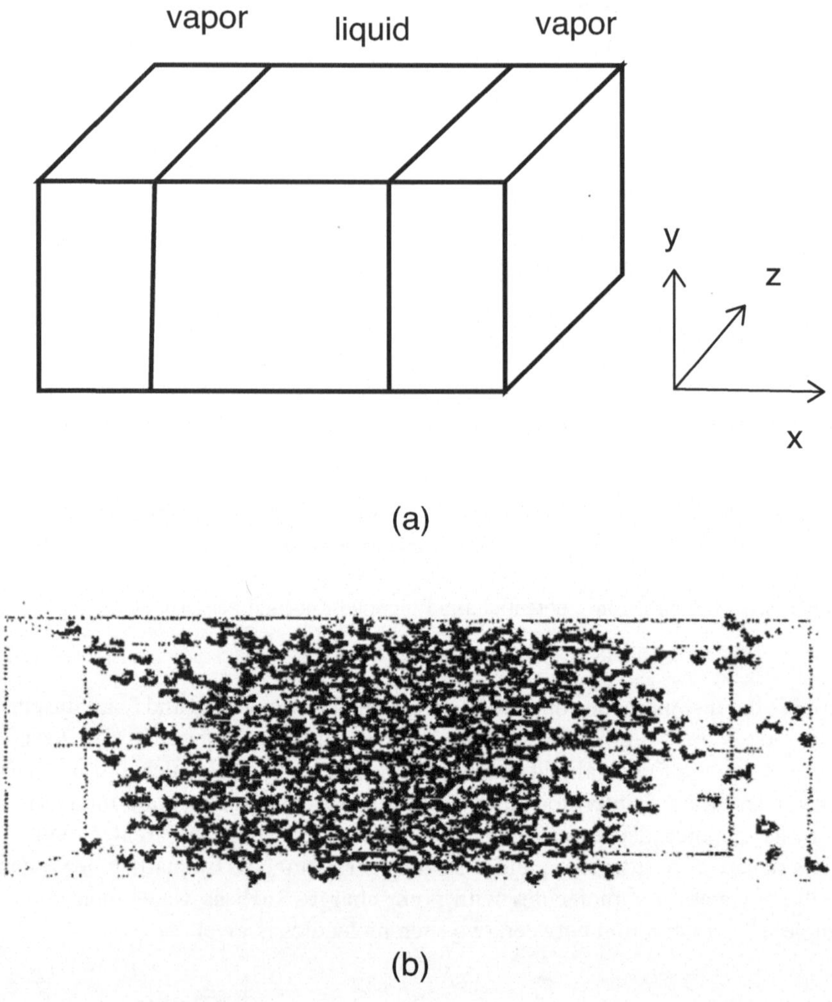

FIGURE 3.8 Computational domain and snapshot of molecular arrangement [12].

Figure 3.8 shows the computational domain and snapshot of molecules at a temperature of 572 K. Surface tension is defined as the integrated difference between normal and tangential pressures from bulk of liquid and vapor phases, or

$$\sigma_{lv} = \int_{-\infty}^{\infty} \left[p_N(z) - p_T(z) \right] dz \tag{3.15}$$

For a plane interface, perpendicular to the *z*-axis, normal and tangential components respectively are:

$$p_N(z) = p_{zz}(z) \tag{3.16}$$

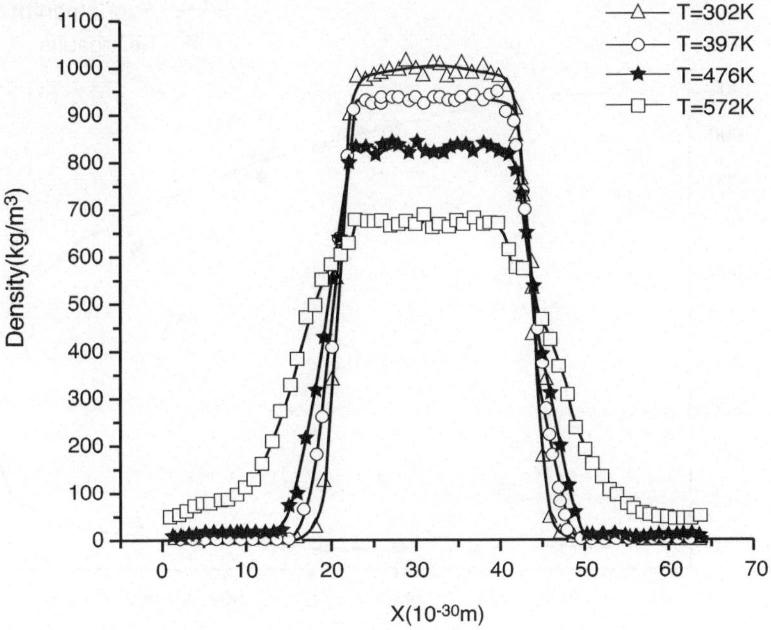

FIGURE 3.9 Density profiles for water at different temperatures [12].

and

$$p_T(z) = \frac{p_{xx}(z) + p_{yy}(z)}{2} \tag{3.17}$$

where $p_{xx}(z)$, $p_{yy}(z)$, and $p_{zz}(z)$ are components of the pressure tensor. The pressure tensor is defined in terms of molecular position, molecular velocities, and interatomic force.

Figure 3.9 shows density profiles for water at different fluid temperatures. The profiles show that a distinct interface between liquid and vapor phases exists on both sides. In Figure 3.10, calculated densities are shown to vary as a function of temperature. Results of molecular dynamic simulation are also compared with data. The solid line is the curve fitted through the experimental data, and the solid squares are the results from molecular dynamic simulation. It is seen that at low temperatures, experimental and simulation results of density agree very well. However, at temperatures near critical, the simulations underpredict water density and overpredict the vapor density. One reason could be the inaccuracy in the space point charge the model used in the Coulomb term.

A comparison of predicted and experimentally obtained surface tension for water as a function of temperature is shown in Figure 3.11. Since there are two interfaces, the calculated surface tension is half of the total integrated value. Filled symbols are the simulation results, and the solid line represents experimental values. Again, the agreement is excellent at low temperatures. However, at higher temperatures, molecular dynamic simulation underpredicts the data.

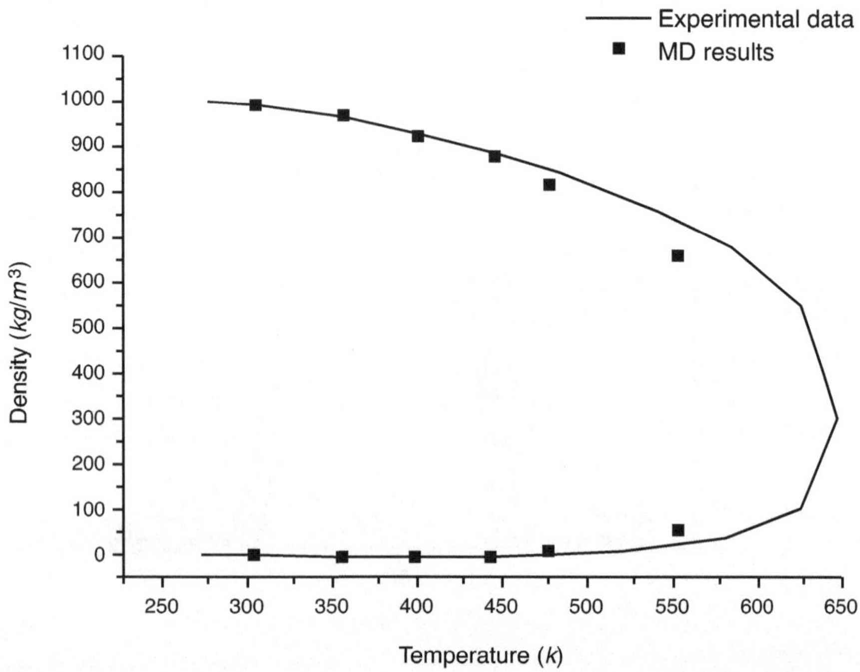

FIGURE 3.10 Liquid and vapor density variation of water with saturation temperature [12].

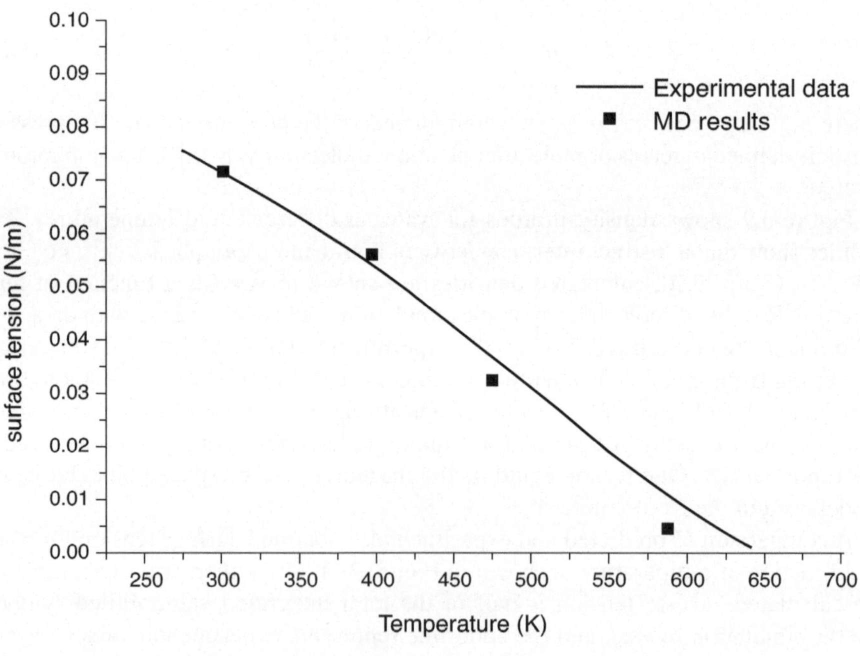

FIGURE 3.11 Comparison with data of surface tension of water obtained from molecular dynamic simulation [12].

Molecular dynamic simulations have been used not only to determine the interfacial tension between the phases but also the contact angle. Following the work of Shi et al. [12], in which for a solid wall made of platinum, the density was taken to be $21.45\times10^3 \frac{\text{kg}}{\text{m}^3}$ and lattice constant was set at 2.776 Å. For the solid–solid interaction, the Lennard-Jones potential was applied and was chosen to be

$$\Phi(r_{ij}) = 4\varepsilon\left[\left(\frac{\ell}{r_{ij}}\right)^{12} - \left(\frac{\ell}{r_{ij}}\right)^{6}\right] \tag{3.18}$$

where $\varepsilon_{ss} = \varepsilon = 8.35\times10^{-20}$ J and $\ell_{ss} = \ell = 2.475$ Å were chosen to model the platinum atomic vibrations of the lattice. Equation (3.18) was also used for the solid–fluid interaction. To model Argon, various parameters used were $\ell_{sf} = 0.8\ell_{ff}$ and $\varepsilon_{sf} = 0.6\varepsilon_{ff}$, where $\ell_{ff} = 3.4$ Å and $\varepsilon_{ff} = 1.65324\times10^{-21}$ J. The ratio $\varepsilon_{\gamma} = \frac{\varepsilon_{sf}}{\varepsilon_{ff}}$ could be varied parametrically because of large density variations of liquid near the solid surface, liquid layer next to the solid surface was chosen to be thicker than 2 nm.

Figure 3.12 shows for Argon the calculated valued of σ_{lv}, σ_{sl}, and σ_{sv} for different dimensionless temperatures, $\left(T/T_c\right)$.

$$T_c = 647.3\text{K}$$

It is noted that at nano-level, all interfacial tensions decrease with temperatures as would be expected at the macrolevel. For temperatures up to critical temperature, all interfacial tensions decrease with temperature. Near the critical temperature

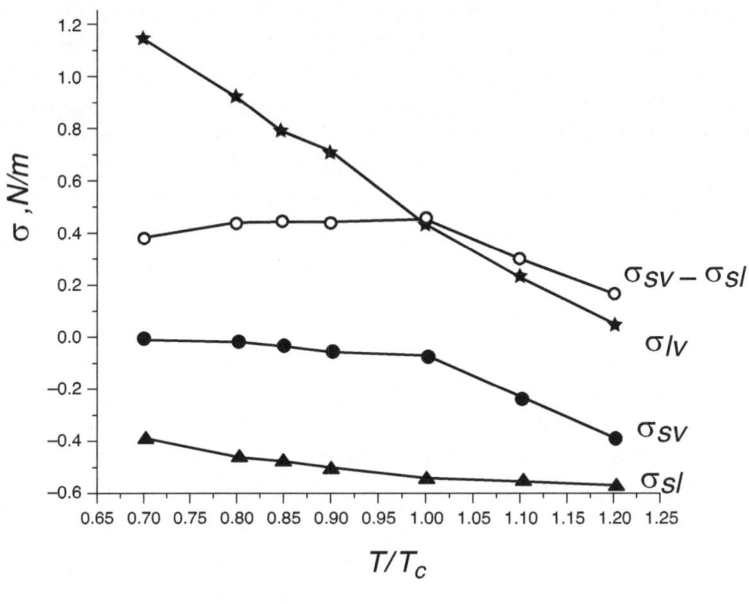

FIGURE 3.12 Dependence for argon of solid–liquid, solid–vapor, and liquid–vapor interfacial tensions on system temperature [12].

TABLE 3.1
σ_{lv}, σ_{sv}, and σ_{sl} (in N/m) Variation with Temperature for Water-Platinum

Temperature $\left(T/T_c\right)$	σ_{lv}	σ_{sv}	σ_{sl}	σ_{sv-sl}
0.465	0.072	−0.0115	−0.073	0.0615
0.543	0.064	−0.0245	−0.081	0.0563
0.620	0.054	−0.0389	−0.089	0.0501
0.695	0.048	−0.0491	−0.095	0.0459
0.775	0.041	−0.0655	−0.106	0.0409
0.825	0.036	−0.0891	−0.127	0.0381
0.887	0.018	−0.1319	−0.127	0.0200

$\left(\frac{T}{T_c} \simeq 1\right)$, difference between solid–vapor and solid–liquid tensions approximately equals interfacial tension between liquid and vapor indicating a complete wetting condition according to Young's eq.

In obtaining the variations of interfacial tensions as a function of temperature for Argon–Platinum combination shown in Figure 3.12, it was found that results of solid–liquid and solid–vapor tensions were dependent on the chosen value of solid–liquid potential. Increasing the potential caused increase in the difference between solid–vapor and solid–liquid potential, with solid–liquid interfacial tension decreasing more than the solid–vapor interfacial tension.

Table 3.1 lists variation of interfacial tensions for water–platinum combination for increasing system temperatures approaching the critical temperatures. In carrying out these calculations, liquid layer thickness was assumed to be 32.5 Å, and four layers of platinum atoms arranged as Face-Centered Cubic lattice were considered. The Particle–Particle, Particle–Mesh (PPPM) method was applied to both the Lennard-Jones and Coulombic terms in carrying out the force calculations. It is noted that all interfacial tensions decrease with increase in system temperature. However, the difference between σ_{sv} and σ_{sl}, decreases slower than σ_{lv}. The two become equal at $T/T_c \simeq 0.775$. According to Young's equation, this will correspond to a contact angle of 0° or well-wetting conditions.

Shi and Dhir [13] have also simulated through molecular dynamics, the contact angles of nano droplets of Argon and water placed on a platinum surface. Figure 3.13 depicts a contour of the liquid droplet of Argon at different temperatures.

It is noted that for $\varepsilon_r = 0.408$ when T^* is increased from 0.7 to 1, the contact angle decreases, consistent with what was noted earlier with change in interfacial tensions. For fixed $T^*\left[=1\right]$ when ε_r is increased from 0.409 to 0.65, contact angle decreases as well. At higher ε_r, droplet spreads on the surface and contact angle is almost zero. From Young's equation, we can deduce that as ε_r is increased, difference between σ_{sv} and σ_{sl} increases. The increase is caused by formation of a monolayer of liquid near the solid surface. Figure 3.14 shows for Argon–Platinum calculated contact angle with increasing ε_r for different T^*. Again increasing ε_r suggests that $\sigma_{sv} - \sigma_{sl}$ increases for a given temperature. At lower temperatures and smaller ε_r, $\sigma_{sv} < \sigma_{sl}$,

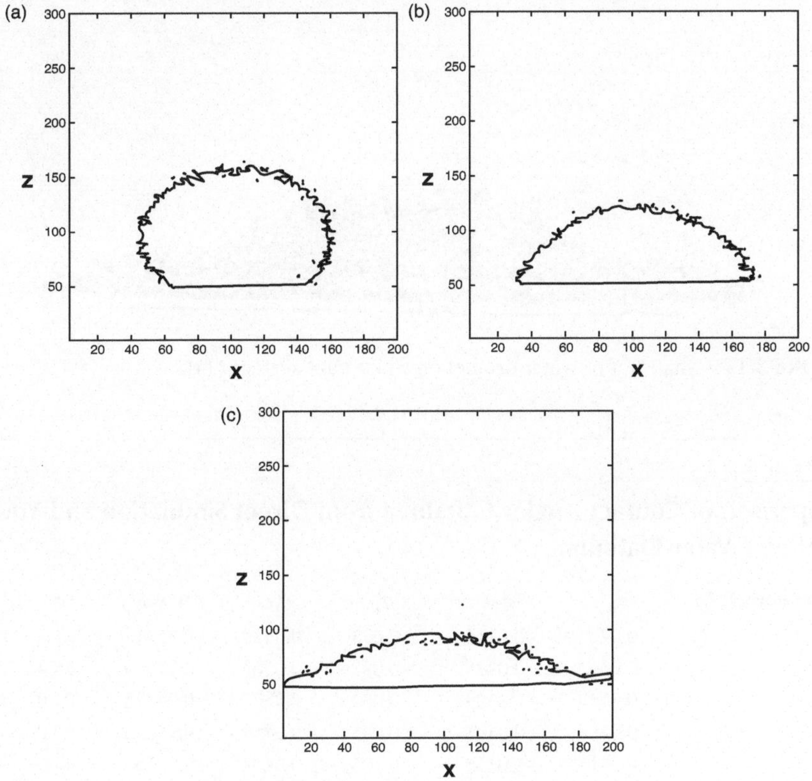

FIGURE 3.13 Argon droplet contours for (a) $\varepsilon_r = 0.408$, $T^* = 0.7$; (b) $\varepsilon_r = 0.408$, $T^* = 1.0$; (c) $\varepsilon_r = 0.65$, $T^* = 1.0$ ($T^* = T/T_c$) [13]. To obtain physical length x and z should be multiplied by 3.4×10^{-10}, m.

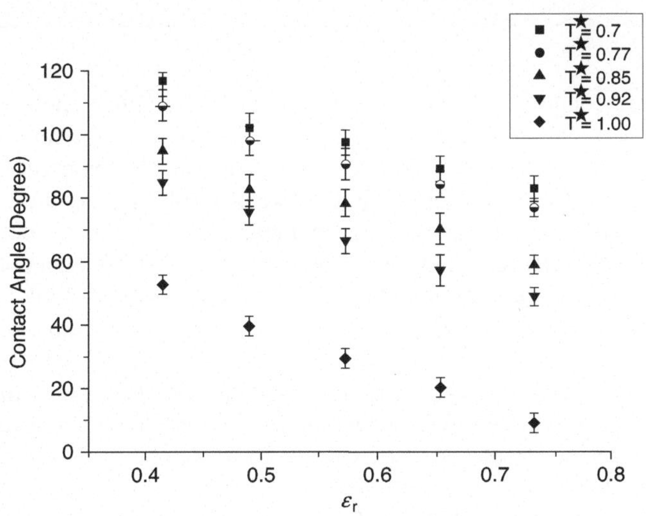

FIGURE 3.14 Variation of the contact angle with ϵ_r (Argon) [13].

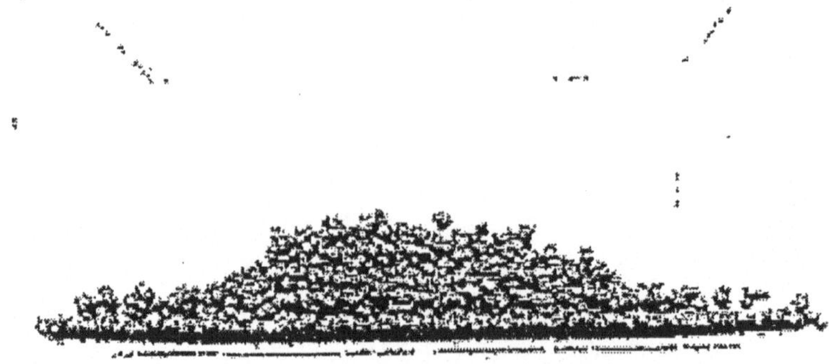

FIGURE 3.15 Snapshot of water droplet on a platinum surface [14].

TABLE 3.2
Comparison of Contact Angles Obtained from Direct Simulation and Young's Equation (Water-Platinum)

Temperature(T/T_c)	σ_{lv}*	σ_{Sv}	σ_{Sl}	σ_{sv-sl}	$\cos\phi_l$	ϕ_y	ϕ_D
0.465	0.072	−0.0115	−0.073	0.0615	0.85417	31.3	25.5
0.543	0.064	−0.0245	−0.081	0.0563	0.87969	28.4	23.0
0.620	0.054	−0.0389	−0.089	0.0501	0.92778	21.9	20.0
0.695	0.048	−0.0491	−0.095	0.0459	0.95625	17.0	11.2
0.775	0.041	−0.0655	−0.106	0.0409	0.99756	4.00	0.00
0.825	0.036	−0.0891	−0.127	0.0381	1.05833	0.00	0.00
0.887	0.018	−0.1319	−0.512	0.0200	1.11111	0.00	0.00

ϕ_Y and ϕ_D are from Young's equation and direct simulation, respectively.
* Interfacial tensions are measured in N/m.

and we observe contact angle to by greater than 90°. A contact angle of 90° corresponds to $\sigma_{sv} = \sigma_{sl}$. Similar calculations have also been carried out for water droplets on platinum. Figure 3.15 shows a simulated water droplet at a temperature of 310 K.

With existence of a monolayer of liquid between droplet and platinum surface, the contact angle of the droplet is about 30°. A comparison of the contact angles obtained from direct simulations (ϕ_D) and those obtained from Young's equation (ϕ_Y) when calculated magnitudes of interfacial tensions are used is made in Table 3.2.

Values greater than unity reflect the inaccuracy in calculated values of interfacial tension.

The chosen parameters for solid–liquid interaction were the same as those described earlier. It is noted that contact angles obtained from molecular dynamic simulation of a nano droplet are comparable to those obtained by using Young's equation when calculated values of interfacial tensions are used. This validates that Young's equation is also valid at the nano/micro scale. It should be stressed that at given temperature, the calculated values of interfacial tensions and contact angle strongly depend on the magnitude of solid–fluid potential. Increasing its values reduces the contact angle.

3.5 EXPERIMENTAL METHODS FOR MEASUREMENTS OF SURFACE TENSION

A number of experimental techniques have been described in the literature to determine surface tension of liquids at various temperatures. In fact, this was a highly investigated topic in late-nineteenth and early-twentieth centuries. See, for example, reference [15].

3.5.1 Capillary Rise Method

The most popular approach has been the *Capillary Rise* method. In this method, a circular capillary tube made of glass and a few mm in diameter is placed vertically in a pool of liquid as shown in Figure 3.16. Because of interfacial tension liquid rises in the capillary tube above its level in the pool. If liquid wets the surface of the capillary completely, i.e., contact angle, $\varphi = 0$, capillary tube is completely vertical and radii of curvature of the interface in the plane of the paper and in the plane normal to it are taken to be the same and equal to the radius of the capillary tube, the pressure difference, ΔP, between the gas/vapor and liquid side of the upward facing concave interface will be

$$\Delta p = \frac{2\sigma}{r} \tag{3.19}$$

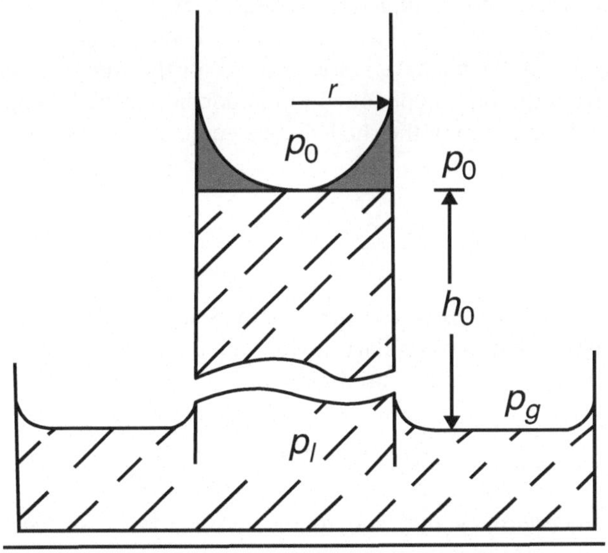

FIGURE 3.16 Capillary rise (capillary tube much magnified in relation to the dish method) [15].

where r is the inner radius of the capillary tube. Pressure p_g at the surface of liquid in the pool can be simply written as

$$p_g = p_0 + \rho_g g h_0 \tag{3.20}$$

where h_0 is the height of the liquid column in the capillary above the flat liquid surface, and p_0 is the pressure in gas at height h_0. This pressure must equal the liquid pressure, p_l, in the capillary at the same level as the pool free surface. Expression for pressure, p_l, with a pressure difference Δp across the interface can be written as

$$p_l = p_0 - \Delta p + \rho_l g h_0 \tag{3.21}$$

Equating, pressure p_g and p_l from eqs. (3.20) and (3.21), we get

$$p_0 + \rho_g g h_0 = p_0 - \Delta p + \rho_l g h_0$$

or

$$\Delta p = (\rho_l - \rho_g) g h_0 \tag{3.22}$$

Substituting for Δp from eq. (3.19) into eq. (3.22), we obtain

$$\sigma = \frac{(\rho_l - \rho_g) g h_0 r}{2} \tag{3.23}$$

If the glass surface is not completely wetted by the liquid, one must multiply right-hand side of eq. (3.19) by $\cos\phi$ to account for contact angle being greater than zero. In writing eq. (3.21), we have not accounted for the weight or pressure difference resulting from the liquid contained in the double-hatched meniscus region. If υ is the volume of the liquid in the shaded area, eq. (3.23) for surface tension would become

$$\sigma = \frac{(\rho_l - \rho_g) g h_0 r}{2} + \frac{(\rho_l - \rho_g) g \upsilon}{2\pi r} \tag{3.24}$$

The correction to surface tension as a result of the second term in eq. (3.24) is small. It has been suggested [16] for capillaries with radius, r, less than 0.2 mm, and water as the test liquid, eq. (3.21) can be used when h_0 is corrected as

$$h_o' = h_o + \frac{r}{3} \tag{3.25}$$

Various other uncertainties that add to the evaluation of surface tension from eq. (3.23) are: non-cylindrical nature of the capillary tube, uncertainty in measurement of contact angle, measurement in height, h_o of the meniscus, and non-sphericity of the concave interface and in evaluation of the contact angle itself.

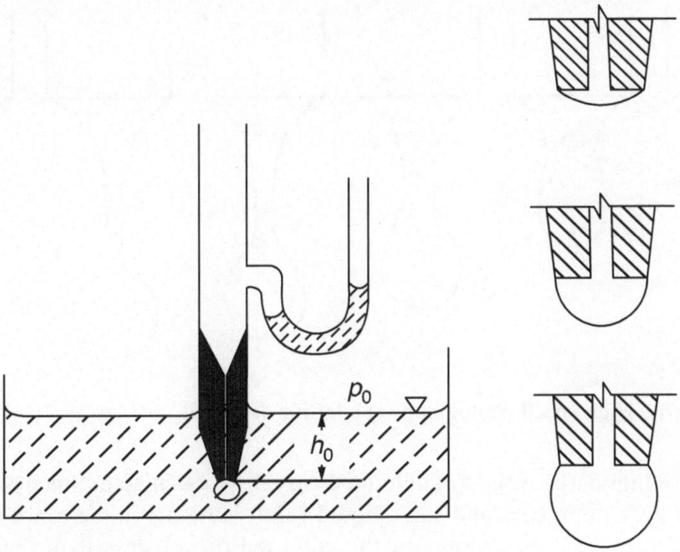

FIGURE 3.17 Maximum bubble pressure method [15].

3.5.2 The Maximum Bubble Pressure Method

In this method as shown in Figure 3.17, a bubble is slowly formed at the tip of a tube submerged in the test liquid.

As the bubble grows, its radius of curvature initially decreases as shown in the sketches depicted on the right-hand side in Figure 3.17. It is minimum when gas bubble becomes hemispherical but increases after it grows beyond a hemisphere. The measured bubble pressure is maximum when bubble is of hemispherical shape. If p_{max} is measured pressure, and h_0 is the submerged depth of the tube below the free surface of liquid pool, Δp_{max}, across the interface is

$$\Delta p_{max} = p_{max} - \left(p_0 + \rho_l g h_0\right) \tag{3.26}$$

For r as the outer radius of the tube, surface tension of the liquid is simply

$$\sigma = \frac{r}{2} \cdot \Delta p_{max} = \frac{r}{2}\left[p_{max} - \left(p_0 + \rho_l g h_0\right)\right] \tag{3.27}$$

In situations where liquid completely wets the surface, bubble base may be limited to the inner radius of the tube. As such, this radius instead of the outer tube radius should be used in eq. (3.27).

3.5.3 Drop Weight Method

Here the approach is to determine the weight of liquid drops falling from the end of a tube. Drops from the end of a tube are allowed to fall in a container. Knowing the number of drops that fall in the container over a given period of time and volume of

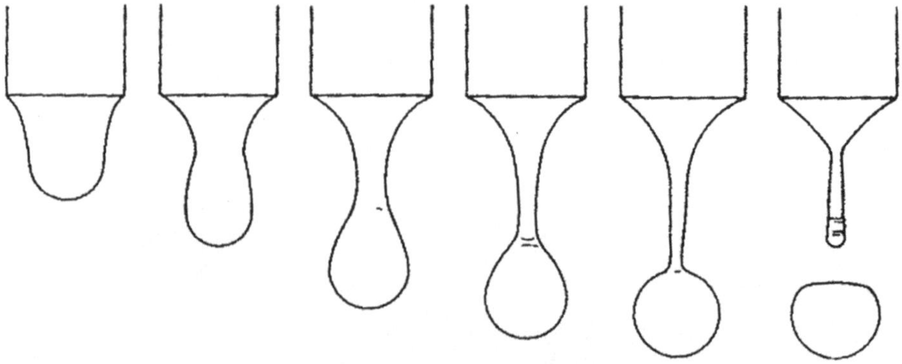

FIGURE 3.18 High-speed photographs of a falling drop [15].

the collected liquid, the weight of a liquid drop can be calculated. Since drop detachment is not a clean process and some liquid is left hanging on the tip of the tube as shown in Figure 3.18, the volume of the collected drop is less than that would be expected if it has clear detachment from the tip of the tube. If r is the radius of the tip, surface tension force is,

$$F = 2\pi r\sigma \tag{3.28}$$

With f being a correction factor for the weight, w' of the collected drop, or the drop being less than the ideal weight of the drop, the surface tension force holding the ideal drop will be

$$F = \frac{w'}{f} \tag{3.29}$$

From eqs. (3.28) and (3.29), we obtain value of the surface tension as

$$\sigma = \frac{w'}{2\pi rf} \tag{3.30}$$

The correction factor has been noted to be a function or r / v where v is the volume of the collected drop. Fluid properties and wetting characteristic of the liquid also affect parameter, f.

3.5.4 The Ring Method

In this method, force required to detach a ring from the free surface of a liquid pool is determined. For a liquid that wets the surface well, the detachment force is equal to the [15] sum of the weight of the ring and the downward force due to surface tension. As shown in Figure 3.19, the expression for the force needed to detach the ring can be written as

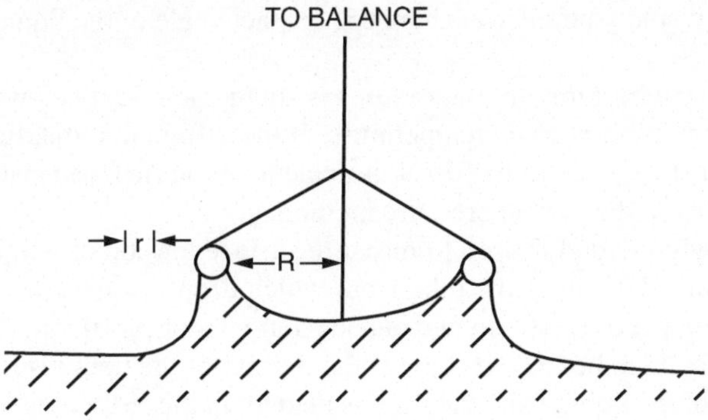

FIGURE 3.19 Ring method [15].

$$F = w + 4\pi R\sigma \tag{3.31}$$

where w is the weight of the ring and R is the radius. From eq. (3.31) knowing F, w, and R, the surface tension, σ, can be determined. Here factor 4 results from the fact that liquid wets the ring on the inside and the outside. As pointed out in [15], surface tension obtained from eq. (3.31) is generally underpredicted. The correction factor is found to depend on the ratios R^3/v and R/r, where v is the volume of the meniscus and r is the radius of the ring wire.

There are several other methods for measurement of surface tension of liquid. For a broader discussion, one is referred to the book by Adamson and Gast [15].

3.6 SUMMARY

In this chapter, we have developed the concept of interfacial tension between the phases. Pressure difference between a vapor bubble and surrounding liquid or a droplet and surrounding vapor is described as a result of interfacial tension. Surface wettability is described and is related through specification of contact angle. Results of molecular dynamic simulations to obtain surface tension and contact angle for water are described, and methods to experimentally determine contact angle and surface tension are discussed.

EXERCISES

1. Obtain the surface tension of Ethylene Glycol at room temperature with $\rho_l = 1{,}111$ kg/m^3, if radius of the capillary tube is 0.5 mm and height of the meniscus is 1.2 cm. Assume that the liquid wets the surface of glass capillary tube well. Compare your result with that reported in the literature.
2. What will be the observed height of the meniscus in Problem 1, if tube radius is increased by 10 or 100 times? Discuss your answer.
3. a. What will be the observed height of the meniscus, if capillary tube in problem 1, is inclined at 60° to the horizontal?

b. How would your answer change if contact angle of the liquid with glass is 60°?

4. A capillary tube 1 mm in inner radius is submerged 5 cm below the surface of a pool of water at room temperature. If the differential maximum bubble pressure is measured to be 635 N/m^2, calculate the surface tension of water obtained from the bubble pressure method.
5. Drop weight method is used to measure surface tension of a liquid at room temperature. The tip of the tube from which drops sequentially fall has an inside diameter of 1×10^{-3} m and outside diameter of 2×10^{-3} m. Weight of 25 drops is found to be 1.6×10^{-4} kg. If the density of the liquid is 1,520 kg/m^3, and 10% of the liquid from the drop is held up in the stem attached to the tip of the tube, calculate the surface tension of the liquid.
6. Plot surface tension of water, ethanol, and sodium as a function of ratio of the saturation temperature to the critical temperature. Discuss your result.

NOMENCLATURE

ds Differential distance
$\vec{F}$ Force vector
f Fractional Area
g Gravitational acceleration
h Height of liquid column
l Characteristic length scale
m Mass
p Pressure
R Gas constant or radius of curvature
r Ratio of actual to projected area, position vector, or radius
T Temperature
x Mole fraction

SUBSCRIPTS

c Composite solid
g Gas
l Liquid
o Above the liquid surface
p Porous
sl Solid–liquid
sv Solid–vapor
y Molecularly smooth surface
w Wall or weight

GREEK LETTERS

ε Characteristic energy
γ Angle with horizontal axis

ρ Density
σ Surface tension
Φ Potential function
ϕ Contact angle measured from liquid side

REFERENCES

1. Vargaftik, N.B., Volkov, B.N., and Voljak, L.D. (1983), "International Tables of the Surface Tension of Water." *J. Phys. Chem. Ref. Data*, Vol. 12, p. 817.
2. Miller, C.A. and Neogi, P. (1985), *Interfacial Phenomena*, Marcel Dekker, New York.
3. Young, T. (1805), "An Essay on the Cohesion of Fluids." *Philos. Trans. R. Soc. Lond.*, Vol. 95, pp. 65–87.
4. Johnson, R.E. Jr, (1959), "Conflicts between Gibbsian and Thermodynamics and Recent Treatments of Interfacial Energies in Solid-Liquid-Vapor Systems." *J. Phys. Chem.*, Vol. 63, pp. 1655–1658.
5. Wenzel, R.N. (1936), "Resistance of Solid Surfaces for Wetting by Water." *Ind. Eng. Chem.*, Vol. 28, pp. 988–994.
6. Cassie, A.B.D. and Baxter, S. (1944), "Wettability of Porous Surfaces." *Trans. Faraday Soc.*, Vol. 40, pp. 546–551.
7. Cassie, A.B.D. (1948), "Contact Angles." *Discuss. Faraday Soc.*, Vol. 3, pp. 11–16.
8. Johnson, R.E. and Dettre, R.H. (1964), "Contact Angle, Wettability and Adhesions." *Adv. Chem. Ser.*, Vol. 43, pp. 112–135.
9. Liaw, S.P. and Dhir, V.K. (1989), "Void Fraction Measurements during Saturated Pool Boiling of Water on Partially Wetted Vertical Surfaces." *ASME J. Heat Transfer*, Vol. 111, pp. 737–738.
10. Burley, R. and Kennedy, B.S. (1976), "An Experimental Study of Air Entrainment at a Solid/Liquid/Gas Interface." *Chem. Eng. Sci.*, Vol. 39, pp. 1357–1372.
11. Huh, C. and Mason, S.G. (1977), "Effect of Surface roughness on Wetting (Theoretical)." *J. Colloid Interface Sci.*, Vol. 60, pp. 11–38.
12. Shi, B., Sinha, S., and Dhir, V.K. (2006), "Molecular Dynamic Simulation of the Density and Surface Tension of Water by Particle-Particle Particle-Mesh Method." *J. Chem. Phys.*, Vol. 124, pp. 1–7.
13. Shi, B. and Dhir, V. K. (2009), "Molecular Dynamic Simulation of the Contact Angle of Liquids on Solid Surfaces." *J. Chem. Phys.*, Vol. 139, pp. 1–5.
14. Shi, B., Sinha, Shashank and Dhir, V. K. (2005), "Molecular Simulation of Contact Angle of Water Droplet on a Platinum Surface." *Proceedings of ASME International Mechanical Engineering Conference*, November 5–11, Orlando, FL.
15. Adamson, A.W. and Gast, A.D. (1997), *Physical Chemistry of Surfaces*, 6th Edition, John Wiley and Sons, Inc.
16. Harkins, W.D. (1949), "Determination of Surface and Interfacial Tension," in *Physical Methods of Organic Chemistry*, edited by A. Weissberger. Interscience Publishers, Inc., New York.

4 Nucleation and Bubble Growth

INTRODUCTION

Nucleation is a process in which finite-size clusters of molecules encompassing properties of the second phase appear in the host liquid. This could be the initiation of vapor bubble formation during boiling, or cavitation, and crystal formation during solidification. Nucleation of vapor bubbles is generally subdivided into two categories: homogeneous nucleation and heterogeneous nucleation. Homogeneous nucleation refers to the first appearance of a vapor bubble in a liquid pool and far away from the bounding walls of a container. Heterogeneous nucleation on the other hand is a process in which bubbles form on defects such as pits, scratches, and grooves on a heated surface submerged in liquid.

Figure 4.1 shows such defects (blind holes) on a polished copper surface. Rate of heat transfer from the solid surface during heterogeneous nucleate boiling depends on a number of subprocesses including number density of active nucleation sites, bubble dynamics, associated heat transfer mechanisms, and thermal response of the solid when heat flux to the surface is controlled. Figure 4.2 shows a schematic of the processes involved. For a predictive model for nucleate boiling heat transfer, one

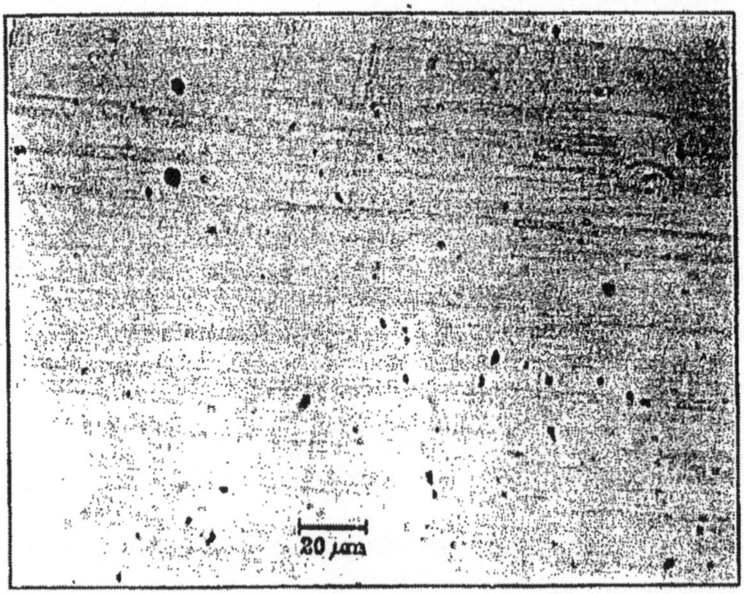

FIGURE 4.1 Optical micrograph of the super smooth surface (500× magnification) [11a,b].

DOI: 10.1201/9781032668437-4

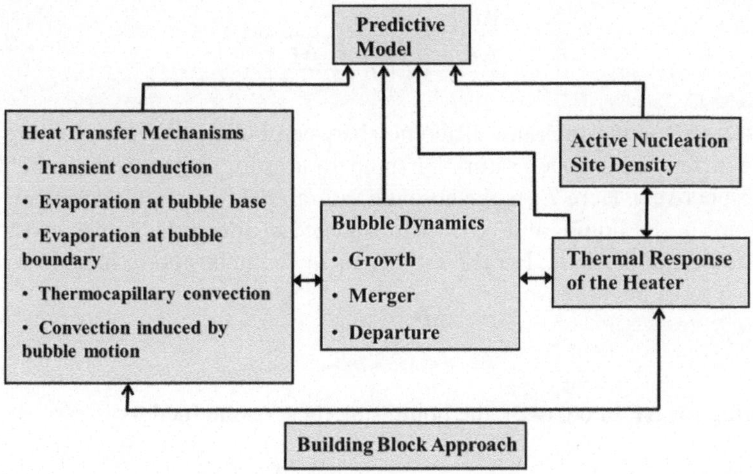

FIGURE 4.2 Subprocesses of nucleate boiling.

must address these subprocesses and their interactions, which tend to be highly nonlinear. In this chapter, we discuss nucleation and bubble dynamics. Before focusing on the detailed description and analyses of these processes, let us first look at the conditions for mechanical and thermal equilibrium of a vapor bubble with the host liquid.

The condition for mechanical equilibrium that was developed earlier in Chapter 3 simply states that the pressure in a perfectly spherical bubble must exceed the ambient by $2\sigma/R$. In a gravitational field, the bubble being lighter than the surrounding liquid will buoy away; however, here we are considering a static equilibrium such that the bubble is relatively small in size and is held in place. For the bubble to be in thermal equilibrium with the host liquid, the temperature of the bubble and the surrounding liquid should be the same, i.e.,

$$T_b = T_l \tag{4.1}$$

Now, if the bubble contains only vapor of the host liquid, from thermodynamic equilibrium consideration as discussed in Chapter 3, the vapor must be at least at the saturation temperature corresponding to pressure in the bubble. We also know from the equation of state that saturation temperature increases with pressure, so the vapor or the liquid temperature should be higher than the saturation temperature corresponding to the pressure in the liquid. Or, the liquid must be superheated.

The degree of superheat can be determined if we combine the mechanical equilibrium condition, $P_b - P_l = \dfrac{2\sigma}{R}$, with the Clausius–Clapeyron equation; which relates the differential change in saturation temperature with change in pressure and described in Chapter 2 as

$$\frac{\Delta P}{\Delta T} = \frac{h_{fg}}{v_{fg}\left(T_{sat} + \frac{\Delta T}{2}\right)} \tag{4.2}$$

In eq. (4.2), ΔP is the pressure difference between bubble and surrounding liquid, ΔT is the difference between saturation temperatures for pressures in the bubble and liquid, respectively. Here T_{sat} is the absolute saturation temperature corresponding to the pressure in the liquid, and v_{fg} is the difference in specific volumes of vapor and liquid at the mean pressure. For $\Delta T << T_{sat}$, eq. (4.2) can be approximately written as

$$\frac{\Delta P}{\Delta T} \approx \frac{h_{fg}}{v_{fg}T_{sat}} \tag{4.3}$$

Substituting for ΔP in eq. (4.3), the liquid superheat is obtained as

$$(T_l - T_{sat}) = (T_b - T_{sat}) = \frac{2\sigma v_{fg} T_{sat}}{Rh_{fg}} \tag{4.4}$$

At low pressures, the specific volume of vapor is much larger than that of the liquid, such that v_{fg} in eq. (4.4), can be replaced by $1/\rho_v$ without making any appreciable error. Similarly, the pressure difference between bubble and liquid is small as such ρ_v can be evaluated with little error at p_l.

With these assumptions, eq. (4.4) becomes

$$(T_l - T_{sat}) = \frac{2\sigma T_{sat}}{R\rho_v h_{fg}} \tag{4.5}$$

Equation (4.5) gives the liquid superheat necessary for a spherical vapor bubble to coexist with the host liquid. Since superheat is not large, σ and h_{fg} can be evaluated at $T_{sat}(p_l)$.

Example:

Calculate water superheat necessary if a steam bubble of radius 10^{-3} cm is to coexist with the liquid at a system of 1 atm pressure.

Solution:

For 1 atm pressure at saturation temperature of 373 K, eq. (4.5) yields

$$(T_l - T_{sat}) = \frac{2 \times 58.9 \times 10^{-3} \times 373}{10^{-5} \times 0.598 \times 2.257 \times 10^{6}} = 3.26 \text{ K}$$

In evaluating eq. (4.5), surface tension, σ, can be obtained at T_{sat} corresponding to pressure in the liquid, although more appropriate value will be at T_l.

An iterative procedure will be required to do so. Regardless of the temperature we choose to evaluate surface tension, the correction is small. Similarly, ρ_v and h_{fg} are obtained at the liquid pressure and corresponding saturation temperature. Thus, we see that for a relatively small bubble, the required liquid superheat is modest. The superheat will tend to decrease as the bubble size is increased and vice versa.

Another observation that can be made about the requirement of thermal equilibrium is that a vapor bubble in the absence of any mass transfer will have to shrink upon introduction into a pool of liquid if the liquid superheat is more than that corresponding to the initial bubble radius. Similarly, if the superheat is smaller than that corresponding to the temperature of vapor in the bubble, the bubble will have to expand. However, mass transfer at the interface will cause the bubble to expand or shrink depending on the bulk temperature.

4.1 HOMOGENEOUS NUCLEATION

The process of creation of a vapor bubble in a superheated liquid is initiated by a cluster of activated molecules. Theoretical formulation of nucleation in superheated liquids has been developed based on the kinetic theory approach [1,2] and statistical mechanical approach (e.g. [3]). Here, we use kinetic theory approach as described by Cole [4]. The activated molecules have energies that are considerably more than the average. Generally, a cluster of activated molecules is constantly growing or shrinking in size as additional activated molecules join or break away from it. When a cluster attains a size such that its excess availability[1] is maximum, it takes on the form of a bubble containing activated molecules. If the activated cluster has a density that is different from that of the surrounding liquid, surface tension can be presumed to exist at the interface between the two to balance the forces acting on molecules in the bulk liquid and cluster side of the interface. For a certain volume of liquid containing an activated cluster of radius r, the availability can be written as

$$A = (n_t - n_a)\,g_l + n_a g_a + 4\pi r^2 \sigma \tag{4.6}$$

Here n_t is the total number of molecules, n_a is the number of activated molecules in the cluster, and g_l and g_a are the specific Gibbs functions of the average and activated molecules, respectively. In eq. (4.6), the last term on the right-hand side represents surface energy. If ρ_a is the fluid density in the activated cluster, M is the molecular weight, and N_A is the Avogadro's number, an expression for the number of activated molecules can be written as

$$n_a = \frac{4}{3}\pi r^3 \rho_a \frac{N_A}{M} \tag{4.7}$$

Substitution of eq. (4.7) in eq. (4.6) yields

$$A = n_t g_l + \frac{4}{3}\pi r^3 \rho_a \frac{N_A}{M}(g_a - g_l) + 4\pi r^2 \sigma \tag{4.8}$$

Since in the absence of a cluster of activated molecules, the availability of the system will be $n_t g_l$ or $A_o = n_t g_l$, the excess availability in the presence of an activated cluster can be written from eq. (4.8) as

$$A - A_o = \Delta A = \frac{4}{3}\pi r^3 \rho_a \frac{N_a}{M}(g_a - g_l) + 4\pi r^2 \sigma \tag{4.9}$$

At the metastable equilibrium, the excess availability of the system with respect to the cluster radius will be maximum. The equilibrium or critical cluster radius, r_c, can be obtained by differentiating ΔA, given by eq. (4.9) with respect to radius, r, of the cluster and setting it equal to zero. If we do so, we obtain

$$\frac{d\Delta A}{dr} = \frac{4\pi r_c^2 \rho_a N_A}{M}(g_a - g_l) + 8\pi r_c \sigma = 0$$

Or critical radius, r_c is

$$r_c = \frac{2\sigma M}{\rho_a N_A (g_l - g_a)} \tag{4.10}$$

Substituting for $(g_l - g_a)$ from eq. (4.10) into eq. (4.9), the excess availability in terms of critical radius is obtained as

$$\Delta A = 4\pi r^2 \sigma \left(1 - \frac{2}{3}\frac{r}{r_c}\right) \tag{4.11}$$

or

$$\frac{\Delta A}{4\pi r_c^2 \sigma} = \left(\frac{r}{r_c}\right)^2 \left(1 - \frac{2}{3}\frac{r}{r_c}\right)$$

Figure 4.3 shows the dimensionless excess availability as a function of dimensionless cluster radius r/r_c. The maximum value of the excess availability ($r = r_c$) is

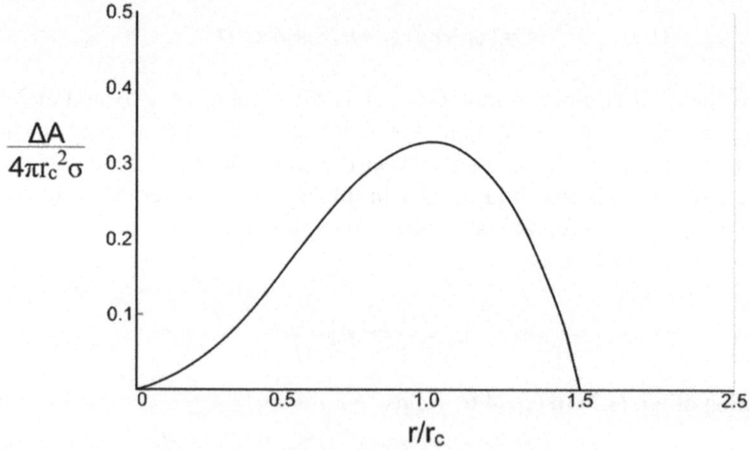

FIGURE 4.3 Excess availability as a function of activated cluster radius.

$$\frac{\Delta A_{max}}{4\pi r_c^2 \sigma} = \frac{1}{3} \tag{4.12}$$

or

$$\Delta A_{max} = \frac{4\pi r_c^2 \sigma}{3}$$

The rate at which activated clusters appear per unit volume of the liquid per unit time can be written as

$$J = fe^{-\left(\frac{\Delta A_{max}}{kT_l}\right)} \tag{4.13}$$

where f is the frequency of molecular collisions per unit volume, and k is Boltzmann's constant. In writing eq. (4.13), it is assumed that nucleation will occur when the excess availability of the cluster is several fold that of the kinetic energy of a molecule, which is characterized by kT_l. As such, the ratio $\Delta A_{max}/kT_l$ represents a dimensionless number rather than number of molecules. From kinetic theory, the frequency of molecular collisions per unit volume per unit time is written as

$$f = \frac{NkT_l}{h} \tag{4.14}$$

In eq. (4.14), N is the number of molecules per m^3, and h is Planck's constant. Substituting eq. (4.14) in eq. (4.13) and taking the logarithms of both sides give

$$\frac{\Delta A_{max}}{\kappa T_l} = \ln\left(\frac{NkT_l}{Jh}\right) \tag{4.15}$$

With ΔA_{max} given by eq. (4.12), a relation for critical cluster radius is obtained from eq. (4.15) as

$$r_c = \left[\frac{3kT_l}{4\pi\sigma}\ln\left(\frac{NkT_l}{Jh}\right)\right]^{1/2} \tag{4.16}$$

The critical radius as given by eq. (4.16) is not very sensitive to the value of J. Any variation in J from 10^{-6} to 10^{6} will yield critical radii, which will differ only by a few percent from each other. However, a number of investigators have attempted to study the effect of J on calculated superheat at different system pressures. In the above expression, the surface tension, σ, and the number density, N, are evaluated at the liquid temperature T_l.

Using the critical radius in the condition for the mechanical equilibrium, the pressure in the cluster of activated molecules, which now appears like a bubble of radius r_c, of the distinct phase can be written as

$$p_b - p_l = \frac{2\sigma}{\left[\dfrac{3kT_l}{4\pi\sigma}\ln\left(\dfrac{NkT_l}{Jh}\right)\right]^{1/2}} \tag{4.17}$$

Equation (4.17) has two unknowns p_b and T_l. However, thermodynamic equilibrium requires T_l, to be the saturation temperature corresponding to pressure p_b. Thus, employing the equation of state, both p_b and T_l can be determined.

Example:

Calculate the superheat necessary for homogeneous nucleation in water at 1 atm pressure.

Solution:

Let us carry out the calculations by assuming $J = 1/(\text{m}^3\text{s})$. We make a table by choosing different values of T_l and calculating p_b from eq. (4.17) and comparing it with that obtained from the Steam Tables when T_l is the saturation temperature corresponding to pressure, p_b. Surface tension and N are calculated at T_l

	p_b, (N/m²)	
T_l (K)	Equation (4.17)	Steam Table
520	27.3×10^6	3.78×10^6
560	12.77×10^6	7.13×10^6
572	8.46×10^6	8.46×10^6

Thus, calculated homogeneous nucleation temperature for water at 1 atm pressure is 572 K. This temperature is about 88% of the critical temperature. Or, water at 1 atm pressure can sustain a superheat of about 200 K before it spontaneously breaks up into two phases—liquid and vapor. Release of excess energy in the form of very rapid creation of vapor upon spontaneous nucleation can lead to mechanical damage in physical systems. This issue has been discussed in the literature, e.g., [5], extensively in connection with chemical and nuclear industry safety. It should be noted that the calculated superheat is the highest needed for homogeneous nucleation. Nucleation in the bulk can occur at much lower superheat in the presence of disturbances, gases dissolved in the host liquid and extraneous materials.

Several efforts have been made to experimentally determine the homogeneous nucleation temperature. These studies generally have utilized superheated liquid drops. Wakeshima and Takata [6] used a vertical column that housed the host liquid. By controlling the heating of the column wall, a stable temperature gradient in the vertical direction was established. As such, temperature of the host liquid at a vertical location was known. The requirements on the host liquid were: it had to be immiscible with the test liquid (droplet), denser than the test liquid, and have a boiling temperature much higher than the test liquid. For example, sulfuric acid was used with droplets of hydrocarbons, such as n-Pentane and n-Hexane. Droplets of the test liquid were introduced at the bottom of the container with the idea that their temperature will increase as they move up in the host liquid. The droplets were 20–200 μm in size. The droplets were expected to move upward slowly enough so that they acquired the local temperature of the host liquid. By noting the location at which droplets exploded as a result of spontaneous or homogeneous nucleation, and in turn the local host liquid temperature, the superheat at homogeneous nucleation could be determined. The measured superheats have been found to be in a reasonable agreement with those predicted by using eq. (4.17). To avoid existence of thermal nonequilibrium as the droplet moves upward in the host liquid because of buoyancy, Apfel [7] used acoustic standing wave to arrest the rising droplet at a particular location. By varying the magnitude of the acoustic pressure and location (temperature of the host liquid) at which the droplet exploded, superheat of the droplet could be determined. The homogeneous nucleation temperature measured in the experiments with n-Hexane droplets in glycerin as the host liquid was found to be in good agreement with the predictions obtained by employing eq. (4.17).

4.2 HETEROGENEOUS NUCLEATION

As mentioned earlier, heterogeneous nucleation is a process in which bubbles nucleate at imperfections on a heated surface submerged in liquid. These imperfections are the scratches, pits, and grooves commonly called cavities, which form on the surface during its preparation and/or manufacture. The wall superheats associated with heterogeneous nucleation are much lower than the liquid superheats for homogeneous nucleation. The early work of Bankoff [8] and Fisher [9] indicated that the surface superheats associated with heterogeneous nucleation are also much lower than those obtained in experiments, in which the test surface was submerged in a pool of liquid maintained at a constant temperature while the system pressure was gradually decreased until a bubble nucleated at the surface. In these experiments, the liquid superheat corresponding to the system pressure at which the first bubble nucleated was noted.

The reason for lower inception superheat on a heated surface can be easily rationalized if we consider that the cavities generally trap vapor, air, or other noncondensables and have radii that are much larger than critical cluster radius given by eq. (4.16). As we will see shortly, the gas-trapping ability of a cavity depends on the surface tension, the contact angle, and the shape of the cavity. With heating or reduction in system pressure, the gas expands and pushes the interface outward. During this process, the radius of curvature decreases and is smallest when the bubble nucleus just covers the mouth of the cavity ($\phi = 90°$). As the bubble grows further, the radius of curvature will begin to increase. Figure 4.4 shows conceptually the evolving interface. Since the liquid superheat required for thermal equilibrium increases with reduction in the radius of a vapor bubble as given by eq. (4.4), the highest superheat will be needed when the bubble radius is equal to the cavity radius. Thus, the superheat at bubble inception will be controlled by the size of the mouth of the cavity. For commercial surfaces, the size of the cavities is

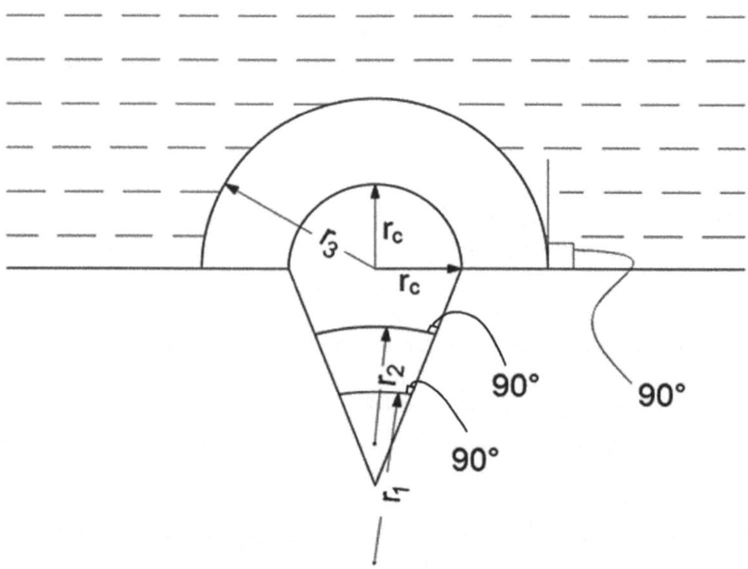

FIGURE 4.4 Conceptualized bubble inception in a nonuniform temperature field ($\varphi = 90°$).

several orders of magnitudes larger than the critical cluster radius, hence lower is the required superheat for inception. The precise value of the nucleation superheat strongly depends on the availability of unflooded cavities with trapped gas/vapor. As these cavities become fewer and fewer and their size decreases, the observed nucleation temperature will increase.

The trapped gas not only promotes nucleation by providing bubble embryos, but it also reduces the vapor pressure in the bubble. Using Dalton's law of partial pressures, the pressure in the bubble nucleus can be written as

$$p_b = p_v + p_g \tag{4.18}$$

Since the vapor pressure difference across the interface is reduced by p_g, the liquid superheat from thermodynamic/thermal equilibrium consideration is also reduced accordingly, as such

$$T_l - T_{\text{sat}} = \frac{\left(\dfrac{2\sigma}{R} - p_g\right)}{\rho_v h_{fg}} T_{\text{sat}} \tag{4.19}$$

Next let us look at the conditions under which cavities will trap gas. Figure 4.5 shows the advance of liquid and gas fronts over wedge-shaped poorly wetted cavities (actually grooves) as originally envisioned by Bankoff [8]. If the contact angle, φ, is larger than the cavity angle, β, the advancing liquid front will fill the upper portion of the cavity while trapping gas underneath. Such a cavity will trigger early and should serve as a favored nucleation site. An advancing gas or vapor front, however, will not

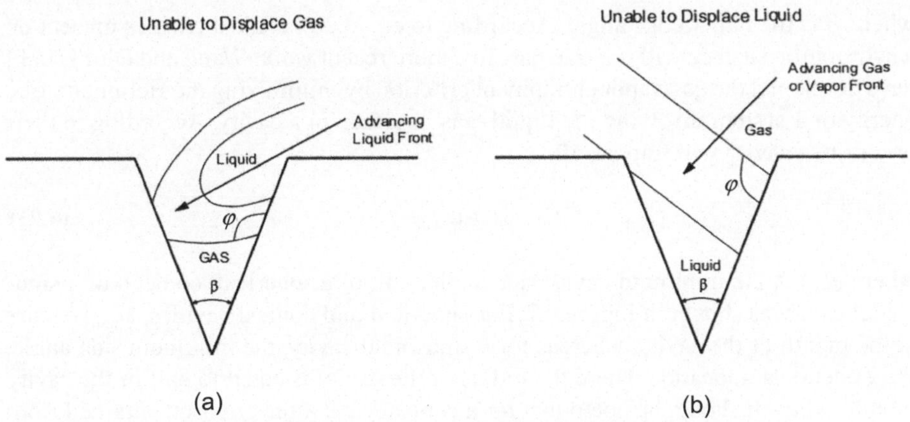

FIGURE 4.5 Vapor–gas-trapping behavior of cavities. (a) If $\beta < \varphi$, the advancing liquid front will hit the further wall of the cavity before it flows completely. (b) If $\pi - \varphi > \beta$, the advancing gas front will strike the farther wall of the cavity before it has advanced completely down the nearer wall. (Adapted from Cole [4].)

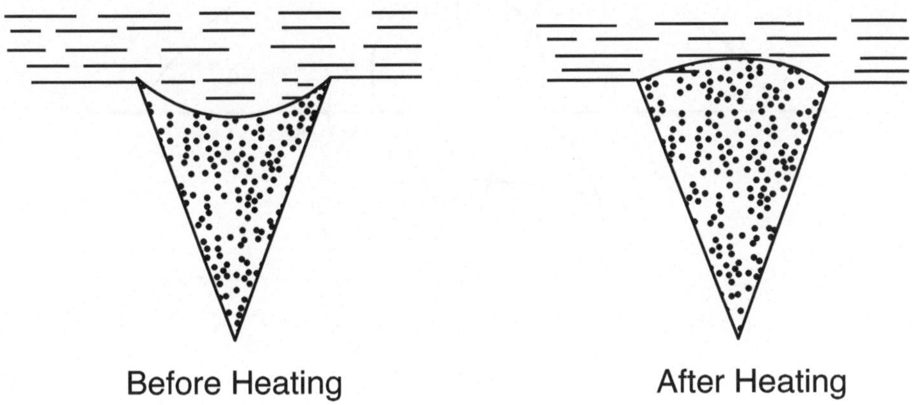

FIGURE 4.6 Liquid–vapor–gas interface shape before and after heating.

be able to displace the liquid if $\pi - \varphi > \beta$. A liquid filled cavity will not be conducive to nucleation. Figure 4.6 shows the interface shape before and after heating of the surface.

Subsequently, Ward and Forest [10] have analyzed the relation between platelet adhesion and roughness of a synthetic material and have shown that the gas/vapor nucleus in a long narrow fissure is stable if:

$$\varphi > \frac{\pi}{2} + \beta \tag{4.20}$$

where β is the half wedge angle. According to eq. (4.20), conical cavities present on a hydrophilic surface will not trap gas. In a more recent work, Wang and Dhir [11a,b] have developed the gas/vapor entrapment criterion by minimizing the Helmholtz free energy of a system involving the liquid–gas interface in a cavity. According to their criterion, a cavity will trap gas if:

$$\varphi > \psi_{min} \tag{4.21}$$

where ψ_{min} is the minimum cavity side angle, 4m, of a spherical, conical, or a sinusoidal cavity as shown in Figure 4.7. For spherical and conical cavities, ψ_{min}, occurs at the mouth of the cavity, whereas for a sinusoidal cavity, the minimum side angle, ψ_{min}, occurs at a location where the radius of the cavity is equal to half of the cavity mouth radius. It should be noted that for a conical cavity, the criterion obtained from eq. (4.21) is the same as that given by eq. (4.20) of Ward and Forest, and liquids with a contact angle less than 90° will not provide a stable gas/vapor embryo for nucleation on a conical cavity.

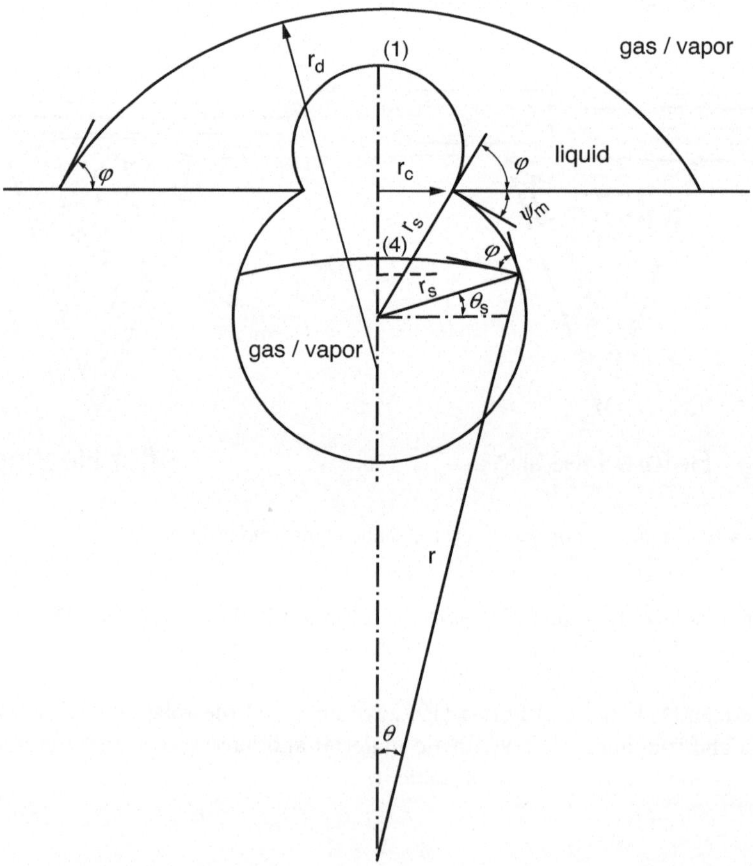

FIGURE 4.7 Liquid front in a spherical cavity [11a].

According to eq. (4.20) or (4.21), hardly any preexisting gas/vapor nuclei in conical cavities are possible for well wetting liquids such as R-113 and FC-72. For these liquids, only spherical cavities with mouth angle smaller than the contact angle, if present, can potentially nucleate. Lack of presence of cavities that trap gas/vapor leads to delayed onset of nucleation (higher superheats) with these liquids. For these liquids, once nucleation occurs, rate of heat transfer to the fluid improves, which in turn leads to reduction of wall superheat at a fixed heat flux. This behavior does not occur if cavities are already active and leads to hysteresis in the boiling curve. It will be discussed further in Chapter 5.

Aside from reduction in wall superheat at nucleation, as we can deduce from eq. (4.19), trapped gas in the defects on the surface plays an important role in initiation of nucleation on a heater surface by providing nuclei for bubble evolution. Because of variation in sizes and shapes of cavities and their gas-trapping ability, number of cavities that become active can differ substantially from surface to surface. Also, experimental conditions especially gas dissolved in the liquid and duration of time a cavity is exposed to liquid that has been degassed before nucleation is attempted can affect the onset nucleation and the number of cavities that nucleate at a given superheat. Consequently, a large scatter in data from different investigations is observed.

4.2.1 Inception Superheat

A number of efforts have been made in the past to predict the incipient superheat for boiling to occur on preexisting gas–vapor-trapping cavities. Griffith and Wallis [12] proposed that superheat at which a cavity nucleates corresponds to the minimum radius of curvature of the vapor–gas/liquid interface. The minimum radius of the interface was assumed to be the radius of the mouth of the cavity. Since pressure in the vapor pocket is higher than that in the liquid by $2\sigma/R$, vapor (neglecting presence of gas) temperature should at least be equal to the saturation temperature corresponding to the pressure in the vapor. Using Clausius–Clapeyron equation, an expression for the well superheat (vapor saturation temperature minus saturation temperature corresponding to pressure in the liquid) at which a cavity of mouth diameter, D_c would nucleate is obtained from eq. (4.15) as:

$$\Delta T_{\text{in}} = \frac{4\sigma T_{\text{sat}}}{\rho_v h_{fg} D_c} \tag{4.22}$$

In eq. (4.22), ΔT_{in} is the difference in wall temperature at nucleation and saturation temperature of the liquid at the imposed pressure. It should be noted that in writing eq. (4.22), no gas in the bubble embryo is assumed to be present. Furthermore, eq. (4.22) neither includes the effect on inception superheat of contact angle nor temperature gradient that exists near a heated surface.

Hsu [13] studied the effect of temperature gradient adjacent to the heated surface on the minimum superheat needed for nucleation. In developing his model, Hsu proposed that liquid at the tip of the bubble embryo must be at saturation temperature corresponding to the pressure in the bubble. Since pressure in the embryo is higher than the liquid by $2\sigma/R$, as discussed earlier, the liquid must be superheated. Because heat is transferred from the wall, liquid temperature decreases with distance from the wall, and the above criterion once satisfied at the tip of the embryo will be satisfied everywhere around the boundary of the embryo. If the liquid temperature at the tip of the embryo is less than the saturation temperature corresponding to the pressure in the embryo, heat transfer to the colder liquid at the top can cause embryo to shrink and eventually collapse. Invariably, it has also been argued that it is the net energy balance around the embryo that determines the superheat at which embryo leads to an active nucleation site.

For prediction of heterogeneous nucleation, the criterion developed by Hsu [13] has been found to be quite successful. This criterion does a fairly good job in predicting the wall superheat necessary for nucleation as will be shown later.

Figure 4.8a shows qualitatively the bubble embryo formed on the heated surface and temperature profile in the thermal layer adjacent to the wall, whereas Figure 4.8b shows as to how Hsu's criterion could be satisfied for a bubble embryo when the wall heat flux is gradually increased while keeping the thermal layer thickness fixed. In plotting the temperature, a fictitious thermal layer of thickness, δ, is defined with a linear temperature profile, which gives the same heat flux at the wall as the actual thermal layer, δ_{th}. Also, it is assumed that δ is independent of the temperature difference across the layer. During natural convection, this is not entirely true since in natural convection, the thermal layer thickness weakly depends on temperature difference, $T_w - T_{\text{sat}}$. The temperature profile traced by the solid line in Figure 4.8b indicates that for a given δ, the wall heat flux or wall temperature at which Hsu's criteria are just satisfied. If the cavity for which the criterion is satisfied is available on the surface (i.e., is not completely filled with liquid), this cavity will be the first one to nucleate. The lowest superheat at which nucleation is possible can be determined from the expression for the temperature distribution in the film and the

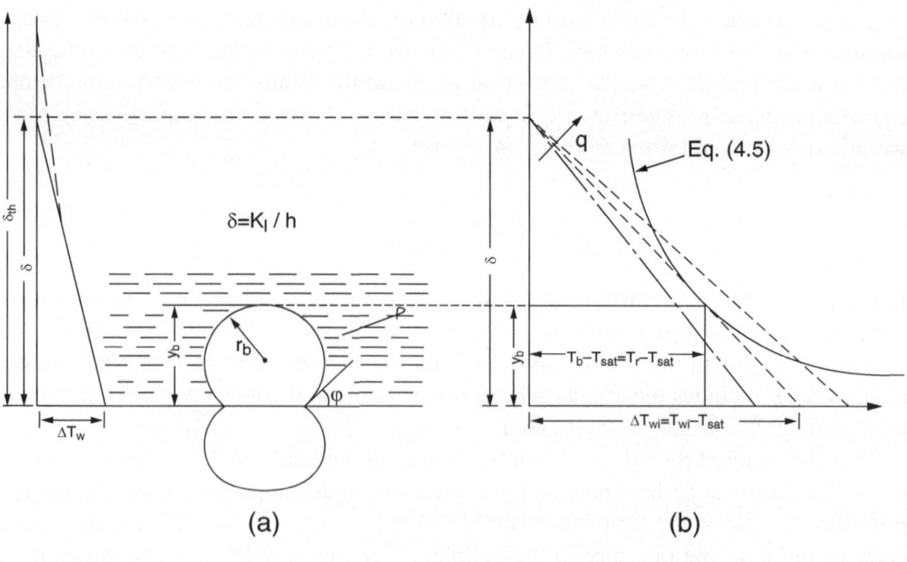

FIGURE 4.8 Inception of nucleation in a nonuniform temperature field (a and b).

liquid temperature dictated by the equilibrium considerations. If the bulk liquid is saturated, the temperature profile in the liquid film is given by

$$T_l - T_{sat} = (T_w - T_{sat})\left(1 - \frac{y}{\delta}\right) \tag{4.23}$$

The vapor bubble thermal equilibrium condition with liquid eq. (4.5) is:

$$T_l - T_{sat} = \frac{2\sigma T_{sat}}{r_b \rho_v h_{fg}} \tag{4.5}$$

For a contact angle, φ, of $\pi/2$, the height of the bubble and the radius of the bubble embryo will be equal to the radius of the mouth of the cavity. For contact angles different than $\pi/2$, the radius and height of a spherical embryo are related to the cavity radius as

$$r_b = \frac{r_c}{C_1} \quad \text{where } C_1 = \sin\varphi \tag{4.24}$$

$$y_b = \frac{r_c C_2}{C_1} \quad \text{where } C_2 = 1 + \cos\varphi \tag{4.25}$$

Differentiating the liquid temperature in eq. (4.23) while replacing r_b by y_b in eq. (4.5) with respect to distance and equating the two derivatives at the tangent condition yields

$$\frac{2C_2\sigma T_{sat}}{y^2 \rho_v h_{fg}} = \frac{(T_w - T_{sat})}{\delta} \tag{4.26}$$

In terms of the radius of the nucleating cavity corresponding to the minimum wall superheat, eq. (4.26) can be written as

$$r_c = \left[\frac{2C_1^2 \sigma T_{sat} \delta}{C_2 \rho_v h_{fg} (T_w - T_{sat})}\right]^{1/2} \tag{4.27}$$

If we replace the wall superheat with wall heat flux as in most applications heat flux is controlled, the expression for r_c becomes

$$r_c = \left[\frac{2C_1^2 \sigma T_{sat} k_l}{C_2 \rho_v h_{fg} q}\right]^{1/2} \tag{4.28}$$

Substituting for r_b in eq. (4.5) by evaluating r_c from eq. (4.28) and using eq. (4.24), the required liquid superheat at the tip of the bubble is

$$(T_l - T_{sat}) = \left[\frac{2C_2 \sigma T_{sat} q}{\rho_v h_{fg} k_l}\right]^{1/2} \tag{4.29}$$

The wall superheat at nucleation is obtained from eq. (4.23) after substituting for $(T_l - T_{sat})$ from eq. (4.29) and y_b from eq. (4.25) while obtaining r_c from eq. (4.28) as

$$(T_w - T_{\text{sat}}) = \left[\frac{8C_2\sigma T_{\text{sat}} q}{\rho_v h_{fg} k_l}\right]^{1/2} \tag{4.30}$$

We find from eq. (4.29) and (4.30) that wall superheat at nucleation is twice the liquid superheat at the tip of the bubble embryo. For a contact angle of $\pi/2$, both C_1 and C_2 are unity or the cavity radius is half of the film thickness.

Example:

Calculate the cavity size and the minimum wall superheat for nucleation from a surface submerged in a pool of saturated water at 1 atm pressure. The wall heat flux is 10^5 w/m^2. Assume the contact angle to be $\pi/2$.

Solution:

From eq. (4.28), the cavity radius is calculated as

$$r_c = \left[\frac{2(58.7)10^{-3}(373.15)(0.681)}{(0.597)(2.257)10^6\left(10^5\right)}\right]^{1/2} = 14.88 \times 10^{-6} \text{ m}$$

The minimum wall superheat for nucleation is found from eq. (4.30) as

$$(T_w - T_{\text{sat}}) = \left[\frac{8(58.7)10^{-3}(373.15)\left(10^5\right)}{(0.597)(2.257)10^6(0.681)}\right]^{1/2} = 4.36°\text{C}$$

The cavity size obtained here is fairly large. It is possible that such a cavity will be filled with a liquid or may not be available to serve as a nucleation site. For smaller cavities to nucleate, heat flux and wall superheat will have to be higher. It is more informative to note that the size of the nucleating cavity is about three orders of magnitude larger than the equilibrium cluster size for homogeneous nucleation. One should also note that this superheat may not be consistent with natural convection existing on the wall prior to nucleation.

Next, let us determine the range of cavities that can nucleate if the wall superheat exceeds the minimum wall superheat for nucleation. Eliminating the liquid temperature from eqs. (4.5) to (4.23) and substituting for bubble radius and bubble height from eq. (4.24) to (4.25) respectively, we obtain

$$(T_w - T_{\text{sat}})\left(1 - \frac{C_2 r_c}{C_1 \delta}\right) = \frac{2C_1\sigma T_{\text{sat}}}{\rho_v h_{fg} r_c} \tag{4.31}$$

or

$$r_c^2 - \frac{C_1}{C_2}\delta r_c + \frac{2C_1^2\sigma T_{sat}\delta}{C_2\rho_v h_{fg}(T_w - T_{sat})} = 0 \quad (4.32)$$

This is a quadratic equation in r_c, the two roots of the equation are found to be

$$r_{c1,2} = \frac{C_1\delta}{2C_2}\left[1 \pm \sqrt{1 - \frac{8C_2\sigma T_{sat}}{\rho_v h_{fg}(T_w - T_{sat})\delta}}\right] \quad (4.33)$$

In Figure 4.9, the intercepts of the straight line representing temperature distribution in the thermal layer with the required liquid superheat predicted from eq. (4.5) represent the range of cavity radii, r_{c1} and r_{c2}, corresponding to the upper and lower limits from eq. (4.33), for which cavities can nucleate if these cavities are available and are not flooded. Now the corresponding wall superheat is higher than the minimum wall superheat given by eq. (4.30). Limited experimental verification of eq. (4.33) of Hsu has been reported by Hsu [13]. In terms of the heat flux, eq. (4.32) can be recast as:

$$q = \frac{C_1 k_l \Delta T}{r_c C_2} - \frac{2C_1^2\sigma T_{sat} k_l}{C_2\rho_v h_{fg} r_c^2} \quad (4.34)$$

For a given r_c, eq. (4.34) represents an equation in two unknowns, namely q and ΔT. To solve explicitly for both, another relation between q and ΔT is

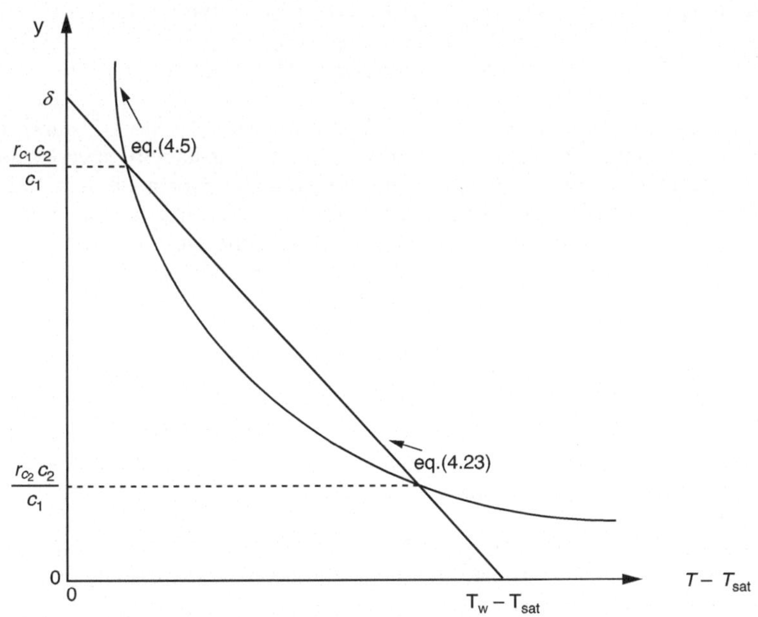

FIGURE 4.9 Range of nucleating cavities.

needed. This relation is provided by the correlations for single-phase natural convection or forced convection that may be present on the heater surface prior to inception. In terms of the wall superheat at inception, eq. (4.31) can also be written as

$$\Delta T = \frac{\dfrac{2C_1\sigma T_{\text{sat}}}{\rho_v h_{fg} r_c}}{\left[1 - \dfrac{C_2 r_c}{C_1 \delta}\right]} \tag{4.35}$$

For r_c much smaller than δ, the denominator in eq. (4.35) is close to unity and the size of the nucleating cavity is found to vary inversely with wall superheat as was the case for eq. (4.22). If the heated surface is subjected to forced flow condition, the thermal layer will thin with flow velocity. This in turn will reduce the film thickness, δ. As can be seen from eq. (4.33), the radii of the nucleating cavities will shrink when δ is reduced. We know from the earlier discussions that smaller cavities require higher superheat or suppress the nucleation process. Another point that should be made here is that in developing flows, the thermal layer thickness is a function of distance from inlet or the leading edge. For example, for laminar flow over a flat plate, the thermal layer thickness increases as square root of distance from the leading edge. Therefore, the farther the location from the leading edge, the larger will be the size of the nucleating cavities and smaller the required wall superheat. Theoretically, the effect of flow velocity and distance from the leading edge or inlet could be included in the evaluation of thermal layer thickness. However, as mentioned earlier, the predictions will be sensitive to the size and shape of the available cavities on the surface and the contact angle. A number of modifications to Hsu's criterion have been proposed by several investigators such as Han and Griffith [14] and Howell and Siegel [15].

As an alternative to Hsu's criterion, Mizukami [16], Nishio [17], and more recently, Wang and Dhir [11a], have proposed that the instability of vapor nuclei in a cavity determines the inception superheat. The vapor nucleus is stable if the curvature of the interface increases with an increase in vapor volume or if it decreases with a decrease in vapor volume. Otherwise, the vapor bubble embryo or nucleus is unstable. If K is the nondimensional modified curvature of the interface, the stability criterion is written as

$$\frac{dK}{dV^*} > 0 \tag{4.36}$$

where V^* is the dimensionless volume of the embryo. The minimum wall superheat required for nucleation will correspond to the value of K for which the interface is unstable.

For a spherical cavity, Wang and Dhir [11a] found that outside of the cavity, maximum value of K corresponds to the condition at which instability of the interface occurs. From geometric considerations, they found that

$$\begin{aligned} K_{max} &= 1 \quad \text{for} \quad \varphi \le 90^\circ \\ &= \sin(\varphi) \quad \text{for} \quad \varphi \ge 90^\circ \end{aligned} \tag{4.37}$$

As such, the incipient wall superheat is obtained as

$$\Delta T = \frac{2\sigma T_{sat}}{\rho_v h_{fg} r_c} K_{max} \tag{4.38}$$

The above expression of wall superheat at nucleation is obtained under the assumption that cavity radius, r_c, is much smaller than the thermal layer thickness, δ. Thus, in this limit, eq. (4.38) suggests that in eq. (4.35) C_1 is unity for $\varphi \le 90°$ and is equal to $\sin\varphi$ for $\varphi \ge 90°$. Or, for non-wetted surfaces, the required superheat is smaller than that given by eq. (4.38) for partially wetted surfaces. Wang and Dhir [11b], from their experiments on surfaces with contact angles less than 90°, have shown the general validity of eq. (4.38).

4.2.2 Number Density of Active Nucleation Sites

As wall superheat is increased, according to eq. (4.38), smaller cavities present on the surface can become active. Thus, with increase in wall superheat number density of active nucleation sites will increase as smaller cavities are added. Several parameters influence the number of cavities that become active at a given wall superheat. These are size and shape of defects that are present in the surface; gas/vapor trapped in the cavities; the extent of dissolved gas present in the host liquid; the duration for which the surface has been exposed to the host liquid prior to experiments and wettability of the surface. Because of the difficulties in control and specification of these parameters in experiments, large variations in data reported in literature have been found. In the very early studies reported by Hsu and Graham [18], it was noted that density of active nucleation sites increases approximately as the square of the heat flux or as the fourth to sixth power of wall superheat. In these studies, no attempt was made to relate either the proportionality constant or the exponent with the surface characteristics or wettability.

Kocamustafagoullari and Ishii [19] have correlated the cumulative active nucleation site density, N_a, reported by various investigators for water boiling on a variety of surfaces at pressures varying from 1 to 198 atm, as

$$N_a^* = \left[D_c^{*-4.4} F(\rho^*)\right]^{1/4.4} \tag{4.39}$$

where

$$N_a^* = N_a D_d^2; \; D_c^* = \frac{D_c}{D_d}$$

and

$$F(\rho^*) = 2.157 \times 10^{-7} \rho^{*-3.2} (1 + 0.0049\rho^*)^{4.13} \tag{4.40}$$

In the above equation, D_c is the cavity diameter, and D_d is the bubble departure diameter, which was obtained by multiplying Fritz's [20] expression for bubble departure diameter, which we will discuss later, as:

$$D_d = 0.0208\varphi\sqrt{\frac{\sigma}{g(\rho_l - \rho_v)}} \tag{4.41}$$

with $0.0012\left(\rho^*\right)^{0.9}$

In eq. (4.41), contact angle φ is measured in degrees. The parameters ρ^* and D_c in eqs. (4.39) and (4.40) are defined as

$$\rho^* = \frac{\rho_l - \rho_v}{\rho_v} \tag{4.42}$$

$$D_c = \frac{4\sigma\left[1+\left(\frac{\rho_l}{\rho_v}\right)\right]}{p_l}\left\{\exp\left(\frac{h_{fg}(T_v - T_{\text{sat}})}{R_v T_v T_{\text{sat}}}\right)-1\right\} \tag{4.43}$$

In eq. (4.43), T_v is the temperature of vapor, p_l is the liquid pressure, and R_v is the gas constant. At low and moderate pressures, eq. (4.43) reduces to eq. (4.35), when cavity diameter is assumed to be much smaller than the thickness of the thermal layer.

Cornwell and Brown [21] have made a systematic study of active nucleation sites on copper surfaces during boiling of water at 1 atm pressure. Their study was limited to low heat fluxes and to surface conditions ranging from a smooth to a scratched rough surface. From their work, it was concluded that the active site density varied with wall superheat as

$$N_a \sim \Delta T^{4.5} \tag{4.44}$$

The proportionality constant in eq. (4.44) was found to increase with surface roughness, but the exponent on ΔT was independent of surface roughness. Mikic and Rohsenow [22] have proposed that on commercial surfaces, the cumulative number of active sites per unit heater area can be assumed to vary in partial nucleate boiling as

$$N_a \sim \left(\frac{D_s}{D_c}\right)^m \tag{4.45}$$

where D_s is the diameter of the largest active cavity present on the surface, and m is an empirical constant. The size, D_c, of a cavity that nucleates at a wall superheat ΔT is obtained from eq. (4.22). Bier et al. [23], on the other hand, have deduced an expression for active site density, N_a, from heat transfer data as

$$\ln(N_a) = \ln(N_{\max})\left[1-\left(\frac{D_c}{D_s}\right)^m\right] \tag{4.46}$$

In eq. (4.46), $N_{\max}$ is the maximum value of N_a, which occurs at $D_c = 0$. The value of m was found to depend on the manner in which a surface was prepared.

With Freon 113 or Freon 11 boiling on a chemically etched copper surface and on a turned surface, values of 0.42 and 0.26, respectively, have been noted for m in eq. (4.46). This observation is not consistent with the conclusion of Cornwell and Brown [21] as described earlier. In the heat transfer experiments, the reduced pressure (system pressure/critical pressure) was varied from 0.0037 to 0.9. It was found that to correlate the data at low and high saturation pressures, some changes in the functional form of eq. (4.46) are necessary.

In order to mechanistically predict the number density of active nucleation sites, one needs to know the size distribution and shape of the defects present on the surface as well as the contact angle so that their gas/vapor trapping ability can be determined. Wang and Dhir [11a,b] essentially did that. By successfully removing a small amount of material through polishing of the surface and taking micrographs of the surface, they found the number density of all types of cavities present on the polished copper surface to correlate as

$$N_s\left(\frac{\text{sites}}{\text{cm}^2}\right) = 9.0\times10^3 D_c^{-2.0} \quad D_c \geq 5.8\ \mu\text{m}$$

$$= 10.3 + 2.4\times10^6 D_c^{-5.2} \quad 3.5 \leq D_c \leq 5.8\ \mu\text{m}$$

$$= 2214 + 1.0\times10^6 D_c^{-5.4} \quad D_c \leq 3.5\ \mu\text{m} \tag{4.47}$$

In eq. (4.47), D_c is measured in μm.

From these expressions, the cumulative number density of reservoir-type cavities (these were cavities that were proposed to nucleate) with mouth angle, ψ_m, less than 90° was correlated as

$$N_{as}\left(\frac{sites}{cm^2}\right) = 5.8\times10^6 D_c^{-5.4} \tag{4.48}$$

Figure 4.7 given earlier shows a spherical cavity with a mouth angle ψ_m. Through analysis, they showed that for cavities to trap gas/vapor, the cavity mouth angle ψ_m should be less than the contact angle. As such, the cumulative number density of cavities that could become active at a given superheat for a particular contact angle, φ, is given as

$$N_a(\varphi, \Delta T) = \frac{1-\cos(\varphi)}{1-\cos(\psi_m^\circ)}\frac{N_{as}(\psi_m^\circ)}{K_{\max}} \tag{4.49}$$

where ψ_m° is the maximum reference mouth angle of the spherical cavities present on the surface. K_{max} is given by eq. (4.37). According to eq. (4.49), when eq. (4.38) is used to relate wall superheat to cavity diameter, the active nucleation site density should vary as as $\Delta T^{5.4}$. Figure 4.10 shows a comparison of the measured and predicted number density of active sites for different contact angles. It is found that on similar surfaces, as contact angle decreases from 90° to 18°, number density of active cavities is reduced by almost a factor of 25.

More recently, Basu et al. [24] have correlated number density of active nucleation sites as a function of wall superheat and contact angle for water at 1 atm pressure as

$$N_a = 0.34(1-\cos\varphi)\Delta T^2 \ \frac{\text{sites}}{\text{cm}^2} \qquad \Delta T < 15°\text{C}$$

$$N_a = 3.4\times10^{-5}(1-\cos\varphi)\Delta T^{5.3} \ \frac{\text{sites}}{\text{cm}^2} \qquad \Delta T \geq 15\ °\text{C} \tag{4.50}$$

It should be pointed out that Pesse et al. [25] have observed that unfilled cavities can get filled with liquid when exposed to degassed host liquid over a long period of time. As such, significant variations in the available gas-trapping cavities can occur depending on the experimental conditions including duration of exposure of the surface to the host liquid and amount of dissolved gas in the liquid. Invariably, the procedure used for surface preparation, wetting characteristics, and other experimental conditions as noted earlier can lead to huge differences in the reported density of active nucleation sites.

4.3 BUBBLE DYNAMICS

The life span of a bubble on an active nucleation site can be divided into two periods: the waiting period, following the departure of a preceding bubble, during which conditions conducive to growth of an embryo are created near the heated surface; and the growth period during which the bubble actually grows to a certain size before departure. With increase in nucleation site density, bubbles may merge laterally before departure. This leads to formation of mushroom-type bubbles with vapor stems attached to the surface and a liquid bridge in between. As the frequency of bubble release from a given site increases, bubble mergers occur normal to the surface as well. We will discuss this further in Chapter 5.

4.3.1 Waiting Period

After inception, a bubble will grow until it encounters cold liquid at the top of the bubble. If the condensation rate at the top of the bubble exceeds the evaporation rate at the base of the bubble, it will shrink. If condensation rate is less than the evaporation rate, the bubble will grow and depart upon reaching a certain size. During the growth period, the bubble will push the surrounding liquid outward causing some mixing of the cold liquid with the warm liquid in the thermal layer next to the heater surface.

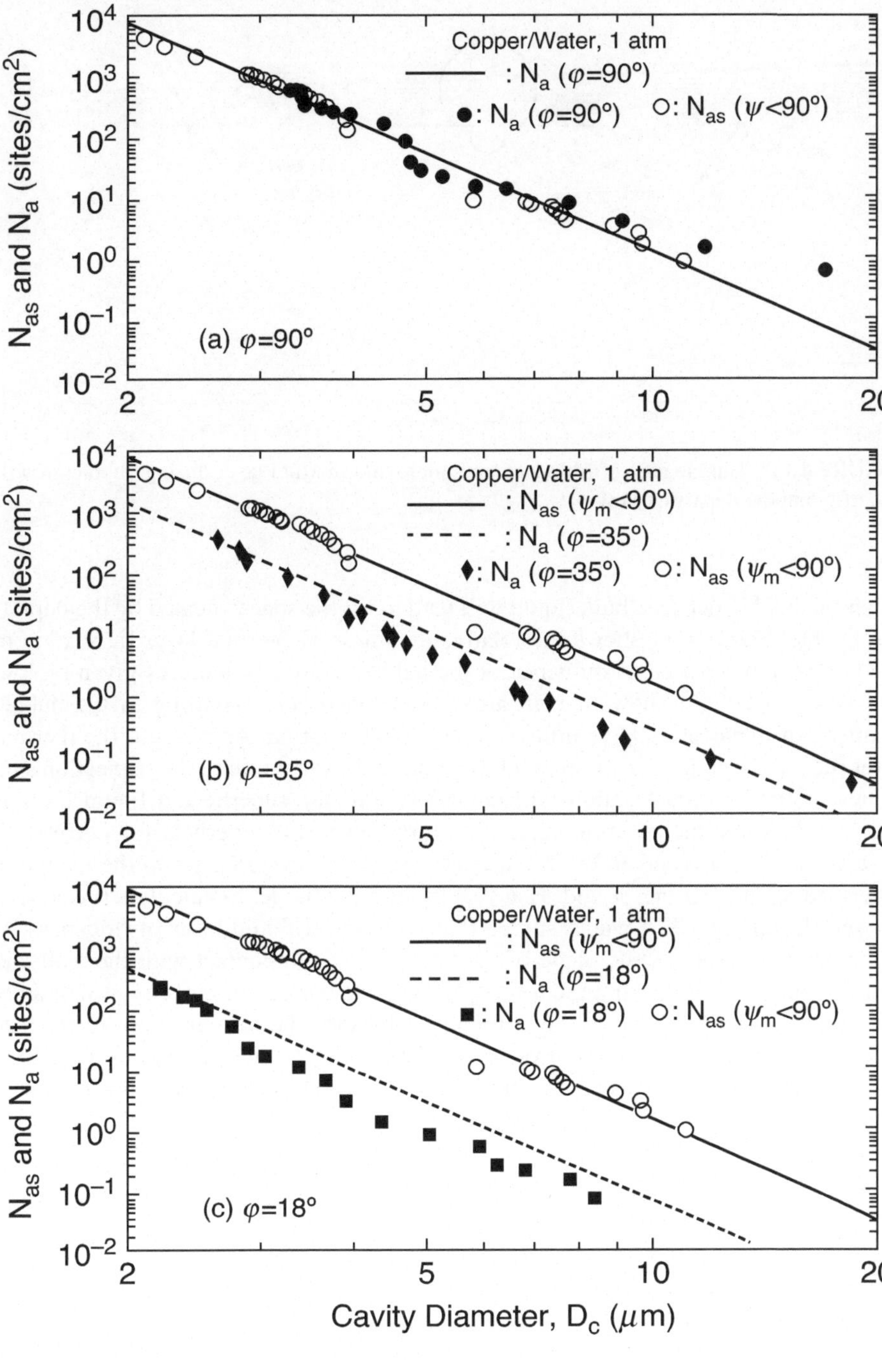

FIGURE 4.10 Number density of cavities present on the surface and those that became active for a given contact angle [11b].

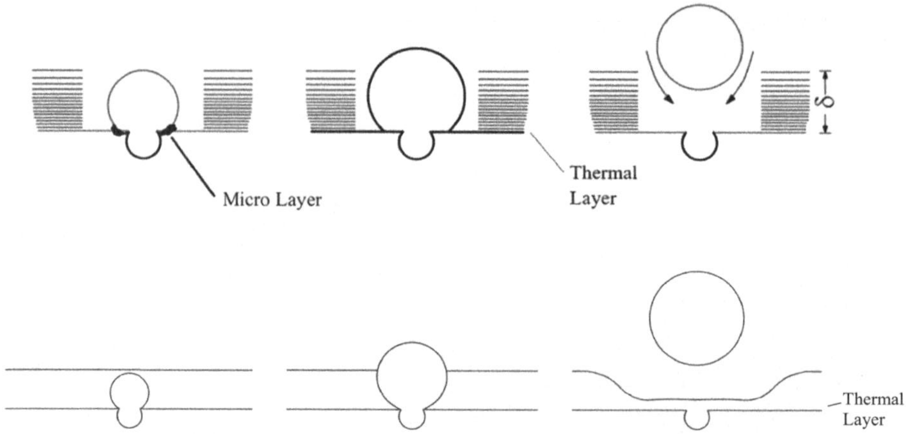

FIGURE 4.11 Bubble area of influence and micro and thermal layer during bubble growth and after bubble departure [18].

After the bubble departs, bulk liquid will tend to fill the space vacated by the departing bubble. Figure 4.11 shows the relative thickness of thermal layer in relation to the bubble and the area of influence of the bubble. Schlieren pictures given by Hsu and Graham [18,26] show that an area about twice corresponding to the bubble diameter of departure, D_d, is influenced by bubble motion. As a result, the thermal layer reforms over an area of circle of diameter $2D_d$, surrounding the nucleation site. A new bubble at this location will not grow until the superheated liquid layer is reestablished and the inception criterion (minimum wall superheat for nucleation) is satisfied. The time taken by the thermal layer to redevelop prior to the inception is termed as the waiting period. Hsu [13] proposed a model to calculate this period as well. In his model, it was assumed that an ambient liquid layer of thickness, δ, extending to infinity in the plane of the heater comes in contact with the wall and is transiently heated. During the heating period, the temperature of the wall and the final thickness of the thermal layer remain constant. The former is not really true when heat flux to the wall is controlled and thermal layer thickness varies with time.

The transient heat diffusion equation for an infinite slab of thickness δ with appropriate initial and boundary conditions can be written as

$$\frac{\partial^2 T}{\partial y^2} = \frac{1}{\alpha_l}\frac{\partial T}{\partial t} \tag{4.51}$$

The initial and boundary conditions are:

Initial Condition:

$$T = T_{\text{sat}} \quad \text{at} \quad t = 0 \quad \text{for all } y \tag{4.52}$$

Boundary Conditions:

$$T = T_w \quad \text{at} \quad y = 0 \quad \text{for all } t \geq 0 \tag{4.53}$$

and

$$T = T_{\text{sat}} \quad \text{at} \quad y = \delta \quad \text{for all } t \geq 0 \tag{4.54}$$

The solution of the problem can easily be written as

$$\frac{T - T_{\text{sat}}}{T_w - T_{\text{sat}}} = \left(1 - \frac{y}{\delta}\right) + \frac{2}{\pi} \sum_{n=1}^{\infty} \frac{(-1)^n}{n} \sin\left[n\pi\left(1 - \frac{y}{\delta}\right)\right] e^{-n^2\pi^2\tau} \tag{4.55}$$

where the dimensionless time, τ, is defined as

$$\tau = \frac{\alpha_l t}{\delta^2} \tag{4.56}$$

The dimensionless temperature profiles obtained from eq. (4.55) for various times, τ, are plotted in Figure 4.12. In this figure, the required liquid superheat determined from eq. (4.5) is also plotted. The intersection of this curve with the temperature profiles gives the waiting time for a particular size cavity (y/δ). Since smaller cavities require higher superheat, waiting time will increase as cavity diameter decreases. However, when cavities are large, the required wall superheat is small, but the thermal layer takes longer to develop and, as such, waiting time increases. It is noted from Figure 4.14 that for a given waiting time, a range of cavities can nucleate.

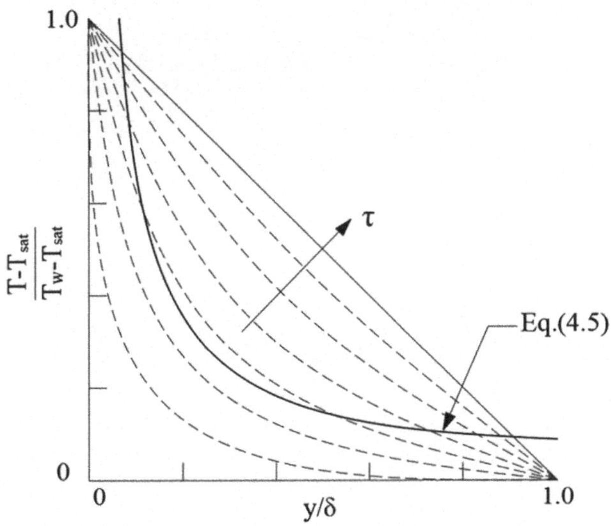

FIGURE 4.12 Waiting period for activation of various size cavities.

As can be seen in Figure 4.14, for a given wall superheat, certain very small and very large cavities may never nucleate.

An analytical expression for the waiting time can be obtained if the cold liquid layer contacting the heated wall is assumed to be semi-infinite. This is exactly the approach that was taken by Han and Griffith [14]. During transient diffusion of heat in a semi-infinite slab, the governing equation, the initial and the first boundary condition, will be the same as we had earlier for a thermal layer of thickness δ. However, the second boundary condition is

$$T = T_{\text{sat}} \quad \text{as} \quad y \to \infty \quad \text{for all } t \geq 0 \tag{4.57}$$

Now the solution will be

$$\frac{T - T_{\text{sat}}}{T_w - T_{\text{sat}}} = erfc\left[\frac{y}{2\sqrt{\alpha_l t}}\right] \tag{4.58}$$

where

$$erfc(\eta) = 1 - \frac{2}{\sqrt{\pi}} \int_0^{\eta} e^{-\eta^2} d\eta \tag{4.59}$$

and

$$\eta = \frac{y}{2\sqrt{\alpha_l t}}$$

Wall heat flux from eq. (4.58) is obtained as

$$q_w = -k \left.\frac{\partial T}{\partial y}\right|_{y=0} = k_l \frac{2}{\sqrt{\pi}} \frac{T_w - T_{\text{sat}}}{2\sqrt{\alpha_l t}} \tag{4.60}$$

we can also write

$$q_w = \frac{k_l (T_w - T_{\text{sat}})}{\delta} \tag{4.61}$$

or

$$\delta = \sqrt{\pi \alpha_l t} \tag{4.62}$$

Elimination of liquid temperature between eqs. (4.23) and (4.5) and upon substituting for bubble height and bubble radius in terms of cavity size, we obtain

$$\frac{C_2 r_c}{C_1 \delta} = \left[1 - \frac{2C_1 \sigma T_{\text{sat}}}{(T_w - T_{\text{sat}}) \rho_v h_{fg} r_c}\right] \tag{4.63}$$

or

$$\delta^2 = \left[\frac{C_2 r_c}{C_1 \left[1 - \dfrac{2C_1 \sigma T_{sat}}{(T_w - T_{sat}) \rho_v h_{fg} r_c} \right]} \right]^2 \quad (4.64)$$

Substituting for δ from eq. (4.62) to (4.64) yields an expression for the waiting time as

$$t_w = \left[\frac{C_2 r_c}{\sqrt{\pi \alpha_l} C_1 \left[1 - \dfrac{2C_1 \sigma T_{sat}}{(T_w - T_{sat}) \rho_v h_{fg} r_c} \right]} \right]^2 \quad (4.65)$$

It can be seen from eq. (4.65) that the waiting time will vary nonlinearly with cavity size. It will first decrease and then increase with cavity size, but it will continue to decrease as the wall superheat is increased.

The waiting times obtained from eq. (4.65) for saturated water at 1 atm pressure are plotted in Figure 4.13 and are based on the assumption that during heating of

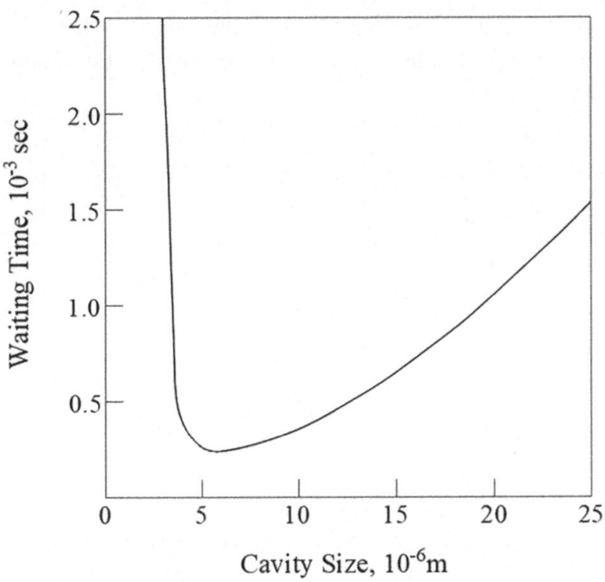

FIGURE 4.13 Waiting time as a function of cavity size for saturated water at 1 atm pressure and $\varphi = 90°$.

the liquid, the heater wall temperature remains constant. This assumption corresponds to a wall heat flux that varies with time. Generally, in laboratory experiments and in many applications, it is the heat flux that is held constant. The latter condition in turn implies that during transient heating, the wall temperature will vary. Also, it is possible that during the bubble growth period, the heat flux associated with evaporation of the microlayer, as we will discuss in more detail later, can exceed the wall heat flux and thereby reduce the wall temperature. Thus, during transient heating of the liquid adjacent to the wall, the wall temperature will have to increase to inception superheat as has also been pointed out by Hatton and Hall [27]. These factors need to be included in the model for a more realistic prediction of the waiting time. However, Figure 4.13 or eq. (4.65) give trends, which are consistent with the experimental data.

Example:

Calculate the waiting time for a 10^{-5} m size cavity on a surface submerged in a pool of saturated water at 1 atm pressure. Assume the wall superheat to be 10°C and the contact angle to be $\pi/2$.

Solution:

For $\varphi = \frac{\pi}{2}$ both constant C_1 and C_2 are unity. Evaluating the various physical properties at the mean temperature of 105°C, the expression for waiting time from eq. (4.65) for $C_1 = C_2 = 1$ is

$$t_w = \left[\frac{r_c}{\sqrt{\pi(1.68)(10^{-7})}\left[1 - \frac{2(57.9)(10^{-3})(373)}{(10)(0.695)(2.245)(10^6)r_c}\right]} \right]^2$$

$$= \left[\frac{1.38 \times 10^3 r_c}{1 - \frac{2.77 \times 10^{-6}}{r_c}} \right]^2 = 3.64 \times 10^{-3}\,\text{s} \quad \text{for} \quad r_c = 10^{-5}\,\text{m}$$

It can be seen from denominator that cavities equal to or smaller than 2.77 μm (knowing that negative value does not have physical significance) will never nucleate at the imposed superheat. Use of mean temperature is an approximation to the liquid temperature at the tip of the bubble embryo.

The growth of a bubble at the wall is influenced by several factors such as the thermal and flow field around and adjacent to the bubble and the extent and thickness of the microlayer underneath the bubble. The bubble shape itself is influenced by the contact angle, the direction and magnitude of gravitational acceleration, and the flow field around the bubble. The quantitative description of growth of such a bubble is quite complex. We can hope to address some features of the growth process of a bubble attached to the solid surface only after developing an understanding of the growth of a totally spherical bubble in an infinite medium under gravity free condition, and we do that next.

4.3.2 Bubble Growth without Heat or Mass Transfer

The growth of a high-pressure gas bubble in an infinite liquid medium was first described by Lord Rayleigh [28]. For an inviscid incompressible liquid, the potential function for radial flow of liquid can be written as

$$\Phi = \frac{R^2\dot{R}}{r} \tag{4.66}$$

where R is the instantaneous radius of the bubble, $\dot{R}$ is first time derivative of the radius, and r is any radius measured from the center of the bubble. From eq. (4.66), the radial velocity is obtained in terms of potential function, Φ, as

$$u = -\frac{\partial \Phi}{\partial r} = \frac{R^2\dot{R}}{r^2} \tag{4.67}$$

For a uniform pressure within the bubble, the momentum equation in radial direction for an inviscid liquid can be written as

$$\frac{\partial u}{\partial t} + u\frac{\partial u}{\partial r} = -\frac{1}{\rho_l}\frac{\partial p}{\partial r} \tag{4.68}$$

After substitution for u from eq. (4.67), eq. (4.68) becomes

$$\frac{R^2\ddot{R}}{r^2} + \frac{2R\dot{R}^2}{r^2} + \frac{1}{2}\frac{\partial}{\partial r}\left(\frac{R^4\dot{R}^2}{r^4}\right) = -\frac{1}{\rho_l}\frac{\partial p}{\partial r} \tag{4.69}$$

In eq. (4.69), $\ddot{R}$ is the second time derivative of R. Integration of (4.69) from far away from the bubble ($r = \infty$) to the outer surface of the bubble ($r = R$) yields

$$-\left[\frac{R^2\ddot{R}}{r} + \frac{2R\dot{R}^2}{r}\right]_{\infty}^{R} + \left[\frac{R^4\dot{R}^2}{2r^4}\right]_{\infty}^{R} = -\frac{1}{\rho_l}[p]_{\infty}^{R} \tag{4.70}$$

or

$$R\ddot{R}+\frac{3}{2}\dot{R}^2=\frac{1}{\rho_l}\left(p_{bo}-p_{l\infty}\right) \tag{4.71}$$

where pressure outside the bubble, p_{bo}, in relation to pressure inside the bubble, p_b, is

$$p_{bo}=p_b-\frac{2\sigma}{R} \tag{4.72}$$

and $p_{l\infty}$ is the pressure in the liquid far away from the bubble. Equation (4.71) is called Rayleigh's equation. This is a second-order differential equation in time, and we need to specify two conditions to solve it, e.g., R and $\dot{R}$ at $t = 0$ when dependence of p_b on R is known.

4.3.3 Bubble Growth with Heat and Mass Transfer

The analysis for the growth of a vapor bubble attached to a heated wall can be extremely involved because of the continuous change in the shape of the bubble, the convective motion around and in the vicinity of the bubble, temperature field in the liquid, presence of microlayer under the bubble, and the effect of the neighboring bubbles. In the literature, several attempts to analytically model the growth of bubbles attached to a wall have been made. All of these efforts have tended to be modifications of the model for growth of a vapor bubble in a uniform temperature field that we do next. Correction factors are generally incorporated to account for the shape of a bubble being different than that of a sphere. One such attempt for growth of a bubble in a uniform temperature field without any interaction with the wall is by Mickic et al. [29] and will be followed here.

Vapor bubble growth in a pool of superheated liquid is marked by two time periods. In the first phase, bubble grows from its metastable state as a result of pressure difference in the bubble and the host liquid in a manner similar to that described in the previous section for growth of a gas bubble in a pool of liquid. This phase of growth is called inertia-controlled growth. This phase lasts for a very short time and most of the bubble growth occurs in the second phase. In this latter phase, bubble grows as a result of evaporation at the bubble interface. Evaporation is supported by heat transfer from the pool to the bubble. This is called heat diffusion growth. If it is assumed that the contribution of surface tension to the pressure in a vapor bubble is small, i.e., $p_{bo} = p_b$ and bubble radius is such that first term in eq. (4.71) is small, eq. (4.71) can be written as

$$p_v-p_l=\frac{3}{2}\rho_l\dot{R}^2 \tag{4.73}$$

where p_v is the pressure of vapor in the bubble. Introducing a constant to account for the shape of the bubble being different than that for a sphere, eq. (4.73) is rewritten as

$$\dot{R}^2 = \frac{b(p_v - p_l)}{\rho_l} \tag{4.74}$$

where b is taken to be 2/3 for a spherical bubble and following eq. (4.71) where the first term on the left-hand side has been ignored. A value of 1/3 for b as has been suggested by Mikic et al. for a bubble attached to an upward facing horizontal surface. Using Clausius–Clapeyron relation, eq. (4.74) can be written as

$$\dot{R}^2 = \frac{b(T_v - T_{\text{sat}})h_{fg}\rho_v}{\rho_l T_{\text{sat}}} \tag{4.75}$$

Where T_v is the saturation temperature corresponding to pressure, p_v in the bubble. Defining liquid superheat as the temperature difference, ΔT, between liquid far away from the bubble and saturation temperature at p_l:

$$\Delta T = T_\infty - T_{\text{sat}}(p_l) \tag{4.76}$$

eq. (4.75) becomes

$$\dot{R}^2 = \frac{A^2(T_v - T_{\text{sat}})}{\Delta T} \tag{4.77}$$

where

$$A^2 = \left(\frac{bh_{fg}\rho_v \Delta T}{\rho_l T_{\text{sat}}}\right) \tag{4.78}$$

and A^2 has dimensions of square of velocity. Equation (4.77) represents inertial growth of the bubble resulting from the expansion due to the pressure difference between the bubble and the liquid while surface tension contribution is neglected.

In a diffusion or thermally controlled phase that follows, the bubble growth occurs as a result of evaporation at the interface. The rate of heat transfer at the interface determines the evaporation rate. Using a plane interface transient conduction analysis as was done earlier for a semi-infinite slab and neglecting evaporative resistance, which we will discuss later, the rate of bubble growth is obtained as

$$\frac{d}{dt}\left(\frac{4}{3}\pi R^3 \rho_v h_{fg}\right) = \frac{k_l}{\sqrt{\pi\alpha_l t}}(T_\infty - T_v)4\pi R^2 \tag{4.79}$$

$$\frac{dR}{dt} = \frac{k_l}{\sqrt{\pi\alpha_l t}}\frac{(T_\infty - T_v)}{\rho_v h_{fg}} \quad \text{or} \tag{4.80}$$

$$\frac{dR}{dt} = \left(\frac{\alpha_l}{\pi t}\right)^{1/2}\frac{\rho_l c_{Pl}(T_\infty - T_v)}{\rho_v h_{fg}} \quad \text{or} \tag{4.81}$$

It should be noted that in writing the right-hand side of eq. (4.79), T_v is implicitly assumed to remain constant with time and the interface is assumed to be plain. Nearly exact solution of Plesset and Zwick [30] in which it is assumed that liquid is simply pushed radially outward as the bubble grows with heat transfer from the superheated liquid suggests that to account for sphericity of the interface the right-hand side of the eq. (4.80) should be multiplied by $\sqrt{3}$. Forster and Zuber [31] also obtained an analytical solution for growth of a spherical vapor bubble. Their numerical constant was about 25% larger than that of Plesset and Zwick. After a correction factor of $\sqrt{3}$ is applied, eq. (4.81) can be written as

$$\frac{dR}{dt} = \frac{1}{2}\frac{B}{\sqrt{t}}\frac{(T_\infty - T_v)}{\Delta T} \tag{4.82}$$

where

$$B = \left(\frac{12\alpha_l}{\pi}\right)^{1/2}\left(\frac{\rho_l c_{Pl}\Delta T}{\rho_v h_{fg}}\right) \tag{4.83}$$

The second term on the right-hand side of eq. (4.83) is called Jakob number, which is a ratio of sensible energy per unit volume of liquid to the latent heat of vaporization per unit volume of vapor.

$$\text{Ja} \equiv \left(\frac{\rho_l c_{Pl}\Delta T}{\rho_v h_{fg}}\right) \tag{4.84}$$

and parameter B has dimensions of $\text{m}/(\text{s})^{1/2}$. Equation (4.82) can be rewritten as

$$\dot{R} = \frac{B}{2\sqrt{t}}\left(1 - \frac{(T_v - T_{\text{sat}})}{\Delta T}\right) \tag{4.85}$$

Eliminating $\dfrac{T_v - T_{\text{sat}}}{\Delta T}$ between eqs. (4.77) and (4.85) yields

$$\dot{R} = \frac{B}{2\sqrt{t}}\left(1 - \frac{\dot{R}^2}{A^2}\right) \tag{4.86}$$

or

$$\frac{1}{A^2}\left(\dot{R}^2\right) + \frac{2\sqrt{t}}{B}\dot{R} - 1 = 0 \tag{4.87}$$

With dimensionless radius and dimensionless time defined as

$$R^* = \frac{AR}{B^2} \tag{4.88}$$

and

$$t^* = \frac{A^2 t}{B^2} \tag{4.89}$$

eq. (4.87) becomes

$$\left(\frac{dR^*}{dt^*}\right)^2 + 2\sqrt{t^*}\frac{dR^*}{dt^*} - 1 = 0 \tag{4.90}$$

The solution of quadratic eq. (4.90) is

$$\left(\frac{dR^*}{dt^*}\right) = \frac{-2\sqrt{t^*} \pm \sqrt{4t^* + 4}}{2}$$

Using only the + sign for bubble growth

$$\frac{dR^*}{dt^*} = \left(t^* + 1\right)^{1/2} - \left(t^*\right)^{1/2} \tag{4.91}$$

If we use the initial condition that $R^* \cong 0$ at $t^* = 0$, the integration of the above equation yields

$$R^* = \frac{2}{3}\left[\left(t^* + 1\right)^{3/2} - \left(t^*\right)^{3/2} - 1\right] \tag{4.92}$$

For $t^* \ll 1$, the above equation reduces to the form

$$R^* = t^* \tag{4.93}$$

On the other hand, for $t^* \gg 1$, it yields

$$R^* = t^{*1/2} \tag{4.94}$$

Thus, for early times, the bubble will grow linearly with time (inertia growth), and in the limit of thermally controlled growth, the radius of the bubble will increase as square root of time. This behavior is generally confirmed by experiments.

4.3.4 Approximate Analysis of Bubble Growth on a Wall

Bubble growth on a heated wall is more complex than that considered in the above for a bubble growing in a sea of superheated liquid. Prior to initiation of bubble growth, thermal layer adjacent to the wall develops and local temperature of the solid reaches nucleation temperature, when heat flux is controlled. This is the waiting period. After inception criterion is met, vapor embryo begins to grow. Initially, the superheated liquid layer adjacent to the wall completely surrounds the bubble embryo; however, as the bubble grows, the thermal layer covers only part of the vapor bubble interface as we saw in Figure 4.11. Also, as the bubble grows, the superheated liquid around the bubble is pushed outward. Now the growth of the bubble is dictated by the rate at which thermal energy can be transferred from the superheated liquid near the heated surface to the interface as well as from the microlayer that forms underneath the bubble. Mikic and Rohsenow [32] also obtained an approximate solution for growth of such a bubble without considering the contribution of microlayer and variation of the flow field around the bubble. In carrying out the somewhat simplistic analysis, they assumed that:

i. Liquid layer adjacent to the wall is semi-infinite
ii. During bubble growth, no relative motion exists between the bubble and the surrounding liquid
iii. Liquid-vapor interface is plane
iv. Interfacial heat flux can be obtained by assuming that vapor temperature remains constant
v. Wall temperature remains constant
vi. Interfacial resistance to vapor production is negligible
vii. Contribution of evaporation from microlayer is neglected

Mikic and Rohsenow [32] decomposed into two periods the energy transfer during bubble growth on the heater surface. During the first phase (waiting period), energy is transferred from the wall to the developing thermal layer adjacent to the heater surface. Whereas during the second phase, energy stored in the thermal layer is utilized for evaporation at the vapor–liquid interface. The transient diffusion of heat into the liquid during the waiting period and out of the superheated liquid during the growth of the bubble is broken up into two one-dimensional transient conduction problems as

$$
\begin{aligned}
&\frac{\partial T}{\partial t}=\alpha_l\frac{\partial^2 T}{\partial y^2}\ \text{for} -t_w\le t\le 0 \quad \text{and} && \frac{\partial T}{\partial t}=\alpha_l\frac{\partial^2 T}{\partial y^2}\ \text{for}\ 0\le t\\
&T=T_l\ \text{for all}\ y\ \text{and}\ t\le -t_w && T=T_l\ \text{for all}\ y\ \text{and}\ t=0\\
&T=T_w\ \text{at}\ y=0\ \text{for}\ 0\ge t\ge -t_w && T=T_l-(T_w-T_v)\ \text{at}\ y=0\ \text{for}\ t\ge 0\\
&T=T_l\ \text{as}\ y\to\infty\ \text{for}\ t\ge -t_w && T=T_l\ \text{as}\ y\to\infty\ \text{for}\ t\ge 0
\end{aligned}
\tag{4.95}
$$

In the above equation, T_l is the liquid temperature away from the wall. It remains constant with time, and it could be at saturation temperature or below saturation. The wall temperature, T_w, is held constant, whereas T_v is the vapor temperature in the bubble and is assumed to remain constant. Figure 4.14 shows the temperature profiles during the thermal layer development period and during the bubble growth period.

Superimposing the solution for the two cases, we obtain

$$
T-T_l=(T_w-T_l)\,erfc\left[\frac{y}{2\sqrt{\alpha_l(t+t_w)}}\right]-(T_w-T_v)\,erfc\left[\frac{y}{2\sqrt{\alpha_l t}}\right] \tag{4.96}
$$

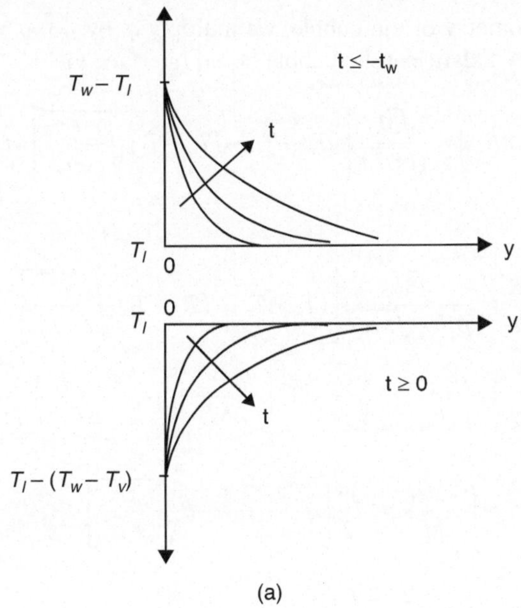

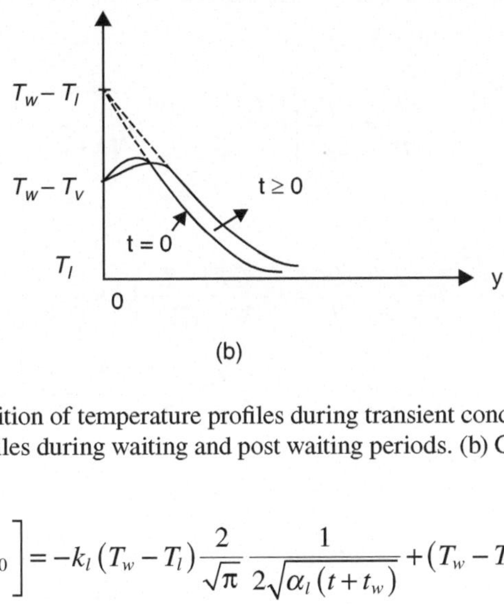

FIGURE 4.14 Superposition of temperature profiles during transient conduction in the liquid [32]. (a) Temperature profiles during waiting and post waiting periods. (b) Combined profile.

$$q_l = -\left[-k_l \frac{\partial T}{\partial y}\bigg|_{y=0}\right] = -k_l\left(T_w - T_l\right)\frac{2}{\sqrt{\pi}}\frac{1}{2\sqrt{\alpha_l\left(t+t_w\right)}} + \left(T_w - T_v\right)\frac{2}{\sqrt{\pi}}\frac{k_l}{2\sqrt{\alpha_l t}}$$

$$= k_l\left[\frac{\left(T_w - T_v\right)}{\sqrt{\pi\alpha_l t}} - \frac{\left(T_w - T_l\right)}{\sqrt{\pi\alpha_l\left(t+t_w\right)}}\right] \quad (4.97)$$

In obtaining eq. (4.97), a tacit assumption has been made that solid surface is replaced by a vapor layer at $y = 0$ during the growth period. Also, a negative sign is added to reflect, transfer during growth period, of energy opposite to increase y. To account for the spherical geometry of the bubble, we multiply q_i by $\sqrt{3}$ as was done earlier. The mass and energy balance at the bubble liquid interface yields

$$\frac{d}{dt}\left(\rho_v h_{fg}\frac{4}{3}\pi R^3\right)=\frac{\sqrt{3}k_l}{\sqrt{\pi\alpha_l t}}\left[(T_w-T_v)-(T_w-T_l)\sqrt{\frac{t}{t+t_w}}\right]4\pi R^2 \tag{4.98}$$

or

$$\begin{aligned}\frac{dR}{dt}&=\frac{\sqrt{3}k_l}{\rho_v h_{fg}\sqrt{\pi\alpha_l t}}\left[(T_w-T_v)-(T_w-T_l)\sqrt{\frac{t}{t+t_w}}\right]\\&=\frac{\rho_l c_{pl}}{\rho_v h_{fg}}\Delta T\sqrt{\frac{\alpha_l}{\pi t}}\sqrt{3}\left[\left(\frac{T_w-T_v}{\Delta T}\right)-\left(\frac{T_w-T_l}{\Delta T}\right)\sqrt{\frac{t}{t+t_w}}\right]\\&=\left(\frac{12\alpha_l}{\pi t}\right)^{1/2}\frac{\text{Ja}}{2}\left[\frac{(T_w-T_v)}{\Delta T}-\theta\sqrt{\frac{t}{t+t_w}}\right]\end{aligned} \tag{4.99}$$

where $\Delta T = T_W - T_{\text{sat}}$ and $\theta = \dfrac{T_w - T_l}{\Delta T}$

In terms of the parameter B defined earlier in eq. (4.83), eq. (4.99) can be written as

$$\frac{dR}{dt}=\frac{B}{2\sqrt{t}}\left[\left(\frac{T_w-T_v}{\Delta T}\right)-\theta\sqrt{\frac{t}{t+t_w}}\right] \tag{4.100}$$

or

$$\begin{aligned}1-\frac{2\sqrt{t}}{B}\frac{dR}{dt}&=1-\left(\frac{T_w-T_v}{\Delta T}\right)+\theta\sqrt{\frac{t}{t+t_w}}\\&=\frac{T_w-T_{sat}-T_w+T_v}{T_w-T_{\text{sat}}}+\theta\sqrt{\frac{t}{t+t_w}}\end{aligned} \tag{4.101}$$

or

$$1-\frac{2\sqrt{t}}{B}\frac{dR}{dt}=\frac{T_v-T_{\text{sat}}}{\Delta T}+\theta\sqrt{\frac{t}{t+t_w}} \tag{4.102}$$

As noted earlier in writing eq. (4.73), Rayleigh equation in an approximate form is written as

$$\left(\frac{dR}{dt}\right)^2=b\left(\frac{p_v-p_l}{\rho_l}\right) \tag{4.103}$$

By using Chausius–Clapeyron relation eq. (4.103) becomes

$$\left(\frac{dR}{dt}\right)^2 \cong b\frac{(T_v - T_{\text{sat}})h_{fg}\rho_v}{T_{\text{sat}}\rho_l \Delta T}\Delta T \tag{4.104}$$

or in terms of parameter A defined in eq. (4.77), eq. (4.104) can be written as

$$\left(\frac{dR}{dt}\right)^2 = A^2\frac{(T_v - T_{\text{sat}})}{\Delta T} \tag{4.105}$$

Substituting for $\frac{(T_v - T_{\text{sat}})}{\Delta T}$ from eq. (4.105) into eq. (4.102) we get

$$1-\frac{2\sqrt{t}}{B}\frac{dR}{dt}=\frac{1}{A^2}\left(\frac{dR}{dt}\right)^2+\theta\sqrt{\frac{t}{t+t_w}} \tag{4.106}$$

Or, in dimensionless form as was done earlier by using eqs. (4.88) and (4.89), the above equation becomes

$$\left(\frac{dR^*}{dt^*}\right)^2 + 2\sqrt{t^*}\frac{dR^*}{dt^*} - \left[1-\theta\sqrt{\frac{t^*}{t^*+t_w^*}}\right] = 0 \tag{4.107}$$

Solution of the quadratic eq. (4.107) while choosing a + sign yields

$$\frac{dR^*}{dt^*} = \left[t^* + 1 - \theta\sqrt{\frac{t^*}{t^*+t_w^*}}\right]^{1/2} - \sqrt{t^*} \tag{4.108}$$

where for a growing bubble, we have used the + sign in the solution of the quadratic eq. (4.107). In the limit of $t_w^* \to \infty$, eq. (4.108) reduces to that for growth of a bubble in a pool of superheated liquid.

Also, we know that bubble growth rate will increase with increase in waiting time, as longer waiting time allows more energy to be deposited in the developing thermal layer. Furthermore, with increase in liquid subcooling, i.e., $\theta > 1$, the bubble growth rate will decrease. Equation (4.108), although obtained with several approximations, captures these features.

As alluded to earlier, during bubble growth temporal and spatial variations in the temperature field over the heater surface and around the bubble occur continuously. Bubble size and shape change as the bubble grows taller. During bubble evolution, a microlayer also forms between the evolving liquid–vapor interface and the solid surface. Figure 4.11 shows a bubble formed on a heated surface. Aside from the heat transfer across the microlayer, the heat transfer from the thermal layer around the bubble also contributes to evaporation at the vapor–liquid interface and consequently to the bubble growth rate. The extent of the microlayer increases as the bubble base expands, but it shrinks as the bubble starts to detach from the surface. The thermal layer around the bubble covers only part of the vapor–liquid interface of the bubble. It is extremely difficult to obtain analytical solution to such a complex process.

As such, we must resort to numerical simulations of the process. We will discuss this in detail in Chapter 13. In Figure 13.9, the growth history of a single bubble on a heated surface obtained from experiments and from numerical simulations is shown.

4.4 BUBBLE DEPARTURE DIAMETER

The size to which a bubble attached to the wall will grow before detaching from an upward facing horizontal surface depends on balance between several forces acting on the bubble. As the bubble grows, its size and shape change. Consequently, the magnitude of the forces acting on it continuously changes. In this section, we identify these forces, but the magnitude of these forces is very approximate because of the complexities arising out of the change of bubble size and shape and results in alteration of the flow field around the bubble. We will also use force balance to determine the terminal velocity of a bubble rising through a pool of inviscid liquid.

4.4.1 Forces Acting on a Bubble Formed on a Wall

In 1935, Fritz [20] proposed that main forces acting on a bubble were the surface tension and buoyancy. Surface tension tends to hold the bubble toward the wall while buoyancy tries to lift it away. For a bubble that is part of a sphere as shown in Figure 4.15a, the surface tension force as depicted in Figure 4.15a is

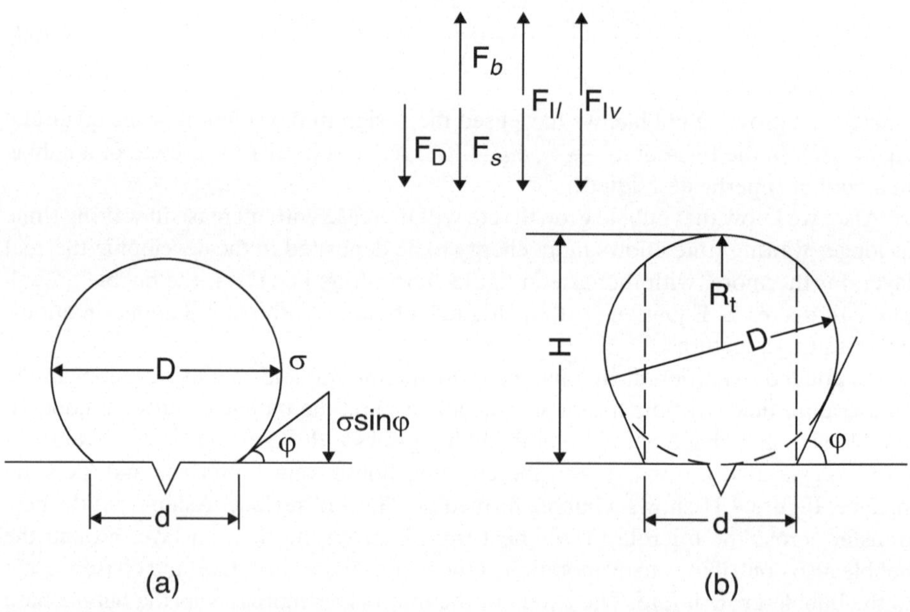

FIGURE 4.15 Bubble shape and forces acting on a bubble attached to wall. (a) Bubble as truncated sphere. (b) Forces acting on a bubble.

$$F_S = \pi D \sigma \sin^2 \varphi \tag{4.109}$$

In the above equation, $D \sin \phi$ is the bubble base diameter, d.

The buoyancy force in terms of volume, V_b, of the bubble is

$$F_B = (\rho_l - \rho_v) g V_b = \frac{\pi D^3}{6} (\rho_l - \rho_v) g \left[\frac{1 + \cos \varphi}{2} \right]^2 [2 - \cos \varphi] \tag{4.110}$$

Equating the surface tension and buoyancy forces from eqs. (4.109) to (4.110) we get

$$D = f(\varphi) \sqrt{\frac{\sigma}{g(\rho_l - \rho_v)}} \tag{4.111}$$

where

$$f(\varphi) = \left[\frac{24 \sin^2 \varphi}{(1 + \cos \varphi)^2 (2 - \cos \varphi)} \right]^{1/2} \tag{4.112}$$

Fritz [20] correlated his experimentally observed bubble diameters at departure as:

$$D = 0.0208 \phi \sqrt{\frac{\sigma}{g(\rho_l - \rho_v)}} \tag{4.113}$$

where φ now is the contact angle measured in degrees. Figure 4.16 shows the dependence on φ of function $f(\varphi)$ obtained in eq. (4.112) and the equivalent function experimentally obtained by Fritz and given in eq. (4.113). The difference between the two expressions is probably reflective of the non-sphericity of the bubbles and non-inclusion of other forces that act on the bubble. However, for contact angles smaller than $\pi/2$, the difference between the two expressions is rather small. As pointed out earlier, for wetting surfaces, the contact angle is much smaller than $\pi/2$.

Bubbles are generally not totally spherical as many investigations such as those of Siegel and Keshock [33], and Cochran and Aydellot [34] have shown. Bubbles not only experience surface tension and buoyancy forces during their growth, but several other forces come in to play. These additional forces are those of inertia and drag. Since these forces act both on volume and surface area, it is difficult to determine a common equivalent diameter of a spherical bubble. If we define an equivalent diameter based on bubble volume as

$$D = \left(\frac{6}{\pi} V \right)^{1/3}$$

The various forces acting on a bubble such as that shown in the Figure 4.15b are:

i. Buoyancy Force, $$F_B = \frac{\pi}{6} D^3 (\rho_l - \rho_v) g \tag{4.114}$$

ii. Surface Tension Force, $$F_s = \pi d \sigma \sin \varphi \tag{4.115}$$

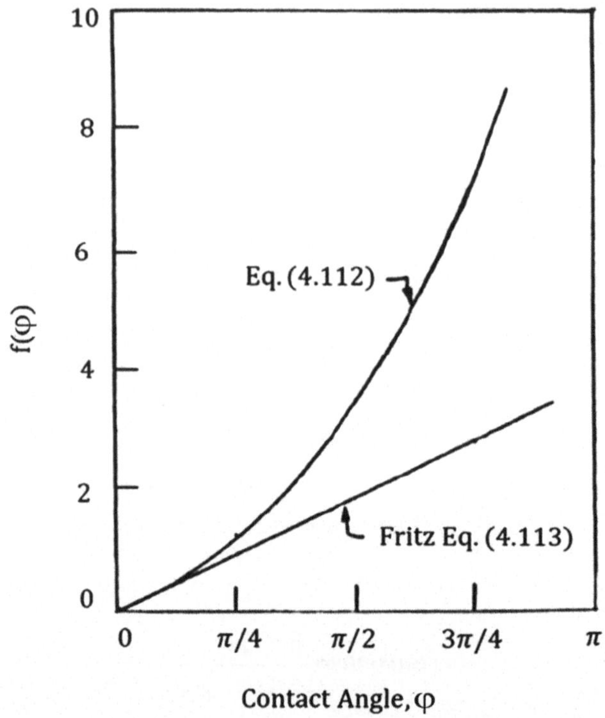

FIGURE 4.16 Comparison of predictions of the effect of contact angle or bubble diameter at departure from Fritz's expression and from Eq. (4.112).

iii. Liquid Inertia Force, $$F_{il} = K\rho_l \frac{d}{dt}\left(\frac{\pi}{12} D^3 \frac{dD}{dt}\right) \tag{4.116}$$

In eq. (4.116), K is the coefficient defining the virtual mass of the liquid surrounding the bubble and $\frac{1}{2}\frac{dD}{dt}$ is taken as the velocity of the center of the bubble mass.

iv. Bubble Inertia Force, $$F_{iv} = \rho_v \frac{d}{dt}\left(\frac{\pi}{12} D^3 \frac{dD}{dt}\right) \tag{4.117}$$

At low pressures, vapor density ρ_v is much smaller than the liquid density ρ_l. Thus, the inertia of the bubble in comparison to that of the liquid can be ignored.

v. Viscous Drag Force, $$F_D = C_D \frac{\rho_l}{2}\frac{\pi}{4} D^2 \left(\frac{dD}{dt}\right)^2 \tag{4.118}$$

In writing eq. (4.118), the velocity of the top surface of the bubble is used instead of the velocity of the center of the bubble. The constant C_D is the drag coefficient. It should be pointed out that evaluation of drag coefficient on an evolving bubble attached to the wall is a difficult proposition. As a first approximation, we can assume C_D to be that for a sphere placed in a stream of liquid.

The buoyancy force written as eq. (4.114) is for a body fully submerged in a liquid. Since the bubble is attached to the wall, the buoyancy force, as has been suggested by Cochran and Aydelott [34], should be corrected such that

$$F_B = (\rho_l - \rho_v) g \left(\frac{\pi}{6} D^3 - \frac{\pi}{4} d^2 H \right) + \frac{\sigma}{R_t} \frac{\pi}{4} d^2 \tag{4.119}$$

In the above equation, the second term accounts for excess pressure in the bubble due to surface tension and is written by assuming that the bubble top has a radius, R_t, and has no curvature in a direction normal to the plane of the figure. The signs of the forces acting on the bubble are shown in Figure 4.15b. Liquid and vapor inertia forces change sign during the growth period. Knowing bubble growth rate from an equation such as (4.108) and the integrated value of bubble diameter as a function of time, one can evaluate various forces. Inertia forces change sign from acting downward when the calculated value is positive and upward when the calculated force is negative. Calculation of the magnitude and direction of these forces and bubble departure diameter is left as an exercise.

4.4.2 Empirical Correlations for Bubble Departure Diameter

In principle, one could use the expressions developed in the previous sections with respect to bubble growth rate and forces that act on the bubble to obtain bubble diameter at departure. However, in a boiling system, such an approach meets with not much success. Some of the reasons for this are:

1. Cavities of different sizes and shapes exist on the surface. Waiting times between cavities may differ significantly.
2. Because of the differences in waiting times, thickness of thermal layers may differ from cavity to cavity. This in turn alters the growth rate of the bubbles. However, this will be true only as long as sufficient distance exists between neighboring cavities.
3. During late phase of bubble growth, thermal layer surrounds only a portion of the bubble interface.
4. Evaporation from liquid layer (microlayer) underneath a bubble can contribute to bubble growth. This process has not been included in the bubble growth model of Mikic, Rohsenow, and Griffith described earlier.
5. Because of the development of flow field and interaction between neighboring sites, thermal layers around bubbles may be distorted.
6. Merger of bubbles at neighboring sites can alter the bubble shape and size, as well as waiting time, and thereby significantly influence the bubble departure diameters and frequency.

As such, several correlations have been suggested in the literature for bubble diameter at departure. Based on bubble departure diameters observed with several liquids, Rohsenow [35] has given a correlation of the type:

For water

$$D_d\sqrt{\frac{g(\rho_l-\rho_v)}{\sigma}}=1.5\times10^{-4}\,\mathrm{Ja}^{*5/4} \tag{4.120}$$

For other liquids

$$D_d\sqrt{\frac{g(\rho_l-\rho_v)}{\sigma}}=4.65\times10^{-4}\,\mathrm{Ja}^{*5/4} \tag{4.121}$$

where

$$\mathrm{Ja}^* = \frac{\rho_l c_{pl} T_{\text{sat}}}{\rho_v h_{fg}} \tag{4.122}$$

As opposed to Fritz's correlation, the correlation of Rohsenow includes the Jakobs number, Ja^*, based on the saturation temperature of the liquid. The above correlations do not account for the effect of the contact angle and the wall superheat.

Gorenflo et al. [36] have proposed an expression for bubble diameter at departure at high heat fluxes as

$$D_d = C_1\left(\frac{\mathrm{Ja}^4\alpha_l^2}{g}\right)^{1/3}\left[1+\left(1+\frac{2\pi}{3\,\mathrm{Ja}}\right)^{1/2}\right]^{4/3} \tag{4.123}$$

Different values of the proportionality constant C_1 have been suggested for different liquids and Ja number is defined as given in eq. (4.84). With the knowledge of the growth rate and the diameter to which a bubble grows before departing, the growth time, t_g, can be calculated. After bubble departure, cooler bulk liquid fills the space vacated by the bubble and a new bubble forms after nucleation criterion is satisfied. It should be stressed that a new bubble at this location will not grow until the superheated liquid layer is reestablished and the inception criterion is satisfied.

4.5 BUBBLE RELEASE FREQUENCY

Bubble release frequency is the reciprocal of the sum of the waiting time, t_w, and growth period, t_g as is given as

$$f=\frac{1}{t_w+t_g} \tag{4.124}$$

The growth period is influenced by the bubble growth rate and diameter at departure (i.e., larger bubbles taking longer to grow). As was noted earlier, waiting time is influenced by the time it takes to redevelop the thermal layer, recovery of wall temperature, and the size of the nucleating cavity. Because of the large variations that can occur in predicted waiting and growth periods, bubble release frequencies can vary greatly. Generally, correlations combining bubble departure diameter and frequency have been proposed in the literature. The general form of such correlations is

$$fD_d^n \equiv C \tag{4.125}$$

where C and n are empirical constants.

Values of n between 1/2 and 3 have been suggested. Jakob [37] proposed a value of unity, while Ivey [38] suggested that $n = 2$ when bubble growth is inertia-controlled and is 1/2 when it is thermally controlled. For inertia-controlled bubble growth, Zuber [39] on the other hand assumed that bubble growth rate was proportional to terminal bubble rise velocity and that waiting time was much smaller than the growth time. Under these assumptions, he suggested that exponent n should have a value of unity and the constant C should be

$$C = 0.59\left[\frac{\sigma g(\rho_l - \rho_v)}{\rho_v^{2}}\right]^{1/4} \tag{4.126}$$

In the above equation, the multiplier 0.59 was obtained by matching the predictions with the data.

Malenkov [40] has suggested a correlation of the type

$$fD_d = \frac{V_d}{\pi\left[1 - \dfrac{1}{1 + (V_d \rho_v h_{fg})/q}\right]} \tag{4.127}$$

where characteristic velocity, V_d, is defined as

$$V_d = \left[\frac{gD_d(\rho_l - \rho_v)}{2(\rho_l + \rho_v)} + \frac{2\sigma}{D_d(\rho_l - \rho_v)}\right]^{1/2} \tag{4.128}$$

From eq. (4.128) for bubbles with large diameter at departure, the characteristic velocity

$$V_d \sim D_d^{1/2} \tag{4.129}$$

whereas for small bubbles

$$V_d \sim D_d^{-1/2} \tag{4.130}$$

Thus, for bubbles with thermally controlled growth rates, Malenkov's expression reduces to a form similar to that of Ivey. Knowing the bubble diameter at departure and frequency, one can determine the rate of energy from the wall that is utilized for evaporation at a particular nucleation site and surrounding area supporting the nucleation site.

In the next chapter, we will see how the correlations and models developed in this section will be helpful in prediction of dependence of nucleate boiling heat flux on wall superheat.

4.6 TERMINAL VELOCITY OF A BUBBLE RISING IN A POOL OF LIQUID

For a spherical cap of a bubble as shown in Figure 4.17 rising in a pool of inviscid liquid at a velocity V_t, the relative velocity with respect to the bubble at location A outside the bubble can be written as

$$V_A = \frac{3}{2} V_t \sin\theta \tag{4.131}$$

where θ is the angular position of A with respect to apex of the bubble. Total pressure at location can be written as

$$P_A = P_o + \rho_l g \Delta H - \frac{1}{2}\rho_l V_A^2 \tag{4.132}$$

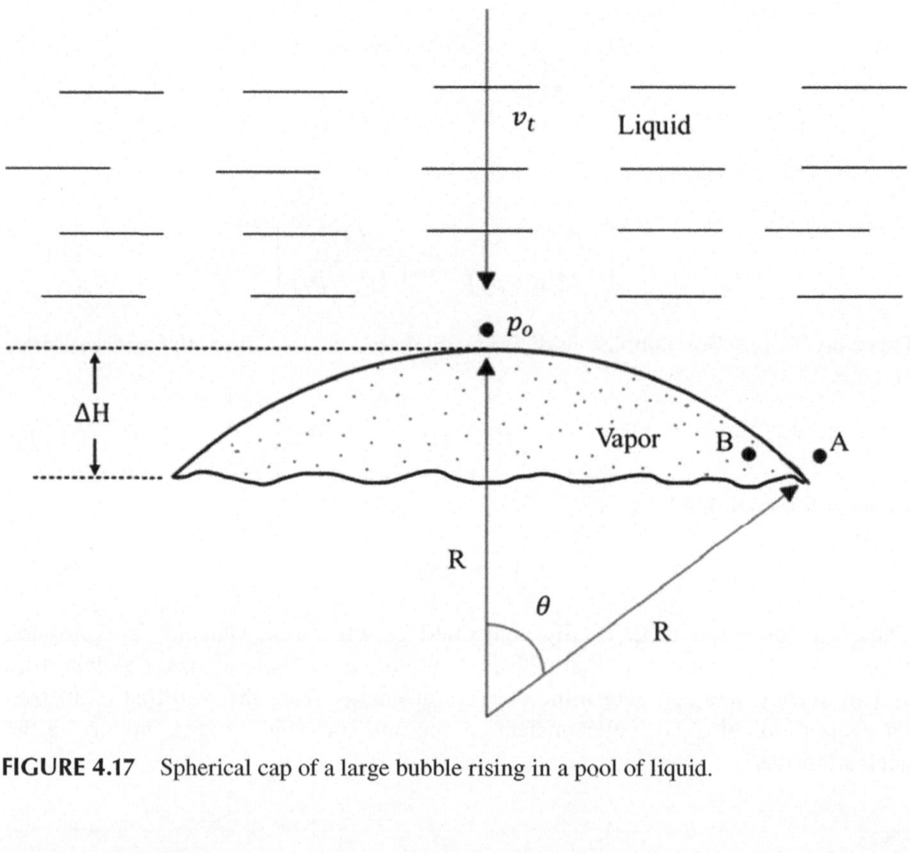

FIGURE 4.17 Spherical cap of a large bubble rising in a pool of liquid.

$$= P_o + \rho_l g \Delta H - \frac{1}{2}\rho_l\left(\frac{3}{2}V_t \sin\theta\right)^2 \tag{4.133}$$

where ΔH is the height of the bubble, and P_o is stagnation pressure at the apex. Similarly, neglecting the contribution of surface tension, total pressure at location B inside the bubble can be written as

$$P_B = P_o + \rho_v g \Delta H \tag{4.134}$$

Equating pressures at A and B from eqs. (4.133) and (4.134), we find

$$\frac{1}{2}\rho_l\left(\frac{3}{2}V_t \sin\theta\right)^2 = (\rho_l - \rho_v) g \Delta H \tag{4.135}$$

If R is the radius of the bubble, the height, ΔH, can be written as

$$\Delta H = R - R\cos\theta = \frac{D}{2}(1 - \cos\theta) \tag{4.136}$$

For small values of θ, $\cos\theta$ and $\sin\theta$ can be approximated as

$$\cos\theta \cong 1 - \frac{\theta^2}{2} \tag{4.137}$$

$$\sin\theta \cong \theta \tag{4.138}$$

Substituting for ΔH in eq. (4.135) from eq. (4.136) and replacing $\cos\theta$ and $\sin\theta$ from eqs. (4.137) to (4.138), we obtain

$$V_t = \frac{1}{3}\sqrt{\frac{2g(\rho_l - \rho_v)D}{\rho_l}} \tag{4.139}$$

4.7 SUMMARY

In this chapter, we have developed expressions for superheat needed for nucleation of a bubble in a pool of superheated liquid and on a heated wall. The former is called homogeneous nucleation, whereas the latter is known as heterogeneous nucleation. We have investigated bubble growth and departure processes, made a force balance to determine the bubble diameter at departure and determine bubble release frequency. Several correlations from literature have also been listed.

EXERCISES

1. Calculate equilibrium radius of a vapor bubble in water and PF-5060 for liquid superheats of 3°C, 5°C, 7°C, 9°C, and 11°C. Assume the pressure to be 1 atm.
2. Calculate equilibrium radius of a vapor bubble in water for system pressures of 10, 50, and 100 atm at a superheat of 5°C.
3. Plot form problems 1 and 2 above, equilibrium bubble size for water as a function of superheat at fixed pressure and as a function of pressure for a given superheat and explain your results.
4. A gas bubble initially at 10 atm pressure and having a diameter of 1mm expends isentropically in water at 1 atm pressure and room temperature, 25°C. Calculate the final diameter of the bubble.
5. Calculate the critical radius and homogeneous nucleation temperature of water at 1 atm pressure. Assume $J = 1$ cluster/cm^3 s. How would your answer change if $J = 1 \times 10^{-2}$ clusters/cm^3 sec or if system pressure was 100 bars? What is the ratio of T_{hom}/T_{crit} for these cases?
6. Calculate the minimum wall superheat and the corresponding cavity size for boiling of saturated water at 1 atm pressure on an upward facing horizontal plate. Wall heat flux is 10^4 W/m^2. If laminar natural connection existed on the plate prior to nucleation, is the predicted wall superheat consistent with the heat flux? Assume contact angle to be 90°.
7. Calculate the wall superheat and the corresponding size of nucleating cavities for saturated water at 1 atm pressure. The wall heat flux is 10^4 W/m^2 and turbulent natural connection exists on the heater prior to nucleation. Assume the contact angle to be 60°.
8. Calculate the waiting time for a cavity of radius 10^{-5} m to nucleate in saturated water at 1 atm pressure. The wall superheat is 8°C and thermal layer has a thickness of 6×10^{-4} m. Assume contact angle to be 90°C.
9. Saturated water at 100°C and 1 atm pressure flows over a flat plate maintained at 108°C. Calculate the distance from the leading edge at which bubble nucleation corresponding to minimum superheat criterion will occur. Flow velocity is 0.5 m/s and laminar forced convection exists on the surface prior to nucleation. Assume a contact angle of 90°. What is the size of the nucleating cavity?
10. A spherical vapor bubble is growing in a pool of superheated water maintained at 110°C at 1 atm pressure. Calculate and plot the bubble radius as a function of time. Sate clearly any assumption you make.
11. A heater is placed in a pool of saturated water at 1 atm pressure. The contact angle is 20°. If at a wall superheat of 8K, the wall heat flux is 6 W/cm^2, calculate the size range of cavities that will nucleate and the corresponding waiting time. Choose two size cavities—one the smallest and one the largest. For the largest cavity, use the model developed in the chapter for a

bubble growing in a nonuniform temperature field to calculate the bubble radius as a function of time. Plot your result.

12. For the largest cavity chosen in Problem 11, plot the forces acting on the bubble as a function of time. Determine the time at which a bubble will detach and the corresponding diameter of the bubble at departure. What will be the bubble release frequency for this case?
13. By just balancing the buoyancy force with surface tension, calculate the bubble departure diameter for water and PF-5060 at 1 atm pressure. Assume the bubble to be a truncated sphere with a contact angle of 10° for both cases.
14. Calculate the terminal velocity of a steam bubble rising in saturated water at 1 atm pressure. Initial diameter of the bubble is 6 mm.

NOMENCLATURE

c_1, c_2	Constants
D	Diameter
F	Force
f	Frequency
G	Gibbs function or gravity
h	Planck's constant
h_{fg}	Latent heat of vaporization
k	Boltzmann's constant or thermal conductivity
Ja	Jakob number
K	Nondimensional modified curvature
M	Molecular weight
N	Number
n	Number of molecules
P	Pressure
q	Heat flux
T	Temperature
t	Time
v_{fg}	Change in specific volume from liquid to vapor
R	Radius of a bubble
r	Radius of a cluster or cavity
V	Volume
y	Vertical distance

SUBSCRIPTS

A	Avogadro
a	Activated
B	Buoyancy
b	Bubble
C	Constant

c	Critical or Cavity
D	Drag
G	Gas
l	Liquid
min, max	Minimum or maximum
v	Vapor
ls	Largest cavity
sat	Saturation
w	Wall or Waiting
m	Constant

GREEK SYMBOLS

α	Thermal diffusivity
β	Half wedge angle
δ	Thermal layer thickness
η	Dimensionless distance
ρ	Density
σ	Surface tension
Φ	Potential function
φ	Contact angle
Ψ	Cavity mouth angle

NOTE

1 It is defined as the maximum possible work the system can do as its state changes from initial to final in equilibrium with its surroundings.

REFERENCES

1. Volmer, M. (1939), *Kinetics of Plate Formation*. Ref. ATI No. 81935(F_TS_7068_RE), Clearinghouse Fed. Schi. Tech. Inform, Springfield, VA.
2. Frankel, J. (1955), *Kinetic Theory of Liquids*, Dover, New York.
3. Reiss, H. (1952), "Theory of the Liquid Drop Model." *Ind. Eng. Chem.*, Vol. 24, pp. 284–288.
4. Cole, R. (1970), "Boiling Nucleation." *Adv. Heat Transfer*, Vol. 10, pp. 86–166.
5. Theofanous, T.G. (1993), "The Study of Stream Explosions in Nuclear Systems." *Proceedings of the International Seminar on the Physics of Vapor Explosions*, Tomakomai, pp. 5–26.
6. Wakeshima, H. and Takata, K. (1958), "On the Limit of Superheat." *J. Phys. Soc. Japan*, Vol. 13, pp. 1398–1403.
7. Apfel, R.E. (1972), "Water Heated to 249.5°C at Atmospheric Pressure." *Nat. Phys. Sci.*, Vol. 238, pp. 63–64.
8. Bankoff, S.G. (1958), "Entrainment of Gas in the Spreading of a Liquid over a Rough Surface." *AIChE J.*, Vol. 4, pp. 24–26.
9. Fisher, J.C. (1948), "The Fracture of Liquids." *J. Appl. Phys.*, Vol. 19, pp. 1062–1067.
10. Ward, C.A. and Forest T.W. (1976), "On the Relation between Platelet Adhesion and the Roughness of a Synthetic Biomaterial." *Ann. Biomed. Eng.*, Vol. 4, pp. 184–207.

11a. Wang, C.H. and Dhir, V.K. (1993), "On the Gas Entrapment and Nucleation Site Density During Pool Boiling of Saturated Water." *J. Heat Transfer*, Vol. 115, pp. 670–679.
11b. Wang, C.H. and Dhir, V.K. (1993), "Effect of Surface Wettability on Active Nucleation Site Density during Pool Boiling of Saturated Water." *J. Heat Transfer*, Vol. 115, pp. 659–669.
12. Griffith, P. and Wallis, G. (1960), "The Role of Surface Conditions in the Nucleate Boiling." *Chem. Eng. Prog. Symp. Ser.*, Vol. 56, No. 30, pp. 49–63.
13. Hsu, Y.Y. (1962), "On the Size Range of Active Nucleation Cavities on a Heating Surface." *J. Heat Transfer*, Vol. 84, pp. 207–216.
14. Han, C.Y. and Griffith, P. (1965), "Bubble Initiation, Growth, and Departure." *Intl. J. Heat Mass Trans.*, Vol. 8, pp. 887–904.
15. Howell, J.R. and Siegel, R. (1967), "Activation, Growth, and Detachment of Boiling Bubbles in Water from Artificial Nucleation Sites of Known geometry and Size." NASA. TN-D-4101.
16. Mizukami, K. (1975), "Entrapment of Vapor in Re-Entrant Cavities." *Lett. Heat Mass Transfer*, Vol. 2, pp. 279–284.
17. Nishio, S. (1985), "Stability of Pre-Existing Vapor Nucleus in Uniform Temperature ield." *Trans. JSME Ser. B.*, Vol. 54-303, pp. 1802–1807.
18. Hsu, Y.Y. and Graham, R.W. (1976), *Transport Processes in Boiling and Two Phase Systems,* Hemisphere Publishing Corporation, Washington, DC.
19. Kocamustafagoullari, G. and Ishii, M. (1983), "Interfacial Area and Nucleation Site Density in Boiling Systems." *Int. J. Heat Mass Transfer*, Vol. 26, pp. 1377–1387.
20. Fritz, W. (1935), "Maximum Volume of Vapor Bubbles." *Physik Zeitschr*, Vol. 36, pp. 379–384.
21. Cornwell, K. and Brown, R.D. (1978), "Boiling Surface Topography, Volume 1." *Proceedings of the 6th International Heat Transfer Conference*, Toronto, ON, Canada, pp. 157–161.
22. Mikic, B.B. and Rohsenow, W.M. (1969a), "A New Correlation of Pool Boiling Data, Including the Effect of Heating Surface Characteristics." *Journal of Heat Transfer*, Vol. 9, pp. 245–250.
23. Bier, K., Gorenflow, D., Salem, M., and Tanes, Y. (1978), "Pool Boiling Heat Transfer and Size of Active Nucleation Centers for Horizontal Plates with Different Surface Roughness, Volume 1." *Proceedings of the 6th International Heat Transfer Conference*, Toronto, ON, Canada, pp. 151–156.
24. Basu, N., Warrier, G.R., and Dhir, V.K. (2005), "Wall Heat Flux Partitioning during Subcooled Flow Boiling: Part I – Model Development." *J. Heat Transfer*, Vol. 127, pp. 131–140.
25. Pesse, Y.V., Warrier, G.R., and Dhir, V.K. (2005), "an experimental Study of the Gas Entrapment Process in Closed End Microchannels." *Int. J. Heat Mass Transfer*, Vol. 48, pp. 5150–5165.
26. Hsu, Y.Y. and Graham, R.W. (1961), "An Analytical and Experimental Study of Thermal Boundary Layer and Ebullition Cycle in Nucleate Boiling." NASA TND-3943.
27. Hatton, A.P. and Hall, I.S. (1966), "Photographic Study of Boiling on Prepared Surfaces." *Proceedings of the 3rd International Heat Transfer Conference*, Chicago, IL, Vol. 4, pp. 24–37.
28. Rayleigh, J.W.S. (1917), "On the Pressure Developed in a Liquid during the Collapse of a Spherical Cavity", *Philos. Mag. Ser.* C., Vol. 94, p. 122.
29. Mikic, B.B., Rohsenow, W.M., and Griffith, P. (1970), "On Bubble Growth Rates." *Int. J. Heat Mass Transfer*, Vol. 13, pp. 657–666.
30. Plesset, M.S. and Zwick, S.A. (1954), "Growth of Vapor Bubbles in Superheated Liquids." *J. Appl. Phys.*, Vol. 25, No. 4, pp. 493–500.

31. Forster, H.K. and Zuber, N. (1954), "Growth of a Vapor Bubble in a Superheated Liquid." *J. Applied Physics*, Vol. 25, No. 4, pp. 474–478.
32. Mikic, B.B., and Rohsenow, W.M. (1969), "Bubble Growth Rates in Non-Uniform Temperature Field." *Prog. Heat Mass Transfer*, Vol. 2, pp. 283–293.
33. Siegel, R. and Keshock, E.G. (1964), "Effects of Reduced Gravity on Nucleate Bubble Dynamics in Water." *AIChEJ.*, Vol. 10, No. 4, pp 509–516.
34. Cochran, T.K. and Aydelott, J.C. (1966), "Effects of Subcooling and Gravity Level on Boiling in the Discrete Bubble Region." NASA TN-D-3449.
35. Rohsenow, W.M. (1985), "Boiling." *Handbook of Heat Transfer Fundamentals*, 2nd edition, W. M. Rohsenow, J. P. Hartnettand, E. N.Garric, eds., McGraw-Hill, New York, pp. 12.1–12.4.
36. Gorenflo, D., Knabe, V., and Bieling, V. (1986), "Bubble Density on Surfaces with Nucleate Boiling – Its Influence on Heat Transfer and Burnout Heat Flux at Elevated Saturation Pressures." *Proceedings of the 8th International Heat Transfer Conference*, Hemisphere Publishing Corporation, Washington, DC, Vol. 4, pp. 1995–2000.
37. Jakob, M. (1949), *Heat Transfer*, Vol. 1, Wiley, New York.
38. Ivey, H.J. (1967), "Relationships between Bubble Frequency, Departure Diameter and Rise Velocity in Nucleate Boiling." *Int. J. Heat Mass Transfer*, Vol. 10, pp. 1023–1040.
39. Zaber, N. (1961), "The Dynamics of Vapor Bubbles in Nun-Uniform Temperature Fields." *Int. J. Heat Mass Trans.*, Vol. 2, pp. 83–98.
40. Malenkov, F.G. (1971), "Detachment Frequency as a Function of Size of Vapor Bubbles." (translated) *Inzh. Fiz. Zhur.*, Vol. 20, p. 99.

5 Nucleate Boiling Heat Transfer

INTRODUCTION

As discussed earlier in Chapter 1, the rate of heat transfer in nucleate boiling is much higher than that with single-phase natural convection or forced convection. Enhancement in heat transfer occurs not only as a result of energy that is used to produce vapor bubbles but also due to repeated disruption and reformation of thermal layer at and around bubble producing locations (nucleation sites) and the agitation created in the liquid by evolving and departing bubbles. Magnitude and rate of heat transfer associated with these processes vary both spatially and temporally, and the evaluation is further complicated by merger of vapor bubbles both laterally and in the vertical direction. Because of the complexity of the processes, only a few mechanistic models to predict nucleate boiling heat fluxes have been developed, and even those have to rely on a number of empirical constants.

We will discuss mechanistic models and correlations that have been developed in the past for nucleate boiling heat transfer. Often, correlations and models developed for partial nucleate boiling are applied to fully developed nucleate boiling as well. However, one should recognize that vapor removal patterns in partial and fully developed nucleate boiling are very different. In partial nucleate boiling, single bubbles leave the heater surface from discretely located sites on the surface, whereas in fully developed nucleate boiling, bubbles merge both laterally and vertically to form mushroom type of bubbles and vapor columns.

5.1 PARTIAL NUCLEATE BOILING

There are a number of mechanisms that contribute to the total rate of heat transfer from the heater surface in partial nucleate boiling. These mechanisms are transient conduction into liquid over the space vacated by the departing bubbles and natural convection over heater areas where no bubble activity exists. During the evolution of a bubble, heat from the wall is also utilized to support evaporation from the microlayer and around the vapor–liquid interface away from the wall, while superheated liquid surrounding the bubble is also pumped away from the wall. It should be stressed that contribution of various mechanisms to total heat transfer rate not only varies spatially and temporally but also with the regime of nucleate boiling (partial vs. fully developed). Attempts at prediction of nucleate boiling heat flux by combining the contribution of transient conduction on and around nucleation sites, microlayer evaporation underneath the bubbles, and natural convection on inactive areas of the heater have been made. In such efforts, tacit assumption is made that energy stored in the thermal layer is utilized for evaporation at the bubble boundary.

DOI: 10.1201/9781032668437-5

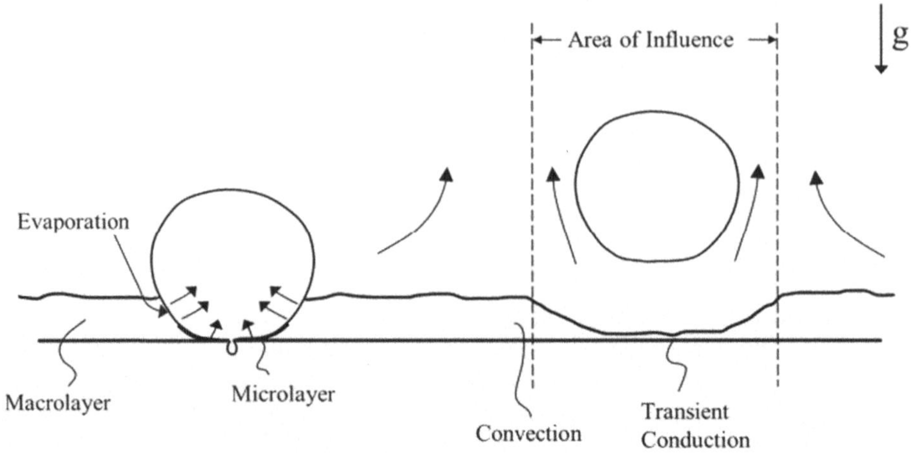

FIGURE 5.1 Various mechanisms of heat transfer during partial nucleate boiling.

Figure 5.1 pictorially shows the various processes. After bubble departure, a new thermal layer forms on the bubble occupied and surrounding area of the wall. Transient conduction into a semi-infinite liquid layer adjacent to the wall is the main mode of heat transfer in the redeveloping thermal layer. This mode of heat transfer occurs during the waiting period over the area of influence of a departed bubble. In the area uninfluenced by the growing and departing bubbles at nucleation sides, heat is removed by natural convection. This is the second mode of heat transfer from portions of the wall. The third mode of heat transfer is the evaporation from the microlayer underneath an evolving bubble. In this formulation, it is implicitly assumed that energy deposited into the thermal layer during the waiting period is utilized for evaporation at the bubble–liquid interface during the bubble growth period.

An expression for the time and area averaged heat flux is written as:

$$q = \frac{K_1^2}{2}\sqrt{\pi\left(k\rho c_p\right)_l f}\, D_d^2 N_a \Delta T + \left(1 - K_1^2 N_a \pi \frac{D_d^2}{4}\right)\bar{h}_{nc}\Delta T + \bar{h}_{ev}\Delta T N_a \frac{\pi}{4} D_d^2 \quad (5.1)$$

The first term accounts for transient conduction into liquid in the developing thermal layer. The second term is contribution of natural convection over areas not influenced by bubble activity at nucleation sites. The third term represents contribution of microlayer evaporation at the base of vapor bubbles. Only the first two terms in the above equation were included in the original model proposed by Mikic and Rohsenow [1]. The evaporation at the bubble boundary as mentioned above is implicitly included in the first term, which represents the transient conduction into the liquid. The addition of the last term on the right-hand side of eq. (5.1) was suggested by Judd and Hwang [2]. Figure 5.1 depicts qualitatively the three modes of heat transfer included in eq. (5.1). In writing the contribution of time average transient conduction in a semi-infinite slab in terms of bubble release frequency in eq. (5.1), growth period is assumed to be much smaller than the waiting period, which is not always true. In eq. (5.1), D_d is the bubble diameter at departure, f is the bubble release frequency, which is the inverse of the sum of waiting period and bubble growth period, and K_1 is the proportionality constant for the bubble area of influence; N_a, is the number of active sites per unit area of the heater; and $\overline{h}_{nc}$ and $\overline{h}_{ev}$ are the average heat transfer coefficients for natural convection and microlayer evaporation, respectively. To evaluate these parameters, one needs to use correlations that themselves have a number of empirical constants embedded in them.

Using empirical correlations for several of these parameters, Mikic and Rohsenow [1] justified the validity of eq. (5.1) when the third term on the right-hand side was not included. Judd and Hwang [2], in matching the heat fluxes, q_P, predicted from eq. (5.1) with those, q_M, observed in experiments in which dichloromethane was boiled on a glass surface, relied on the measured values of microlayer thickness to evaluate $\overline{h}_{ev}$ and on the assumption that K_1^2 has a value of 1.8. Additionally, experimentally measured values of active nucleation site density and bubble release frequency were used in the model. Figure 5.2 shows their data and predictions (as solid lines). It is seen that at the total measured heat flux of 6 w/cm^2, about one-third of the energy is dissipated through evaporation at the bubble base.

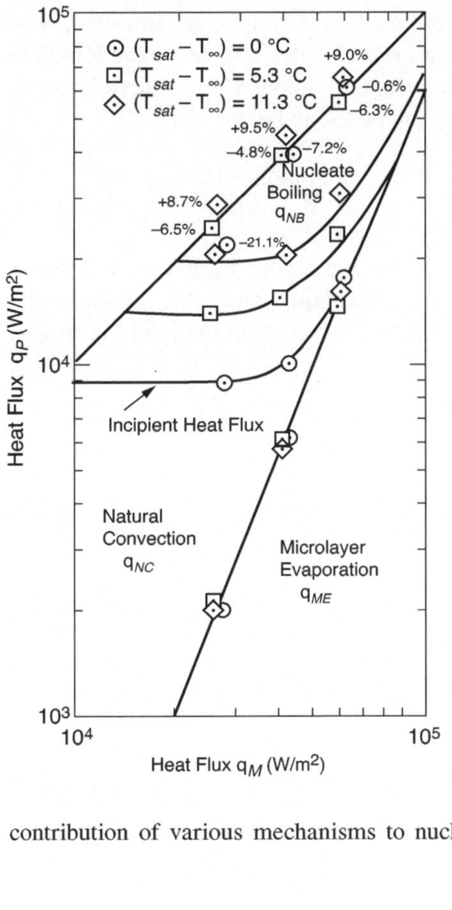

FIGURE 5.2 Relative contribution of various mechanisms to nucleate boiling heat flux (Judd and Hwang) [2].

The data plotted in Figure 5.2 shows that at high heat fluxes or in fully developed nucleate boiling, most of the energy from the heater is removed by evaporation. A similar observation was made by Paul and Abdel-Khalik [3] and by Ammerman et al. [4] on horizontal cylinders. The results of all these studies are in general agreement with the findings of Gaertner [5]—that after the first transition (from partial to fully developed nucleate boiling), evaporation or vapor production is the dominant mode of heat transfer.

5.2 EVAPORATIVE RESISTANCE

During transfer of heat through microlayer underneath a bubble, two resistances contribute: the conductive resistance of the microlayer and the evaporative resistance at the interface. The evaporative resistance results from the fact that net efflux of vapor molecules from the interface takes place. To accomplish this, the temperature of the interface must be higher than the vapor temperature. In obtaining the evaporative resistance, we assume that a very thin layer known as Knudsen layer exists just outside the liquid layer. In the Knudsen layer, exchange of molecules from liquid side and vapor side of the interface occurs. In the absence of any heat transfer, an equilibrium will exist with respect vapor molecules that enter and leave the Knudsen layer. Same will be true for liquid molecules. However, when there is heat transfer from the wall, more vapor molecules will leave the Knudsen layer than entering it. Figure 5.3 shows the conceptual model. From kinetic theory, the number density, n_v, of molecules in the vapor space can be written as

$$n_v = \frac{\text{molecules}}{\text{volume}} = \frac{p_v N_A}{R T_v M} \tag{5.2}$$

where P_v is the pressure in the vapor, N_A is the Avogadro's number, R is the gas constant, T_v is the vapor temperature, and M is the molecular weight. Similarly, the number density of liquid molecules, n_l, in the liquid layer is written as

$$n_l = \frac{p_l N_A}{R T_l M} \tag{5.3}$$

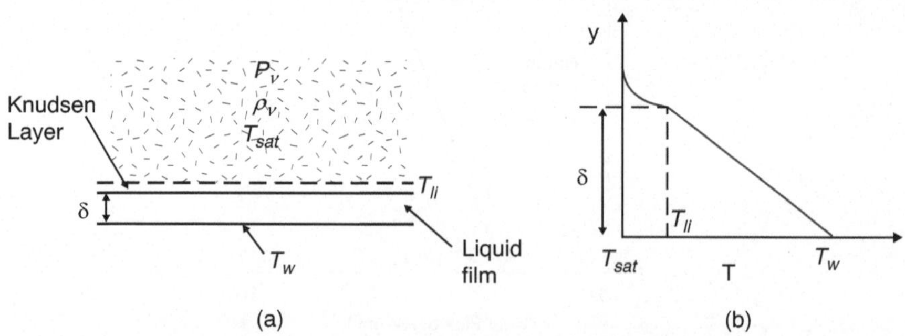

FIGURE 5.3 Model for evaporative resistance (a), and temperature profile (b).

The average velocity of molecules on either side can be written as

$$\overline{c_v} = \left[\frac{8RT_v}{\pi}\right]^{1/2} \tag{5.4}$$

$$\overline{c_l} = \left[\frac{8RT_l}{\pi}\right]^{1/2} \tag{5.5}$$

For evaporation, net flux of molecules from the liquid side of the Knudsen layer is

$$J_m\left(\frac{\text{molecules}}{\text{area time}}\right) = \frac{n_l\overline{c_l}}{4} - \frac{n_v\overline{c_v}}{4} \tag{5.6}$$

$$= \frac{p_l N_A}{4RT_l M}\left[\frac{8RT_l}{\pi}\right]^{1/2} - \frac{p_v N_A}{4RT_v M}\left[\frac{8RT_v}{\pi}\right]^{1/2}$$

$$= \frac{p_l N_A}{M}\left[\frac{1}{2\pi RT_l}\right]^{1/2} - \frac{p_v N_A}{M}\left[\frac{1}{2\pi RT_v}\right]^{1/2} \tag{5.7}$$

Difference between T_l and T_v is small, and as such, we can assume that absolute temperature $T_v \simeq T_l \simeq T_{\text{sat}}$, with this assumption eq. (5.7) becomes

$$J_m = \frac{(p_l - p_v)N_A}{M}\left[\frac{1}{2\pi RT_{\text{sat}}}\right]^{1/2} \tag{5.8}$$

or heat flux at the interface is

$$q = \frac{J_m M}{N_A}h_{fg} = (p_l - p_v)h_{fg}\left[\frac{1}{2\pi RT_{\text{sat}}}\right]^{1/2} \tag{5.9}$$

Using Clausius–Clapeyron relation

$$p_l - p_v = \frac{(T_l - T_v)h_{fg}}{T_{\text{sat}}v_{fg}} \tag{5.10}$$

in eq. (5.9), we obtain

$$q = \frac{(T_l - T_v)h_{fg}^2}{T_{\text{sat}}v_{fg}}\left[\frac{1}{2\pi RT_{\text{sat}}}\right]^{1/2} \tag{5.11}$$

With temperature at the interface being T_{li}, expression for the evaporative resistance at the interface is obtained as

$$\frac{T_{li} - T_v}{q} = \frac{T_{\text{sat}}v_{fg}(2\pi RT_{\text{sat}})^{1/2}}{h_{fg}^2} \tag{5.12}$$

For low pressures, $v_{fg} \simeq \frac{1}{\rho_v}$, and as such, eq. (5.12) for evaporative resistance becomes

$$\frac{T_{li} - T_v}{q} = \frac{T_{sat}\left(2\pi R T_{sat}\right)^{1/2}}{\rho_v h_{fg}^{\ 2}} \tag{5.13}$$

or corresponding heat transfer coefficients, h_i, s_i

$$h_i = \frac{\rho_v h_{fg}^{\ 2}}{T_{sat}\left(2\pi R T_{sat}\right)^{1/2}} \tag{5.14}$$

Example:

Calculate the evaporative resistance for a microlayer of water with a thickness of 10^{-6}m. The layer is supported by a substrate at 105°C, whereas the microlayer is exposed to saturated vapor at 1 atm pressure. What is the interface temperature and what is the evaporative heat flux?

Solution:

For transfer of heat from the substrate to vapor, there are two resistances that act in series: the conductive resistance, $\frac{\delta}{k_l}$, of the microlayer and the evaporative resistance $\frac{1}{h_i}$. As such, heat flux from the wall can be written as

$$q = \frac{(T_w - T_{sat})}{\frac{\delta}{k_l} + \frac{1}{h_i}}$$

From eq. (5.14), h_i is obtained as

$$h_i = \frac{\rho_v h_{fg}^{\ 2}}{T_{sat}\left(2\pi R T_{sat}\right)^{1/2}}$$

$$= \frac{0.596 \times \left(2.257 \times 10^6\right)^2}{373(2\pi \times 461.95 \times 373)^{1/2}} = 7.81 \times 10^6 \ \text{W/m}^2\text{K}$$

or

$$q = \frac{(5)}{\frac{10^{-6}}{0.69} + \frac{1}{7.81 \times 10^6}} = 3.11 \times 10^6 \ \text{W/m}^2$$

Knowing q, we can determine the interface temperature as

$$T_{li} = T_w - \frac{q\delta}{k_l} = 105 - 4.51$$
$$= 100.49°C$$

Thus, we see that in this case interface temperature is 0.49°C higher than the saturation temperature.

5.3 FULLY DEVELOPED NUCLEATE BOILING

As wall superheat is increased, number density of nucleation sites and bubble release frequency at a given site increase. With increase in nucleation site density, bubbles nucleating at neighboring sites begin to merge before departure. Lateral bubble merger leads to formation of mushroom type of bubbles as shown in the photographs in Figure 5.4 [6]. Mushroom-type bubbles are attached to the solid surface through vapor stems with a liquid bridge in between. Prior to departure from the surface merged vapor mass tries to become spherical and goes through size and shape change in the plane of the page and normal to it. According to Gaertner [5], most evaporation occurs at the periphery of vapor stems. Energy for the phase change is supplied by the superheated liquid layer in which the vapor stems are implanted. With increase in frequency of bubble release at a particular site, bubbles begin to merge in the vertical direction to form vapor columns. Photographs in Figure 5.5 show the merger process [6].

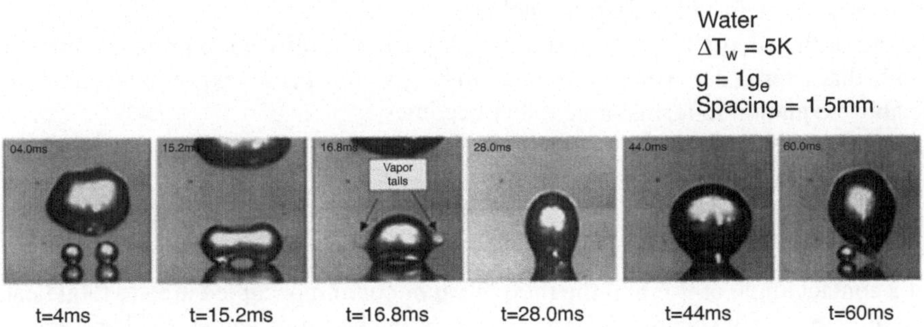

FIGURE 5.4 Sequential photographs of lateral merger of bubbles at two neighboring sites [6].

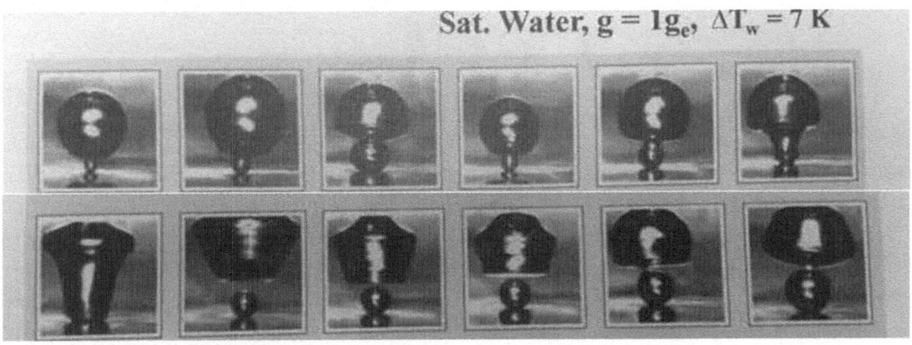

FIGURE 5.5 Sequential photographs of vertical merger of vapor bubbles at a nucleation site [6].

Vertical bubble merger occurs when the upper interface of the succeeding bubble catches up with the rate at which lower interface of the proceeding bubble moves away from the solid wall. After merger, a cylindrical bubble is formed. As the bubble lifts off from the solid surface, a strong inverse wall jet is formed, and the merger process is interpreted for a short while before it starts all over again. Formation of mushroom type of bubbles and vapor columns signals the onset of fully developed nucleate boiling. Although no significant change in the slope of the boiling curve is observed after onset of fully developed nucleate boiling, larger fraction of energy from the wall is transferred by evaporation at the bubble base and boundary.

Moissis and Berenson [7] identified from experiments on a flat plate that vapor columns start to appear on the heater surface when

$$\frac{\dot{V}_v}{A_v V_t} = 9 \tag{5.15}$$

where $\dot{V}_v$ is the vapor volume flow rate over a vapor producing area A_v, and V_t is the terminal velocity of a bubble rising in a pool of liquid.

Use of eq. (4.139) from Chapter 4 for V_t in eq. (5.15) yields

$$\dot{V}_v = 6A_v\left(\frac{g(\rho_l - \rho_v)R}{\rho_l}\right)^{1/2} \tag{5.16}$$

where R is the radius of the rising bubble.

If all of the energy from the heater is assumed to be utilized to produce vapor and bubble diameter at departure is assumed to be given by Fritz's expression, eq. (4.111), the above equation in terms of heat flux becomes

$$q = 0.61\rho_v h_{fg}\frac{A_v}{A_t}\sqrt{\varphi}\left(\frac{\sigma g(\rho_l - \rho_v)}{\rho_l^2}\right)^{1/4} \tag{5.17}$$

For a contact angle of 90° and the ratio of vapor column occupied area to total heater area of $\pi/16$, which corresponds to creation of bubbles on a square grid and spacing equal to twice the bubble departure diameter, eq. (5.17) for heat flux at transition from partial to fully developed nucleate boiling becomes

$$q = 1.14\rho_v h_{fg}\left(\frac{\sigma g(\rho_l - \rho_v)}{\rho_l^2}\right)^{1/4} \tag{5.18}$$

A number of attempts have been made in the past to describe the process of heat transfer in fully developed nucleate boiling. Gaertner [5] proposed that in fully developed nucleate boiling, evaporation is the dominant mode of heat transfer. Evaporation occurs at the periphery of the vapor stems supporting mushroom-type bubbles. Energy for phase change is supplied by the superheated liquid layer in which the stems are implanted. Thus, the boiling heat flux can be calculated if heater area and averaged fractional area occupied by the vapor stems and thickness of the thermal layer are known.

With increase in wall superheat, waiting time between successive nucleations decreases, as a result for much larger fraction of time during a nucleation cycle, bubbles are present at the nucleation sites. Thus, in the limit of zero waiting time in a quasi-static model with mushroom type of bubbles present on the heater surface, the heater area fraction occupied by vapor stems is equal to the product of the number density of stems and the area occupied by one stem. Experimental observations of Iida and Kobayasi [8] and Liaw and Dhir [9] show that vapor occupied fractional area of the heater increases with wall superheat or heat flux. Gaertner and Westwater [10], however, found the stem diameter to decrease with wall heat flux. Because of the rapid increase in number density of nucleation sites, reduction in stem diameter is accompanied by an increase in heater fractional area occupied by vapor. Gaertner and Westwater [10] and Iida and Kobayasi [8] determined that the thickness of the thermal layer varied spatially and the average thermal layer thickness decreased with wall heat flux or superheat. The functional dependence of film thickness on wall heat flux was found to be about the same as of the diameter of the vapor stems. Bobst and Colver [11] also measured the thickness of the thermal layer adjacent to the wall and found the thickness δ_{th}, to vary with wall superheat, ΔT, as

$$\delta_{th} \sim \Delta T^{-6} \tag{5.19}$$

Only mechanistic model for fully developed nucleate boiling is that of Dhir and Liaw [12]. In their model, it was assumed that vapor stems locate on a square grid with spacing L, being $Na^{-1/2}$. The vapor stem diameter was deduced from measured void fractions near the wall and was found to vary as $\Delta T^{-1/4}$. Figure 5.6 shows a sketch of cross-sectional area of stems and superheated liquid layer. Energy from the wall was assumed to be conducted into the liquid micro/macro layer and was used for evaporation at the stationary liquid–vapor interface. The heat transfer rate into the thermal layer and the temperature distribution in it were determined by solving a two-dimensional steady-state conduction equation in the liquid-occupied region. Conduction equation was solved for each ray while including the evaporative resistance at the vapor liquid interface. Figure 5.6 shows their conceptual model. The heater surface area over which stems existed was assumed to be dry. Using average heat transfer coefficients, $\bar{h}_l$ and $\bar{h}_v$ over the liquid and vapor occupied regions, respectively, an expression for the wall heat flux in fully developed nucleate boiling was obtained as

$$q = \left\{ \bar{h}_l \left[1 - \frac{\frac{\pi}{4} D^2}{L^2} \right] + \bar{h}_v \frac{\frac{\pi}{4} D^2}{L^2} \right\} \Delta T \tag{5.20}$$

where D is the diameter of a stem, and L is the spacing between stems. In nucleate boiling, the contribution of the second term in the curly brackets is much smaller than the first term; hence, it can be neglected. The predictions from eq. (5.20) as shown in Figure 5.7 were found to be in good agreement with high heat flux data for a vertical surface at difference contact angles. Equation (5.20) is also valid in transition boiling. At high wall superheats, the contribution of the second term starts to become important, and it cannot be neglected.

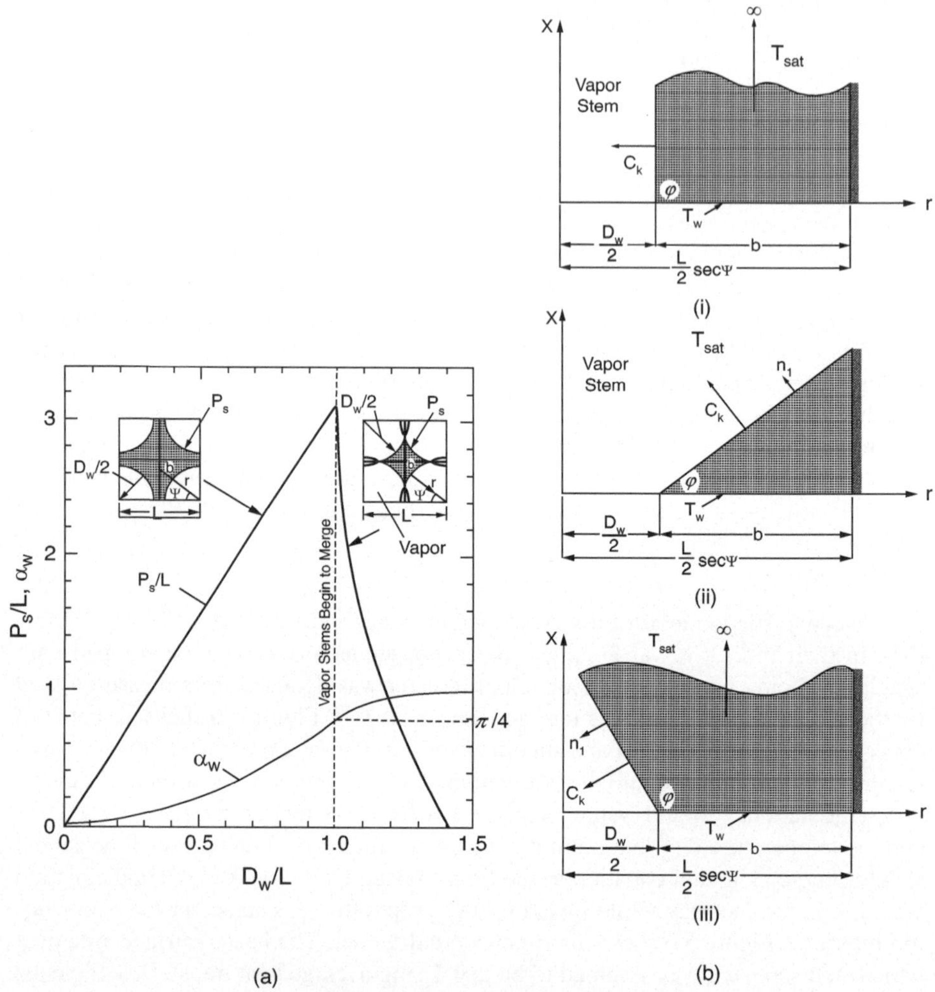

FIGURE 5.6 (a) Dependence of stem periphery and void fraction on stem diameter, and (b) Liquid layer configuration for different contact angles: (i) contact angle equal to 90°; (ii) contact angle less than 90°; (iii) contact angle greater than 90° [12].

5.4 CORRELATIONS FOR NUCLEATE BOILING HEAT TRANSFER

Earliest attempt to develop a predicative correlation for partial nucleate boiling (without consideration of bubble merger) was made by Rohsenow [13].

In developing the correlation, Rohsenow assumed that heat transfer from the surface with boiling could be reduced to a problem similar to that for single-phase forced convection. The convective motion was created by the liquid that flowed downward to the heater to compensate for that used for vapor generation. Hence, a formation similar to that for forced convection could be used. The characteristic length was assumed to be equal to the diameter of a spherical bubble rising through the pool.

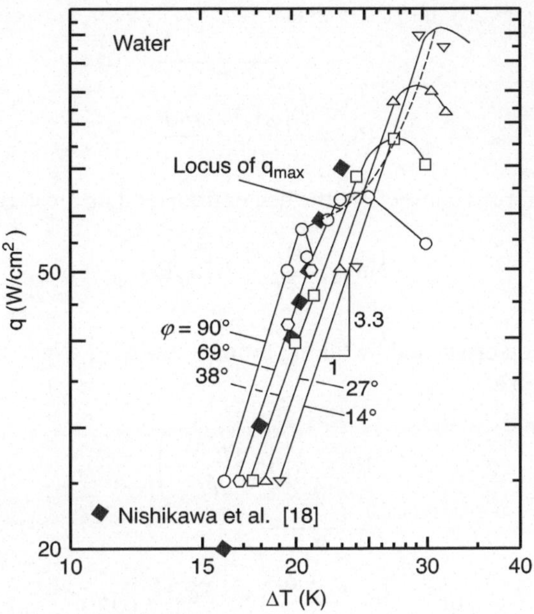

FIGURE 5.7 Effect of surface wettability on high nucleate boiling heat fluxes [12].

The superficial velocity of liquid was obtained by knowing the mass flow rate of liquid needed to compensate for evaporation rate. Assuming that all of the energy lost from the heater was utilized in vapor production, the mass flux of vapor and liquid per unit heater area can be related to the heat flux as

$$\dot{m}_v = \dot{m}_l = \frac{q}{h_{fg}} \tag{5.21}$$

Or, the superficial velocity of liquid (area averaged) can be written as

$$V_l = \frac{q}{\rho_l h_{fg}} \tag{5.22}$$

Or, Reynolds number characterizing the flow of liquid toward the surface is obtained as

$$\text{Re} = \frac{\rho_l V_l D_d}{\mu_l} = \frac{q D_d}{\mu_l h_{fg}} \tag{5.23}$$

where D_d is the bubble diameter at departure. From Fritz's expression, bubble diameter at departure is written as

$$D_d \sim \sqrt{\frac{\sigma}{g(\rho_l - \rho_v)}} \tag{5.24}$$

Substitution of eq. (5.24) in (5.23) yields

$$\mathrm{Re} = \frac{q\sqrt{\dfrac{\sigma}{g(\rho_l - \rho_v)}}}{\mu_l h_{fg}} \tag{5.25}$$

For single-phase forced convection, the heater transfer data are correlated in the form

$$\mathrm{Nu} = \frac{qD_d}{k_l \Delta T} = f(\mathrm{Re}, \mathrm{Pr}) \tag{5.26}$$

Using the definition of Nu and by taking product of $\mathrm{Re} \cdot \mathrm{Pr} \cdot \mathrm{Nu}^{-1}$, the above relationship can be written as

$$\mathrm{Re} = f_1\left(\mathrm{Pr}, \frac{cp_l \Delta T}{h_{fg}}\right)$$

or

$$\frac{q\sqrt{\dfrac{\sigma}{g(\rho_l - \rho_v)}}}{\mu_l h_{fg}} = \left[\frac{c_{pl}\Delta T}{C_s h_{fg}}\right]^n \mathrm{Pr}_l^{-m} \tag{5.27}$$

In eq. (5.27), C_s, n, and m are empirical constants. On commercial surfaces, values of n generally lie between 3 and 4. Constant, m, was found to have a value of 3 for water and 5.1 for other organic liquids. The values of C_S depend on liquid–solid combination. Table 5.1 gives the typical values for C_S. It should be noted that the magnitude of C_S depends on the surface wettability and roughness. Liaw and Dhir [14] systematically studied the effect of surface wettability on vertical copper surfaces placed

TABLE 5.1
Selected Values of the Surface Correlation Factor for Use with eq. (5.27)

Surface-Fluid Combination	C_s
Water-nickel	0.006
Water-platinum	0.013
Water-copper	0.013
Water-brass	0.006
CCl_4-copper	0.013
Benzene-chromium	0.010
n-Pentane-chromium	0.015
Ethyl alcohol-chromium	0.0027
Isopropyl alcohol-copper	0.0025
35% K_2CO_3-copper	0.0054
50% K_2CO_3-copper	0.0027
n-Butyl alcohol-copper	0.0030

TABLE 5.2
Values of C_s for Different Contact Angles with Water as Test Liquid [14]

φ (°)	C_s
14	0.0209
27	0.0202
38	0.0194
69	0.0186
90	0.0172

in a pool of saturated water. They used static contact angle as a measure of wettability. Table 5.2 lists their reported values of C_S with contact angle. The value of C_S increases as heater wettability improves or boiling curve shifts to the right.

Example:

Using Rohsenow's correlation eq. (5.27), calculate nucleate boiling heat flux on a nickel surface for saturated water at 1 atm pressure when the wall superheat is 8°C.

$$q = \frac{\mu_l h_{fg}}{\sqrt{\dfrac{\sigma}{g(\rho_l - \rho_v)}}} \cdot \left[\frac{c_{pl}\Delta T}{C_s h_{fg}}\right]^n \mathrm{Pr}^{-m}$$

Here we use liquid properties at the mean temperature of 377 K.

$$\rho_l = 955 \text{ kg/m}^3, \quad \mathrm{Pr}_l = 1.64$$

$$\mu_l = 267 \times 10^{-6} \text{ Ns/m}^2$$

$$c_{pl} = 4{,}223 \text{ J/kgK}$$

The following properties are obtained at T_{sat} (373 K)

$$\sigma = 58.9 \times 10^{-3} \text{ N/m}$$

$$h_{fg} = 2.257 \times 10^6 \text{ J/kgK}$$

$$\rho_v = 0.597 \text{ kg/m}^3$$

However, since temperature difference is small, little error is made if all the properties are evaluated at the saturation temperature.

From Table 5.1, $C_s = 0.006$ and with $n = 3$ and $m = 3$ we obtain

$$q = \frac{267\times10^{-6}\times2.257\times10^{6}}{\sqrt{\dfrac{58.9\times10^{-3}}{9.81\times(954.4)}}}\cdot\left[\frac{4223\times8}{0.006\times2.257\times10^{6}}\right]^{3}(1.64)^{-3}$$

$$q = 8.47\times10^{5}\ \text{W/m}^2$$

Rohsenow's correlation, though developed for partial nucleate boiling, has been generally found to work for both partial and fully developed nucleate boiling. Although the physics underlying this correlation is not correct and is not in agreement with the mechanistic approach described earlier, the empirical constants used in this correlation generally yield on the average observed dependence of heat flux on wall superheat. Hence, the correlation has found wide acceptance in literature. It should be noted that the magnitude of heat flux at a given wall superheat is very sensitive to the values of C_s. A factor of two change in C_s can cause almost an order of magnitude change in heat flux at a given wall superheat. As such, the values given in Table 5.1 must be considered as a guide, and for a particular application, it is always advisable to establish the value of C_s for conditions of particular interest. Subsequent to Rohsenow's correlation, a number of other correlations for dependence of pool nucleate boiling heat flux on wall superheat have been reported in the literature.

Stephan and Abdelsalam [15] have developed a comprehensive correlation for saturated nucleate pool boiling of different liquids. In developing their correlation, they have divided the liquids into four groups: (i) water, (ii) hydrocarbons, (iii) cryogenic liquids, and (iv) refrigerants. In these correlations, dimensionless heat transfer coefficients (Nusselt numbers) were written in terms of several dimensionless parameters that depend on fluid and solid properties. The important fluid property groups were identified through regression analysis, and the values of exponents of the property groups were obtained by matching predictions with the data. In developing the correlations, data from different heater geometries (such as flat plate, horizontal cylinders, and vertical cylinders) were used. Also, a mean surface roughness of 1 μm was assumed for the heaters. Their correlations for different liquids and contact angels are:

Water:

$$\text{Nu} = 2.46\times10^{6}\cdot X_1^{0.673}\cdot X_4^{-1.58}\cdot X_3^{1.26}\cdot X_8^{5.22}$$

$$10^{-4} \le \frac{p}{p_c} \le 0.886 \quad \text{and} \quad \varphi = 45° \tag{5.28}$$

Hydrocarbons:

$$\text{Nu} = 0.0546\left(X_5^{0.5}\cdot X_1\right)^{0.67}\cdot X_8^{4.33}X_4^{0.248}$$

$$5.7\times10^{-3} \le \frac{p}{p_c} \le 0.9 \quad \text{and} \quad \varphi = 35° \tag{5.29}$$

Cryogenic Liquids:

$$\mathrm{Nu} = 4.82X_1^{0.624} \cdot X_7^{0.117} \cdot X_5^{0.257} \cdot X_3^{0.374} \cdot X_4^{-0.329}$$

$$4.0\times10^{-3} \le \frac{p}{p_c} \le 0.97 \quad \text{and} \quad \varphi = 1° \tag{5.30}$$

Refrigerants:

$$Nu = 207X_1^{0.745} \cdot X_5^{0.581} \cdot X_6^{0.533}$$

$$3.0\times10^{-3} \le \frac{p}{p_c} \le 0.78 \quad \text{and} \quad \varphi = 35° \tag{5.31}$$

All of the above equations correlate the data within ±15% mean absolute error. Nusselt number and various dimensionless groups $X_1,\ldots X_8$ are defined as

$$\mathrm{Nu} = \frac{qD_d}{\Delta T \cdot k_l}$$

$$X_1 = \frac{qD_d}{k_l \cdot T_{\text{sat}}} \qquad X_5 = \frac{\rho_v}{\rho_l}$$

$$X_2 = \frac{\alpha_l^2 \rho_l}{\sigma D_d} \qquad X_6 = \frac{v_l}{\alpha_l}$$

$$X_3 = \frac{c_{pl} T_{\text{sat}} D_d^2}{\alpha_l^2} \qquad X_7 = \frac{(\rho c_p k)_s}{(\rho c_p k)_l}$$

$$X_4 = \frac{h_{fg} D_d^2}{\alpha_l^2} \qquad X_8 = \frac{\rho_l - \rho_v}{\rho_l}$$

Stephan and Abdelsalam have also given a generalized correlation that is applicable to all liquids, but has a larger mean absolute error:

$$\mathrm{Nu} = 0.23X_1^{0.674} \cdot X_5^{0.297} \cdot X_4^{0.371} \cdot X_8^{-1.73} \cdot X_2^{0.35} \tag{5.32}$$

It should be noted that all these correlations suggest that $q \sim \Delta T^{3-4}$ and are valid for earth normal gravity.

Cooper [16] has proposed a much simpler correlation for saturated nucleate pool boiling. His correlation employs reduced pressure, molecular weight, and surface roughness as the correlating parameters. His correlation for a flat plate is:

$$\frac{(q)^{1/3}}{\Delta T} = 55.0\left(\frac{p}{p_c}\right)^{0.12-0.21\log_{10} R_p}\left(-\log_{10}\frac{p}{p_c}\right)^{-0.55} M^{-0.50} \tag{5.33}$$

In the above equation, the roughness, R_p, is measured in microns, and M is the molecular weight in Kg/mole, ΔT is measured in degrees C, and q is given in W/m^2.

Cooper suggests that for application of the correlation to horizontal cylinders, the lead constant on the right-hand side of eq. (5.20) should be increased to 95. It should be noted that correlation eq. (5.20) accounts for roughness but does not account for the variations in degree of surface wettability. Similarly, eqs. (5.28)–(5.31) are for a specific contact angle and for an assumed value of roughness. Also, while eq. (5.33) accounts for geometry of the heater, eqs. (5.28)–(5.31) are independent of the heater geometry. Equation (5.27) is easy to use and is recommended. However, any of the equations should be used with caution, as large deviations between actual data and predictions from these equations can occur when the conditions under which the data used in developing the correlation are not duplicated.

Example:

Using Cooper's correlation eq. (5.33) and $R_p = 1\,\mu\text{m}$, calculate the wall superheat for a heat flux of 10^6 W/m^2 for saturated water at 1 atm pressure.

Solution:

$$\frac{\left(10^6\right)^{1/3}}{\Delta T} = 55.0\left(\frac{1}{220.9}\right)^{0.12-0.21\log_{10}1}\left(-\log_{10}\frac{1}{220.9}\right)^{-0.55} 18^{-0.50}$$

$$\frac{10^2}{\Delta T} = 55.0\left(\frac{1}{220.9}\right)^{0.12}(2.34)^{-0.55}\cdot 0.236$$

$$= 55.0(0.523\times 0.626\times 0.236)$$

$$= 4.25$$

$$\Delta T = 23.5°\text{C}$$

5.5 EFFECT OF VARIOUS SYSTEM VARIABLES ON NUCLEATE BOILING HEAT TRANSFER

Rate of nucleate boiling heat transfer is affected by a number of system variables. Although correlation of Stephan and Abdelsalam includes the effect of thermophysical properties of the solid, some effort has also been made in the past to numerically investigate the effect of thermophysical properties of the substrate, thickness of the substrate, and contact angle on single bubble dynamics. However, no such study exists on the effect of these independent variables on overall heat transfer rate when numerous cavities nucleate on the surface. There are several variables that affect the relationship between heat flux and wall superheat, the effect of which has been only investigated experimentally. We will discuss the influence of several of these variables next.

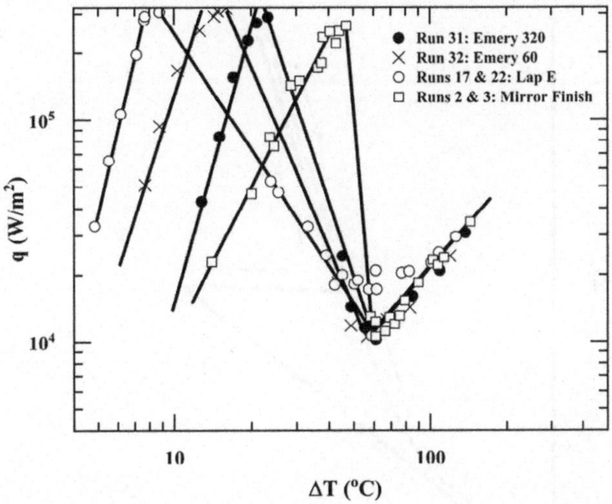

FIGURE 5.8 Effect of surface roughness on boiling transition. (From Berenson [17].)

5.5.1 Surface Finish

On a rough surface, there are many more defects or cavities on the surface than on a smooth polished surface. As such, on a rough surface, potentially more cavities nucleate at a given wall superheat. This in turn leads to higher overall heat transfer rate and the nucleate boiling curve shifts to the left. Figure 5.8 shows Berenson's [17] data on surfaces with different roughnesses. More recently, several studies have been performed on surfaces with protrusions and several types of surface modifications. Again surface modifications as with roughness shift the nucleate boiling curve to the left.

5.5.2 Surface Wettability

Contact angle, which is an indicator of surface wettability, has been shown to affect single bubble dynamics. However, it also affects in a significant way the number of cavities that become active at a given superheat. Liaw and Dhir [14] have systematically studied the effect on nucleate boiling heat transfer rate of static contact angle on vertical surfaces with saturated water at 1 atm pressure as the test liquid. Figure 5.7 given earlier shows their data as well as that of Nishikawa et al. [18] on a vertical surface. With increase in wettability (decrease in contact angle) of the surface, the nucleate boiling curve shifts to the right or the nucleate boiling process becomes less efficient. However, the dependence of heat flux on wall superheat remains about the same. It is interesting to note that maximum heat flux in nucleate boiling on the other hand increases as the surface wettability improves. This is discussed further later.

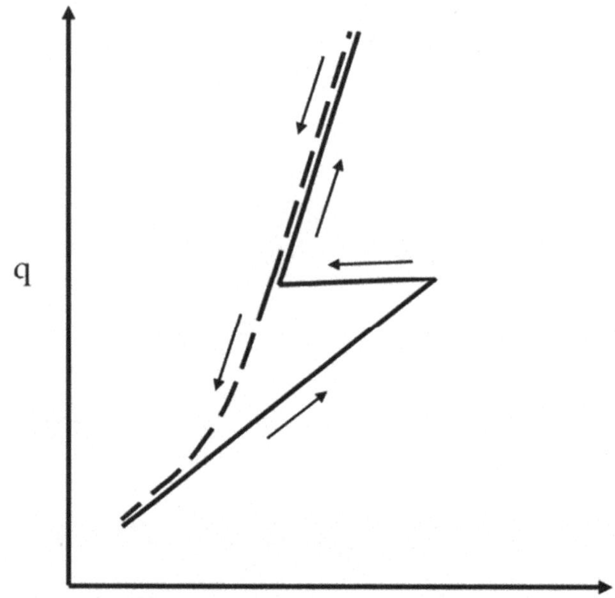

FIGURE 5.9 Hysteresis in boiling curve for well wetting surfaces.

For liquids that wet the surface well (contact angle approaching zero), onset of nucleation may be delayed. For these liquids, single-phase convection will continue to persist up to higher superheats. Upon nucleation of a large number of smaller cavities, a sudden drop in temperature occurs as rate of heat transfer improves from a surface on which heat flux applied to the surface is controlled. Thereafter, the wall superheat will increase with increase in heat flux as that with liquids that partially wet the surface. This behavior does not occur when heat flux is reduced from heigh end of nucleate boiling. As a result, boiling curve shows hysteresis for well-wetting liquids, Figure 5.9 shows qualitatively the process.

5.5.3 Heater Orientation and Geometry

With change in the orientation of the flat surface in relation to the gravitation acceleration vector, bubble dynamics changes and so does the contribution of various mechanisms to total heat transfer from the wall. As the inclination of the surface changes from facing upward to facing downward, bubbles slide along the heater wall before moving away from the wall. These bubbles are generally larger in size when they depart and disrupt the thermal boundary layer adjacent to the heated wall as they slide. Disruption of thermal layer enhances heat transfer by transient conduction and convection from the wall. Nishikawa et al. [18] studied nucleate boiling on flat plates inclined at different angels to the horizontal. In the experiments, saturated water at 1 atm. pressure was used as the test liquid. Figure 5.10 shows their data. It is found that in partial nucleate boiling, heat transfer rates increase as the angle of inclination

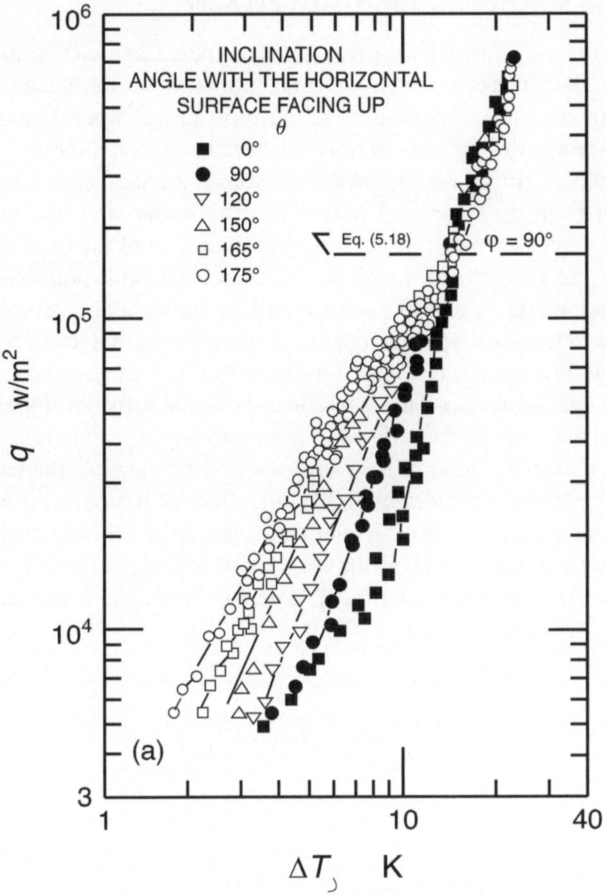

FIGURE 5.10 Nucleate boiling data of Nishikawa et al. [18] on plates oriented at different angles to the horizontal.

of the surface is increased. However, at upper end of nucleate boiling (fully developed), all the data appear to collapse on a single line indicating little effect of surface orientation. Main reason for this behavior is that in fully developed nucleate boiling most of the energy from the heated surface is utilized in evaporation, which is not sensitive to orientation of the plate. In partial nucleate boiling the wall heat flux, q, is found to vary with wall superheat as ΔT^n, where n varies from 2 to 4. However, in fully developed nucleate boiling, value of n is found to lie between 3.3 and 4.3. On a horizontal cylinder, data of Cornwell and Einarsson [19] show that in partial nucleate boiling, the heat transfer coefficient is highest at the lowest stagnation point and is lowest at the top. However, the heat transfer coefficient remains fairly constant over most of the periphery of the cylinder. In summary, we can state that heater geometry does play a role in partial nucleate boiling, but its role diminishes in fully developed nucleate boiling.

5.5.4 Liquid Subcooling and System Pressure

Contribution of the convective heat transfer increases with liquid subcooling. Depending on the thickness of the thermal layer next to the heater surface, liquid subcooling can lead to condensation at the protruding vapor–liquid interface and as a result increase the growth periods of bubbles and reduce the bubble release frequency. Bubbles sitting on the heater for longer periods act as heat pipes while removing heat from the near wall region by evaporation and depositing it into the bulk liquid through condensation on the upper interface of the bubble.

Because of the increase of convective heat transfer with liquid subcooling, heat fluxes at inception and in partial nucleate boiling increase at a given wall superheat in comparison to nucleate boiling at saturation temperature. However, at high nucleate boiling heat fluxes (fully developed nucleate boiling), the saturated and subcooled boiling curves nearly overlap. This is a reflection that in fully developed nucleate boiling, more and more energy from the wall goes into vapor production.

The effect of system pressure is to shift the boiling curve to the left. Correlations described earlier either include implicitly the effect of pressure through changes in thermophysical properties in the correlating groups or have been developed explicitly in terms of system pressure such as the correlation of Cooper [16]. Gorenflo et al. [20] have correlated heat transfer coefficient data for Freon-12, Freon-22, and propane obtained at pressure, $p^* = \left(\frac{p}{p_c}\right)$ up to 0.93 as

$$\frac{h}{h_0} = \left(\frac{q}{q_0}\right)^{n(p^*)} \cdot F(p^*) \tag{5.34}$$

where

$$F(p^*) = 2.1p^{*0.27} + \left(4.4 + \frac{1.8}{1-p^*}\right)p^*$$

and

$$n(p^*) = 0.9 - 0.3p^{*0.3}$$

The values of the reference heat transfer coefficient, h_0, at reference heat flux, $q_0 = 2\times10^4$ W/m^2, and the reference pressure, $p_o^* = 0.03$, for these liquids are:

Liquid	R-12	R-22	C_3H_8
h_2 (W/m^{2}K)	2,300	2,200	2,100

5.5.5 Surface Contamination

Physical and chemical changes at the boiling surface can take place in several ways: deposition of inert matter contained in the host liquid, chemical reaction of the surface with the gases dissolved in the liquid or with the vapor, and chemical reaction of the surface with a concentrated electrolytic solution. During boiling of liquid metals, reaction of the host liquid with oxides on the heated surface can also occur. These processes, if continued over a long period of time, can lead to loss of material or corrosion of the surface. Surface degradation can also occur due to collapse of vapor bubbles in subcooled liquids and repeated collisions with the surface of bubbles entrained in liquid. The presence of foreign material on the surface can affect nucleation behavior by deactivating some cavities, by providing new nucleation sites, and by changing the wettability of the surface. With increase in surface wettability, nucleate boiling curve shifts to the right or the heat transfer coefficient decreases.

The presence of oxygen in the host liquid leads to contamination of the surface via slow chemical reaction. The process, if continued over a long period, can lead to the aging of the surface and a resulting permanent change in its boiling characteristics. The metal ions combine with oxygen to form metal oxides that precipitate around the mouth of the cavities or at the base of bubbles, where high heat and mass transfer gradients exist. The effect of contamination of the surface is similar to that of improved wettability, i.e., to push the nucleate boiling curve to the right or to increase the wall superheat for a given heat flux. In liquid metals, the effect may be opposite as liquid metals remove oxide layer present on the heated surface.

From corrosion studies performed during long-term boiling on iron, copper, and aluminum surfaces, Ross [21] has noted that mass loss rate increases with an increase in the concentration of oxygen in water. At low heat fluxes, the mass loss rate is independent of heat flux, but at high heat fluxes, a reduction in mass loss rate is observed as the heat flux is increased. When sodium chloride (electrolyte) was dissolved in water, the mass loss rate curves show local minima, because at low heat fluxes, the mass loss rate decrease with heat flux, whereas at high heat fluxes, an opposite trend was observed. It has been suggested that minima in mass loss rate may be due to the competing effects of oxygen and chloride ions at the surface.

5.5.6 Magnitude of Gravity

In pool boiling, gravity provides the main force (buoyancy) that lifts the vapor bubbles away from the heater surface and subsequently the processes of bubble inception, growth, merger, and departure repeat. As such, both the magnitude and the components of the gravitational acceleration normal to and along the heater surface influence bubble dynamics and associated heat transfer. On an upward-facing horizontal surface, bubble diameter at departure varies, according to Fritz's expression, as $g^{-1/2}$. Numerical simulation results as summarized by Warrier et al. [22] and described later show that predicted bubble diameter at departure is approximately consistent with Fritz's expression. Figure 5.11 shows predictions from numerical simulations of bubble departure diameter and growth period normalized with that at earth normal gravity as a function of level of gravity for water and PF-5060. For water with a contact angle of 54°, bubble departure diameter is predicted to vary as $g^{-1/2}$, whereas for PF-5060 with contact angle 10°, bubble departure diameter varies as $g^{-0.42}$. The data obtained by Qiu et al. [23] in the parabolic flights and that by Siegel and Keshock [24] in drop tower are consistent with the prediction. Single

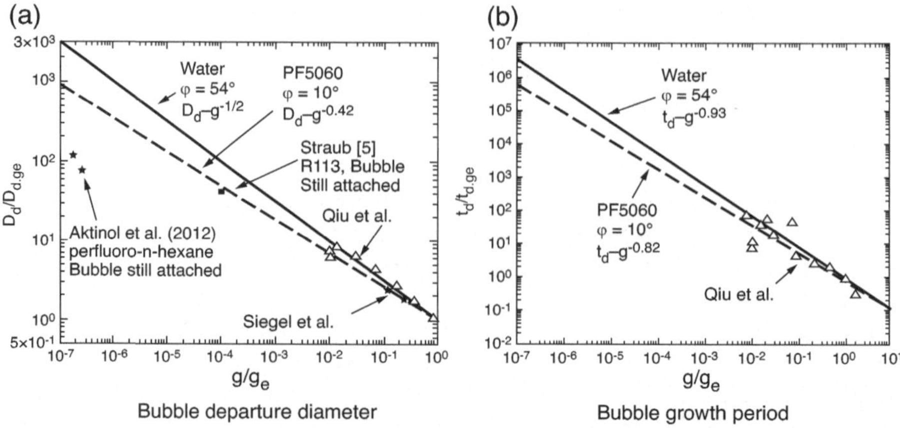

FIGURE 5.11 Variation of normalized single bubble diameter at departure (a) and growth period (b) with level of gravity [22].

bubble data of Straub [25] obtained on space shuttle $\left(\frac{g}{g_e} \cong 10^{-4}\right)$ and that of Aktinol et al. [26] on the International Space Station (ISS) with $\frac{g}{g_e} \cong 10^{-7}$ are also shown. In both cases, bubble diameters are below departure diameters predicted from numerical simulations, and at those diameters, bubbles were still attached to the heater surface—thus confirming the validation of the simulation results. Bubble growth periods are seen to increase more strongly with reduction of gravity. The negative exponent on gravity is found to be weakly dependent on fluid properties and contact angle and is about twice that for bubble departure diameter.

Growth period data obtained in KC-135 aircraft are somewhat overpredicted from the numerical simulations. Thus, based on the limited data for single bubbles, the effect of reduced gravity is to stretch both the length scale and time scale of the boiling process.

At earth normal gravity and the reduced gravity environment of KC-135, vertical and lateral mergers of bubbles lead to formation of columns and mushrooms type of bubbles respectively as described earlier. After mergers, the vapor mass departed from the heater surface. However, in microgravity environment of ISS $\left(\frac{g}{g_e} \cong 10^{-7}\right)$, after initial merger of neighboring bubbles and departure of a merged bubble, a large bubble was seen to form in the middle of the test surface. This bubble served as a vapor sink. Smaller bubbles growing on the surface moved radially inward and merged with the large bubble. Figure 5.12 shows a photograph of the nucleate boiling process obtained from experiments on ISS. At higher superheats, the large bubble may lift off from the surface as a result of the push created by merging small bubbles but continue to hover near the surface while growing in size.

A number of investigations of nucleate boiling heat transfer at different levels of gravity have been reported in the literature and are reviewed by Warrier et al. [22].

FIGURE 5.12 Vapor bubble formed in nucleate boiling on the heated surface on ISS. Bubble shown in picture is about 6 cm in diameter and still sitting on the surface. The test liquid is perfluorohexane.

These studies have been inconclusive. Some have reported higher nucleate boiling heat transfer rates at reduced gravity, some lower and some insensitive to the magnitude of gravity. There are several reasons for these conflicting observations: transient nature of the experiments; the duration of experiments for which reduced gravity persisted, size of the heater in relation to the size of the chamber, and liquid subcooling. With the stretching of the characteristic length for the boiling phenomenon as gravity is reduced, it is extremely important to recognize the relative size of the heater in relation to bubble size as well as the size of the container housing the heater. From experiments conducted at earth normal gravity, KC-135 aircraft and on ISS, Warrier et al. [22] have noted that nucleate boiling heat fluxes decrease as the level of gravity is reduced. From limited available experimental data, which have a significant scatter, Warrier et al. [22] have found that nucleate boiling heat transfer coefficient has a very weak dependence on gravity and correlates as $g^{\frac{1}{8}}$ during seven orders of variation in the gravity level. However, it should be stressed again that size of the heater and container size can play an important role as capillary or the boiling length stretches with reduced gravity. We will discuss this further in Chapter 8.

5.6 SUMMARY

In this chapter, we have developed expressions for partial and fully developed nucleate boiling using mechanistic approach. We have also listed several correlations that apply to both partial and fully developed nucleate boiling. Effect of a number of experimental variables that affect the nucleate boiling heat transfer data has been described.

EXERCISES

1. Calculate the evaporative resistance for water at 10 and 100 atm. pressures, respectively. Compare the evaporative resistance with conductive resistance of a 10×10^{-6} m thick layer of water.
2. Working with the mechanistic model, plot for saturated water at 1 atm. pressure the nucleate boiling heat flux as a function of wall superheat. In the model, use Fritz' expression for bubble diameter at departure and Malenkov's correlation for the product of bubble diameter at departure and frequency. For number density of active sites, use Basu's correlation given in chapter 4. Neglect microlayer contribution. Assume contact angle to be 45°. Discuss your result.
3. If area and time averaged microlayer has a thickness of 10^{-5} m and occupies ½ of the bubble base diameter prior to departure, how would your result from Problem 2 change?
4. Employing Rohsenow's empirical model, plot the wall heat flux as a function of wall superheat for saturated water and FC-72 at 1 atm. pressure. Heater surface material is copper. C_s for water and PF-5060 has a value of 0.013 and 0.006, respectively. Compare your results with those obtained from the correlation of Stephan and Abdelsalam.

NOMENCLATURE

A	Area
C_k	Evaporative heat transfer coefficient
c_s	Empirical constant
c_p	Specific heat at constant pressure
D	Bubble diameter
f	Frequency
g	Gravitational acceleration
h	Heat transfer coefficient
$\bar{h}$	Time or spatially averaged
h_{fg}	Latent heat of vaporization
K_l	Empirical constant
K	Thermal conductivity
L	Spacing of stems
M	Molecular weight
m, n	Empirical constants
$\dot{m}$	Mass flux
N	Number density
p	Pressure
Pr	Prandtl number
R	Radius
Re	Reynolds number

q Heat flux
T Temperature
V Volume or velocity
$\dot{v}$ Volume flow rate

SUBSCRIPTS

a Active
d Departure
e Earth
ev Evaporation
l Liquid
nc Natural convection
t Terminal or total
th Thermal
v Vapor
w Wall

GREEK LETTERS

α Void fraction
ρ Density
σ Surface tension
δ Film thickness
μ Molecular density

REFERENCES

1. Mikic, B.B. and Rohsenow, W.M. (1969), "A New Correlation of Pool Boiling Data, Including the Effect of Heating Surface Characteristics." *J. Heat Transfer*, Vol. 91, No. 2, pp. 245–250.
2. Judd, R.L. and Hwang, K.S. (1976), "A Comprehensive Model for Nucleate Boiling Heat Transfer Including Microlayer Evaporation." *J. Heat Transfer*, Vol. 98, pp. 623–629.
3. Paul, D.D. and Abdel-Khalik, S.I. (1983), "A Statistical Analysis of Saturated Nucleate Boiling Along a Pleated Wire." *Int. J. Heat Mass Transfer*, Vol. 26, pp. 509–519.
4. Ammerman, C.N., You, S.S., and Hong, Y.S. (1996), "Identification of Pool Boiling Heat Transfer from a Wire Immersed in Saturated FC-72 Using a Simple/LDA Method." *J. Heat Transfer*, Vol. 117, pp. 117–123.
5. Gaertner, R.F. (1965), "Photographic Study of Nucleate Pool Boiling on a Horizontal Surface." *ASME J. Heat Transfer*, Vol. 87, pp. 17–29.
6. Son, G., and Dhir, V.K. (2008), "Numerical Simulation of Nucleate Boiling at High Heat Fluxes." *Int. J. Heat Mass Transfer*, Vol. 51, pp. 2566–2582.
7. Moissis, R., and Berenson, P.J. (1963), "On the Hydrodynamic Transitions in Nucleate Boiling." *ASME J. Heat Mass Transfer*, Vol. 27, pp. 221–229.
8. Iida, Y., and Kobayasi, K. (1969), "Distribution of Void Fraction above a Horizontal Heating Surface in Pool Boiling." *Bull. JSME*, Vol. 12, pp. 283–290.

9. Liaw, S.P. and Dhir, V.K. (1989), "Void Fraction Measurement during Saturated Pool Boiling of Water on Partially Wetted Vertical Surfaces." *ASME Journal of Heat Transfer*, Vol. 111, pp 731–738.
10. Gaertner, R.F. and Westwater, J.W. (1960), "Population of Active Sites in Nucleate Boiling Heat Transfer." *Chem. Eng. Prog. Symp. Ser.*, Vol. 56, pp. 39–48.
11. Bobst, R.W. and Colver, C.P. (1968), "Temperature Profiles up to Burnout Adjacent to a Horizontal Heating Surface in Nucleate Pool Boiling of Water." *Chem. Eng. Prog. Symp. Ser.*, Vol. 64, pp. 26–32.
12. Dhir, V.K. and Liaw, S.P. (1989), "Framework for a Unified Model for Nucleate and Transition Boiling." *ASME J. Heat Transfer*, Vol. 111, pp. 739–746.
13. Rohsenow, W.M. (1952), "A Method of Correlating Heat Transfer Data for Surface Boiling of Liquids." *Trans. ASME*, Vol. 74, pp. 969–976.
14. Liaw, S.P. and Dhir, V.K. (1986), "Effect of Surface Wettability on Transition Boiling Heat Flux from a Vertical Surface", *Proceedings of International Heat Transfer Conference*, Vol. 4, San Francisco, pp. 2031–2036.
15. Stephan, K. and Abdelsalam, M. (1980), "Heat Transfer Correlation for Natural Convection Boiling." *Intl. J. Heat Mass Transfer*, Vol. 23, pp. 73–87.
16. Cooper, M.G. (1984), "Saturation Nucleate Pool Boiling – A Simple Correlation." *IChemE Symp. Ser.*, Vol. 86, pp. 786–793.
17. Berenson, P. (1962), "Experiments in Pool Boiling Heat Transfer." *Intl. J. Heat Mass Transfer*, Vol. 5, pp. 985–999.
18. Nishikawa, K., Fujita, Y., and Ohta, H. (1984), "Effect of Surface Configuration on Nucleate Boiling Heat Transfer." *International Journal of Heat and Mass Transfer*, Vol. 27, pp. 1559–1571.
19. Cornwell, R. and Einarsson, J.G. (1989), "The influence of Fluid Flow on Nucleate Boiling from a Tube." *Proceedings of European Seminar, No. 8: On Advances in Pool Boiling Heat Transfer*, University of Paderborn, May 11–12, 1989, Paderborn, Germany, pp. 28–41.
20. Gorenflo, D., Knabe, V., and Bieling, V. (1986), "Bubble Density on Surfaces with Nucleate Boiling – Its Influence on Heat Transfer and Burnout Heat Flux at Elevated Saturation Pressures." *Proceedings of the 8th International Heat Transfer Conference*, Vol. 4, August 1986, San Francisco, CA, pp. 1995–2000.
21. Ross, T.K. (1967), "Corrosion and Heat Transfer Review." Ed. Corros. J., Vol. 2, pp. 131–142.
22. Warrier, G.R., Dhir, V.K., and Chao, D.F. (2015), "Nucleate Pool Boiling Experiment (NPBX) in Microgravity." *Int. J. Heat Mass Transfer*, Vol. 83, pp. 781–798.
23. Qiu, D.M., Dhir V.K., Chao, D., Hasan, M., Neumann, E., Yee, G., and Birchenough, A. (2002), "Single Bubble Dynamics during Pool Boiling Under Low Gravity Conditions." *J. Thermophys. Heat Transfer*, Vol. 16, pp. 336–345.
24. Siegel, R. and Keshock, E.G. (1964), "Effect of Reduced Gravity on Nucleate Boiling Bubble Dynamics in Saturated Water." *AIChE J.*, Vol. 10, pp. 509–517.
25. Straub, J. (2001), "Boiling Heat Transfer and Bubble Dynamics in Microgravity." *Adv. Heat Transfer*, Vol. 35, pp. 57–172.
26. Aktinol, E., Warrier, G., and Dhir, V.K. (2014), "Single Bubble Dynamics Under Microgravity Conditions in the Presence of Dissolved Gas in the Liquid." *Int. J. Heat Mass Transfer*, Vol. 79, pp. 251–268.

6 Hydrodynamic Analysis of Maximum and Minimum Heat Fluxes

INTRODUCTION

As noted in the introductory chapter, maximum heat flux represents the upper limit of nucleate boiling or efficient heat removal. Minimum heat flux, on the other hand, marks the lowest heat flux at which the surface is fully covered with vapor and no liquid–solid contact occurs. Thus, maximum and minimum heat fluxes represent two extremes of the transition boiling portion of the boiling curve. In a heat flux-controlled process, the occurrence of maximum heat flux condition leads to severe degradation in the rate of heat transfer from the surface. The resulting rapid rise in temperature can lead to local melting and in turn damage to the surface on which boiling occurs. Because of the practical importance of the knowledge of the magnitude of the maximum heat flux, numerous studies involving correlations and mechanistic models have been developed over the past eight decades. Bonilla and Perry [1] in 1941 proposed that the process of occurrence of maximum heat flux on a horizontal plate was similar to flooding of columns in the chemical industry. By modifying the dimensionless groups used to correlate the flooding data, they were reasonably successful in correlating the maximum heat flux data for binary mixtures. However, they did not pursue this approach further. Kutateladze [2,3], as noted by Zuber [4], proposed, "essential feature of the theory of the phenomenon can be derived if one assumes that the crisis in the boiling process is purely a hydrodynamic phenomenon: the destruction of stability of two phase flow existing close to the heating surface." Accordingly, the maximum heat flux occurs when the velocity of vapor leaving the heater surface reaches a critical value. Developing dimensionless groups from the momentum and energy equations, Kutateladze obtained an expression for the maximum heat flux on a horizontal surface that required an empirical constant to correlate the data. The expression obtained by Kutateladze involved simply the square root of the main dimensionless groups identified by Bonilla and Perry [1].

Zuber [4] proposed that maximum and minimum heat fluxes representing two ends of the transition boiling curve were governed by hydrodynamic processes present in transition boiling. In transition boiling, an irregular film covers portions of the heater surface. The film is inherently unstable. At heat fluxes at and above the minimum heat flux, film covers the entire surface uniformly and instability of the interface determines the bubble release from the interface. Near maximum heat flux, liquid contacts most of the surface, rate of evaporation is high, and release of bubbles resembles small explosions or burst of vapor. Spacing of these vapor bursts is similar

DOI: 10.1201/9781032668437-6

to that of release of vapor bubbles in film boiling. However, the occurrence of the critical velocity in these vapor bursts (columns) leads to accumulation of vapor on the surface and onset of transition from nucleate boiling. The formalization by Zuber [4] of the hydrodynamic theory for maximum and minimum heat fluxes was based on the instability of the vapor film in film boiling and critical velocity of vapor columns at which they became unstable and interrupted the vapor outflow in nucleate boiling. This was a major step forward in developing analytical expressions for highly complex phenomena behind occurrence of minimum and maximum heat fluxes. In recent years, a number of alternative approaches to describe the occurrence of maximum heat flux have been proposed. For the sake of completeness, we will briefly describe these at the end of the chapter. Before we proceed with the analytical approach used by Zuber [4], we carry out analysis of two-dimensional Taylor and Helmholtz instabilities at the vapor–liquid interfaces, which are the corner stones of Zuber's analyses.

6.1 TAYLOR INSTABILITY

Taylor instability is the name given to instability of an interface between two fluid layers when the heavier fluid lies over a lighter fluid and acceleration is directed from the heavier to the lighter fluid and normal to the interface. The instability of a lighter fluid rising in a heavier fluid was first considered by Rayleigh as given in Lamb [5]. G.I. Taylor [6] analyzed this instability between two incompressible and inviscid fluid layers of infinite extent in the manner we have come to recognize it but without including surface tension. Bellman and Pennington [7] extended Taylor's analysis by incorporating the effect of surface tension and viscosity. Figure 6.1 shows a typical configuration. Under the assumptions that fluids are inviscid, incompressible; fluid layers are stationary and are of infinite depth, wave amplitude is much smaller than the wavelength; no heat or mass transfer occurs; the two layers are in thermal equilibrium, and the nonlinear terms are small, the continuity and momentum equations in two dimensions (x, y) for each phase with the liquid at the top and vapor underneath can be written as

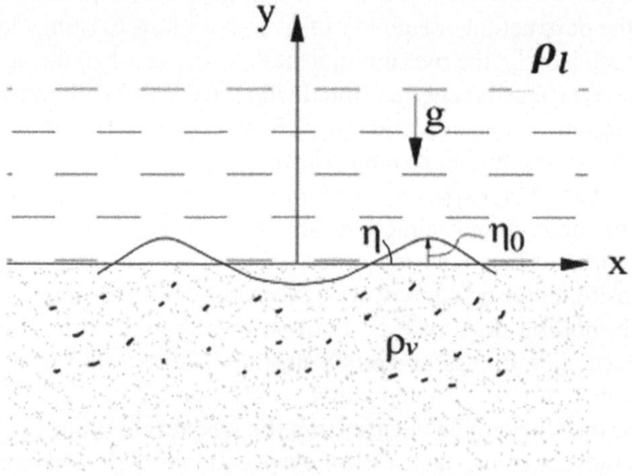

FIGURE 6.1 Configuration for two-dimensional Taylor instability analysis.

$$\frac{\partial u}{\partial x}+\frac{\partial v}{\partial y}=0 \tag{6.1}$$

$$\frac{\partial u}{\partial t}=-\frac{1}{\rho}\frac{\partial p}{\partial x} \tag{6.2}$$

$$\frac{\partial v}{\partial t}=-\frac{1}{\rho}\frac{\partial p}{\partial y}-g \tag{6.3}$$

In writing the conservation equations, we also assume that perturbation in pressure does not lead to evaporation or condensation at the interface and vapor behaves like a gas. Let us perturb the plane interface such that

$$\eta=\eta_0 e^{i(\omega_r+i\omega_i)t}\cos kx \tag{6.4}$$

where ω_r and ω_i are the real and imaginary components of the wave frequency, and k is the wave number. For a stationary wave, $\omega_r=0$ or

$$\eta=\eta_0 e^{-\omega_i t}\cos kx \tag{6.5}$$

If ω_i is a negative number, the interface will grow with time and the interface is unstable. If on the other hand, ω_i is positive, the perturbation will die with time and the interface is stable. For $\omega_i=0$, magnitude of the perturbation remains constant with time. Such a disturbance is called neutral disturbance.

The kinematic condition at the vapor–liquid interface can be written as

$$\frac{d\eta}{dt}\approx\frac{\partial \eta}{\partial t}=v \tag{6.6}$$

or

$$-\omega_i\eta_0 e^{-\omega_i t}\cos kx=v$$

or

$$v=-\eta\omega_i \tag{6.7}$$

Next, we define a potential function, Φ, such that

$$u=\frac{\partial \Phi}{\partial x} \tag{6.8}$$

$$v=\frac{\partial \Phi}{\partial y}. \tag{6.9}$$

Noting that far away from the interface, perturbation components of velocity should vanish and eliminating v between eqs. (6.7) and (6.9), we find for the liquid phase

$$\Phi_l = +\eta_0 \frac{\omega_i}{\kappa} e^{-ky} e^{-\omega_i t} \cos kx \tag{6.10}$$

and for the vapor phase

$$\Phi_v = -\eta_0 \frac{\omega_i}{\kappa} e^{+ky} e^{-\omega_i t} coskx \tag{6.11}$$

The chosen potential functions go to zero as $y \rightarrow +\infty$ or $-\infty$ and satisfy the continuity equations.

In terms of potential function for the liquid phase, the momentum equations in the x and y directions respectively become

$$\frac{\partial^2 \Phi_l}{\partial x \partial t} = -\frac{1}{\rho_l} \frac{\partial p_l}{\partial x} \tag{6.12}$$

and

$$\frac{\partial^2 \Phi_l}{\partial y \partial t} = -\frac{1}{\rho_l} \frac{\partial p_l}{\partial y} - g \tag{6.13}$$

Integrating eq. (6.13) with respect to y, we find

$$\frac{\partial \Phi_l}{\partial t} = -\frac{1}{\rho_l} p_l - gy + f(x) \tag{6.14}$$

The constant of integration $f(x)$ can, without any loss of generality, be assumed to be a constant independent of x. It can be easily verified that eq. (6.14) will satisfy eq. (6.12) as well. The constant of integration can be determined if we evaluate eq. (6.14) in the limit of $y = 0$ while recognizing that at the unperturbed interface, $\frac{\partial \Phi}{\partial t} = 0$, therefore

$$f(x) = C = \frac{p_{0l}}{\rho_l} \tag{6.15}$$

where p_{0l} is the pressure on the liquid side of the unperturbed interface. After substitution for $f(x)$, eq. (6.14) can be written as

$$\frac{\partial \Phi_l}{\partial t} = \frac{(p_{0l} - p_l)}{\rho_l} - gy \tag{6.16}$$

An equation similar to eq. (6.16) is obtained for the vapor phase as

$$\frac{\partial \Phi_v}{\partial t} = \frac{(p_{ov} - p_v)}{\rho_v} - gy \tag{6.17}$$

At the undisturbed interface pressure, p_{ol} on the liquid side will be equal to pressure, p_{ov}, on the vapor side. Eliminating p_{ol} and p_{ov} between eqs. (6.16) and (6.17), we obtain

$$\rho_l \frac{\partial \Phi_l}{\partial t} - \rho_v \frac{\partial \Phi_v}{\partial t} = (p_v - p_l) - (\rho_l - \rho_v) g y \tag{6.18}$$

Substituting for $\frac{\partial \Phi_l}{\partial t}$ and $\frac{\partial \Phi_v}{\partial t}$ from eqs. (6.10) to (6.11) respectively and evaluating eq. (6.18) at $y = \eta$, we obtain

$$-\left[\rho_l \frac{\omega_i^2}{\kappa}\eta + \rho_v \frac{\omega_i^2}{\kappa}\eta\right] = (p_v - p_l)\,|_\eta - (\rho_l - \rho_v) g \eta \tag{6.19}$$

For an interface with curvature only in the x-y plane, the pressure difference between vapor and liquid can be written as

$$(p_v - p_l)\,|_\eta = \frac{\sigma}{R_1} \tag{6.20}$$

The curvature, $\frac{1}{R_1}$, is related to the interface shape as

$$\frac{1}{R_1} = -\frac{\frac{\partial^2 \eta}{\partial x^2}}{\left[1 + \left(\frac{\partial \eta}{\partial x}\right)^2\right]^{\frac{3}{2}}} \approx -\frac{\partial^2 \eta}{\partial x^2} = \kappa^2 \eta \tag{6.21}$$

where we have assumed that $\frac{\partial \eta}{\partial x}$ is much smaller than 1.

After substitution for $(p_v - p_l)$ from eq. (6.20) and R_1 from eq. (6.21), eq. (6.19) becomes

$$(\rho_l + \rho_v)\omega_i^2 \frac{\eta}{k} = -k^2 \eta \sigma + (\rho_l - \rho_v) g \eta \tag{6.22}$$

or

$$\omega_i^2 = \left[-\frac{\sigma k^3}{(\rho_l + \rho_v)} + \frac{(\rho_l - \rho_v) g k}{(\rho_l + \rho_v)}\right]$$

or

$$\omega_i = \pm\left[-\frac{\sigma k^3}{(\rho_l + \rho_v)} + \frac{(\rho_l - \rho_v) g k}{(\rho_l + \rho_v)}\right]^{1/2}$$

Choosing the negative sign for an unstable interface, we find

$$-\omega_i = \left[\frac{(\rho_l - \rho_v)gk}{(\rho_l + \rho_v)} - \frac{\sigma k^3}{(\rho_l + \rho_v)}\right]^{1/2} \tag{6.23}$$

From eq. (6.23) we see that buoyancy will tend to destabilize the interface, whereas surface tension will have a stabilizing effect. For two-dimensional neutral wave, $\omega_i = 0$, or

$$\frac{(\rho_l - \rho_v)gk}{(\rho_l + \rho_v)} = \frac{\sigma k^3}{(\rho_l + \rho_v)}$$

or, critical wave number k_{c2}, for onset of instability is

$$k_{c2}^2 = \left(\frac{\lambda_{c2}}{2\pi}\right)^2 = \frac{(\rho_l - \rho_v)g}{\sigma}$$

or, the critical wavelength, λ_{c2}, is

$$\lambda_{c2} = 2\pi\sqrt{\frac{\sigma}{g(\rho_l - \rho_v)}} \tag{6.24}$$

Subscript 2 denotes that it is for a two-dimensional case. Interface will be unstable for all wavelengths greater than the critical wavelength, λ_{c2}, and stable for perturbations with wavelengths less than λ_{c2}. The "most dangerous" wavelength, λ_{d2}, is the wavelength for which the growth rate is maximum. Differentiating w_i with respect to k and setting $\frac{d\omega_i}{dk} = 0$, we obtain

$$\frac{(\rho_l - \rho_v)gk}{(\rho_l + \rho_v)} - \frac{3\sigma k^2}{(\rho_l + \rho_v)} = 0$$

or

$$k_{d2}^2 = \frac{(\rho_l - \rho_v)g}{3\sigma} \tag{6.25}$$

or, the "most dangerous" wavelength λ_{d2}, is

$$\lambda_{d2} = 2\pi\sqrt{3}\sqrt{\frac{\sigma}{(\rho_l - \rho_v)g}} \tag{6.26}$$

Thus, we see that the "most dangerous" wavelength is $\sqrt{3}$ times longer than the critical wavelength. Substituting for κ_{d2} from eq. (6.25) into eq. (6.23), we obtain the frequency corresponding to the "most dangerous" wavelength as

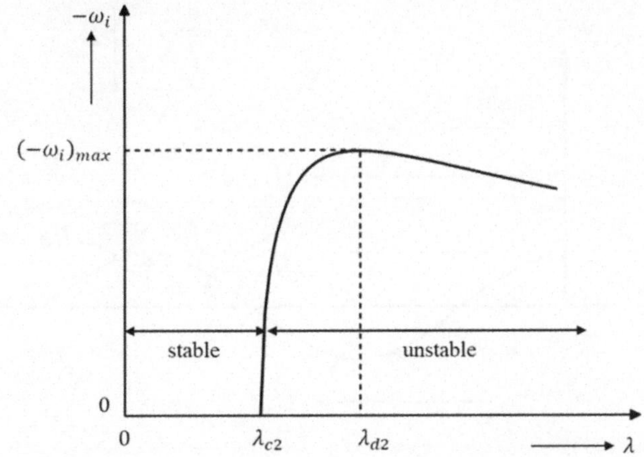

FIGURE 6.2 Dispersion relation for two-dimensional Taylor instability.

$$(-\omega_i)_{\max} = \sqrt{\frac{2}{3\sqrt{3}} \frac{g(\rho_l - \rho_v)}{(\rho_l + \rho_v)} \sqrt{\frac{g(\rho_l - \rho_v)}{\sigma}}} \tag{6.27}$$

Figure 6.2 shows a plot of the dispersion relation, eq. (6.23), between frequency and wavelength. Three- dimensioal analysis [10] will yield three-dimensional "most dangerous" wavelength to be $\sqrt{2}$ times longer than the two-dimensional wavelength. Next, we look at the instability of an interface between two parallel flowing fluid streams.

6.2 INSTABILITY OF THE PLANE INTERFACE BETWEEN TWO PARALLEL STREAMS

Figure 6.3 shows the typical configuration in which streams of liquid and vapor at velocities U_l and U_v move parallel to each other. The interface is disturbed so that only two-dimensional disturbances are considered.

In carrying out the analysis, we assume that:

1. Fluids are inviscid and incompressible
2. Second-order nonlinear terms can be neglected
3. Wave amplitude is much smaller than the wavelength
4. Fluid layers are of infinite extent
5. Liquid and gas/vapor layers move at a free stream velocity of U_l and U_v respectively
6. Vapor and liquid are in thermal equilibrium, and no heat and mass transfer occurs across the interface.

If the initial plane interface between the two streams is disturbed such that displacement is given by

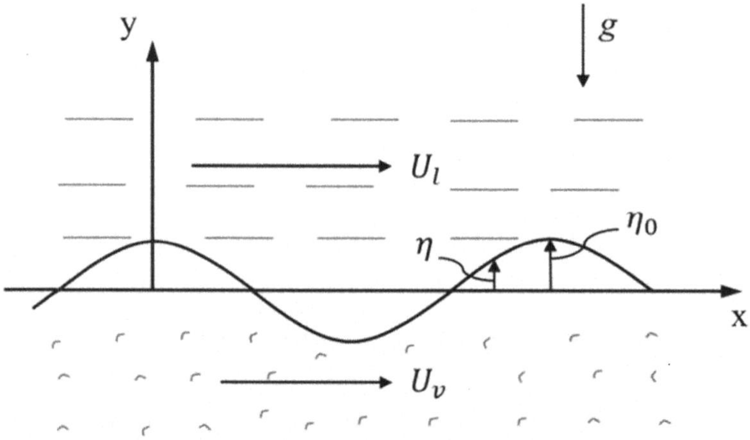

FIGURE 6.3 Instability of plane interface between two parallel streams.

$$\eta = \eta_0 e^{i(\omega t - kx)} \tag{6.28}$$

where ω is a complex frequency defined as

$$\omega = \omega_r + i\omega_i \tag{6.29}$$

Potential functions satisfying the continuity equations for liquid and vapor phases can be written as

$$\Phi_l = U_l x + A_l e^{-ky} e^{i(\omega t - kx)} \tag{6.30}$$

$$\Phi_v = U_v x + A_v e^{ky} e^{i(\omega t - kx)} \tag{6.31}$$

where U_l and U_v are the free stream velocities of liquid and vapor streams respectively and A_l and A_v are constants.

A relation between constants A_l, A_v and η_0 is obtained by using the kinematic condition at the interface as

$$v \equiv \frac{\partial \Phi}{\partial y} = \frac{d\eta}{dt} = \frac{\partial \eta}{\partial t} + U\frac{\partial \eta}{\partial x} \tag{6.32}$$

where U is the fluid velocity in the x-direction.

Substitution in eq. (6.32) for Φ from eq. (6.30) and η from eq. (6.28) yields for the liquid phase

$$A_l = -i\eta_0 (c - U_l) \tag{6.33}$$

In eq. (6.33), c is the wave velocity and is a complex number defined as

$$c = \frac{\omega}{k} = \frac{\omega_r + i\omega_i}{k} \tag{6.34}$$

where k is the wave number.

Similarly, for the vapor/gas phase

$$A_v = i\eta_0 (c - U_v) \tag{6.35}$$

After A_l and A_v are substituted from eqs. (6.33) and (6.35) respectively into eqs. (6.30) and (6.31), the expressions for the potential functions for liquid and vapor respectively become

$$\Phi_l = U_l x - i\eta_0 (c - U_l) e^{-ky} e^{i(\omega t - kx)} \tag{6.30a}$$

$$\Phi_v = U_v x + i\eta_0 (c - U_v) e^{ky} e^{i(\omega t - kx)} \tag{6.31a}$$

Neglecting the second-order terms, the momentum equations in the x and y direction for either of the phases can be written as

$$\frac{\partial u}{\partial t} + U\frac{\partial u}{\partial x} = -\frac{1}{\rho}\frac{\partial p}{\partial x} \tag{6.36}$$

$$\frac{\partial v}{\partial t} + U\frac{\partial v}{\partial x} = -\frac{1}{\rho}\frac{\partial p}{\partial y} - g \tag{6.37}$$

In terms of the potential function, eqs. (6.36) and (6.37) are written as

$$\frac{\partial^2 \Phi}{\partial x\, \partial t} + U\frac{\partial^2 \Phi}{\partial x^2} = -\frac{1}{\rho}\frac{\partial p}{\partial x} \tag{6.38}$$

$$\frac{\partial^2 \Phi}{\partial y\, \partial t} + U\frac{\partial^2 \Phi}{\partial x\, \partial y} = -\frac{1}{\rho}\frac{\partial p}{\partial y} - g \tag{6.39}$$

Integration of eq. (6.39) with respect to y yields

$$\frac{\partial \Phi}{\partial t} + U\frac{\partial \Phi}{\partial x} = -\frac{p}{\rho} - gy + f(x) \tag{6.40}$$

Without any loss of generality, the function $f(x)$ can be assumed to be a constant. As such, eq. (6.40) can be shown to satisfy eq. (6.38) as well. The constant of integration can be obtained by setting the limits of integration on eq. (6.40) as

$$\left[\frac{\partial\Phi}{\partial t}\right]_0^y + \left[U\frac{\partial\Phi}{\partial x}\right]_0^y = -\left[\frac{p}{\rho}\right]_0^y - gy \tag{6.41}$$

At the plane undisturbed interface ($y=0$), $\dfrac{\partial\Phi}{\partial t}=0$ and $\dfrac{\partial\Phi}{\partial x}=U$. With these substitutions, eq. (6.41) becomes for any position of the interface as

$$\frac{\partial\Phi}{\partial t} + U\frac{\partial\Phi}{\partial x} = -\frac{1}{\rho}(p-p_0) - gy + U^2 \tag{6.42}$$

In eq. (6.42), p_0 is the static pressure at the undisturbed interface. A comparison of eq. (6.42) with eq. (6.40) shows that the constant of integration in eq. (6.40) is

$$\frac{p_0}{\rho} + U^2 \tag{6.43}$$

After substitution for Φ from eq. (6.30), eq. (6.42) when evaluated at $y=\eta$ yields for the liquid phase

$$\rho_l\left[\eta_0(c-U_l)\omega - U_l\eta_0(c-U_l)k\right]e^{i(\omega t-kx)} = -p_l + p_0 - \rho_l g\eta \tag{6.44}$$

Similarly, for the vapor phase

$$\rho_v\left[-\eta_0(c-U_v)\omega + U_v\eta_0(c-U_v)k\right]e^{i(wt-kx)} = -p_v + p_0 - \rho_v g\eta \tag{6.45}$$

Elimination of p_0 between eqs. (6.44) and (6.45) yields

$$\begin{aligned}(\rho_l+\rho_v)c^2 - 2(\rho_v U_v + \rho_l U_l)c + \rho_l U_l^2 + \rho_v U_v^2 &= \frac{p_v - p_l}{\eta k} - \frac{(\rho_l-\rho_v)g}{k} \\ &= \frac{\sigma}{R_1\eta k} - \frac{(\rho_l-\rho_v)g}{k}\end{aligned} \tag{6.46}$$

Where R_1 is the radius of curvature of the interface. As noted earlier for a two-dimensional interface in the x-y plane, R_1^{-1}, can be approximated as

$$R_1^{-1} \approx -\frac{\partial^2\eta}{\partial x^2} = k^2\eta \tag{6.47}$$

After substitution for R_1^{-1} from eq. (6.47), eq. (6.46) becomes

$$(\rho_l+\rho_v)c^2 - 2(\rho_v U_v + \rho_l U_l)c + \rho_l U_l^2 + \rho_v U_v^2 = \sigma k - \frac{(\rho_l-\rho_v)g}{k} \tag{6.48}$$

Equation (6.48) is a quadratic equation in c. The solution for this equation is written as

$$c = \left[\frac{\rho_l U_l + \rho_v U_v}{(\rho_l + \rho_v)}\right] \pm i\sqrt{\frac{\rho_v \rho_l (U_v - U_l)^2}{(\rho_l + \rho_v)^2} - \frac{\sigma k}{(\rho_l + \rho_v)} + \frac{(\rho_l - \rho_v) g}{(\rho_l + \rho_v) k}} \tag{6.49}$$

For $U_v = U_l = 0$, eq. (6.49) reduces to the dispersion relation derived earlier for stationary fluid layers. The first term under the square root sign represents the contribution of inertia, the second term that of surface tension, and the last term represents the contribution of buoyancy. It is seen that inertia and buoyancy tend to destabilize the interface, whereas surface tension stabilizes the interface. The real part in eq. (6.49) represents mass-mean velocity of the interface. From instability considerations, our interest is in the imaginary part of the wave velocity. Critical condition will occur when the wave velocity or the growth rate given by the imaginary component is zero. For such a case, we find

$$\frac{\rho_v \rho_l (U_v - U_l)^2}{(\rho_l + \rho_v)} = \sigma k - \frac{(\rho_l - \rho_v) g}{k}$$

If gravity is zero or the streams move parallel to gravitational acceleration vector, we find

$$\left|(U_v - U_l)\right|_c = \sqrt{\frac{\sigma k (\rho_l + \rho_v)}{\rho_l \rho_v}} \tag{6.50}$$

To obtain the critical velocity difference between the two streams before the interface becomes unstable, we need to know the magnitude of the wave number corresponding to the dominant wavelength on the interface. To obtain the dominant wavelength, we need another relation.

Example:

Calculate the critical and most dangerous two-dimensional wavelengths for a stationary interface between saturated water and steam at 1 atm pressure. What will be the critical velocity difference for flowing streams if the dominant wavelength is taken to be equal to the two-dimensional "most dangerous" wavelength between the two stationary layers.

Solution:

$$\lambda_{c2} = 2\pi\sqrt{\frac{\sigma}{g(\rho_l - \rho_v)}}$$

$$= 2\pi\sqrt{\frac{58.9 \times 10^{-3}}{9.81(957.9 - 0.596)}}$$

$$= 1.57 \text{ cm}$$

$$\lambda_{d_2} = 2\pi\sqrt{3}\sqrt{\frac{\sigma}{g(\rho_l - \rho_v)}}$$

$$= 2\pi\sqrt{3}\sqrt{\frac{58.9\times10^{-3}}{9.81(957.9 - 0.596)}}$$

$$= 2.72 \text{ cm}$$

$$(U_v - U_c)_c = \sqrt{\frac{2\pi\sigma(\rho_l + \rho_v)}{\lambda_{d_2}\rho_v\rho_l}}$$

$$= \sqrt{\frac{2\pi\times58.9\times10^{-3}(957.9 + 0.596)}{2.72\times10^{-2}\times0.596\times957.9}}$$

$$= 4.778 \text{ m/s}$$

6.3 MAXIMUM HEAT FLUX

As noted earlier, Kutateladze [2,3] suggested that occurrence of maximum heat flux condition was purely a hydrodynamic phenomenon. Based on similarity considerations related to the equations of motion and energy, he identified two dimensionless groups:

$$\frac{q_{max}}{\rho_v h_{fg} U_v} \quad \text{and} \quad \frac{\sigma(\rho_l - \rho_v)g}{U_v^4 \rho_v^2}$$

where q_{max} is the maximum heat flux, and U_v is the maximum superficial velocity at which vapor can escape from the surface in fully developed nucleate boiling. By eliminating U_v contained in the two dimensionless groups, Kuteladze obtained a correlation for the maximum heat fluxes as

$$q_{max} = 0.16\rho_v h_{fg}\left[\frac{\sigma g(\rho_l - \rho_v)}{\rho_v^2}\right]^{1/4} \tag{6.51}$$

The numerical constant on the right-hand side of eq. (6.51) was obtained empirically by comparing the predictions with the data available at that time. Subsequently, Zuber [4] argued that onset of maximum heat flux on infinite flat plates was a consequence of hydrodynamic instability of vapor columns. Vapor leaves the heater in the form of jets. The maximum heat flux condition occurs when vapor velocity in the jets reaches a critical velocity at which vapor jets become unstable. As a result of instability, excess vapor starts to accumulate on the heater surface. The instability of the jets thus places an upper limit on the vapor removal rate from the heater surface.

Accumulation of vapor on the heater surface leads to a rapid increase in the fraction of the heater area that is dry. The insulating effect of vapor causes a reduction in heat removal rate from the surface. The difference in the heat input rate and the heat removal rate further amplifies the increase in the wall superheat with continued degradation of rate of the heat transfer as the fractional wall area covered with a vapor layer increases.

If U_{vc} is the vapor velocity at which vapor jets become unstable, an expression for the maximum heat flux is obtained by making an energy balance at the heater surface as

$$q_{\max} = \rho_v h_{fg} U_{vc} A_v / A \tag{6.52}$$

In the eq. (6.52), A_v / A is the fraction of the heater area that is occupied by vapor jets. In writing eq. (6.52), it is assumed that all of the energy released from the heater is utilized in production of vapor at and near the heater surface, although a small fraction may be removed by convection.

The determination of critical velocity of the jets requires knowledge of dominant wavelength at the interface. Earlier, we obtained an expression, eq. (6.50), for inviscid fluid streams flowing parallel to each other, the critical relative velocity between the two streams in terms of the dominant wavelength as

$$\left(U_v - U_l\right)_c = \sqrt{\frac{2\pi\sigma\left(\rho_l + \rho_v\right)}{\lambda_{Hc}\rho_v\rho_l}} \tag{6.53}$$

In eq. (6.53), λ_{Hc} is the dominant wavelength on the interface. Equation (6.53) as shown earlier (eq. (6.50)) is obtained from initially plane interface analysis. However, to obtain λ_{Hc}, Zuber used results of two-dimensional analysis for instability of a vapor column.

Zuber argued that on a large horizontal plate, the jets locate themselves on a square grid with spacing given by two-dimensional Taylor wavelength. Furthermore, he assumed, rightly so, that no vapor jets located in the valleys of the Taylor wave, although as we will see later, in film boiling vapor bubbles are released alternatively from peak and valley regions, respectively. Figure 6.4 obtained from Lienhard and Dhir [8] shows the conceptual configuration. The diameter of the jets was assumed to be equal to half of the jet spacing. The support for these assumptions was based on the spacing and bubble diameter relationship observed in film boiling. As such, $\frac{A_v}{A}$ was taken to be $\frac{\pi}{16}$. Since liquid must flow in the downward direction to keep up with the vapor outflow, a mass balance over one square cell of width λ yields

$$\rho_v U_v \left[\pi\left(\frac{\lambda}{4}\right)^2\right] = -\rho_l U_l \left[\lambda^2 - \pi\left(\frac{\lambda}{4}\right)^2\right] \tag{6.54}$$

or

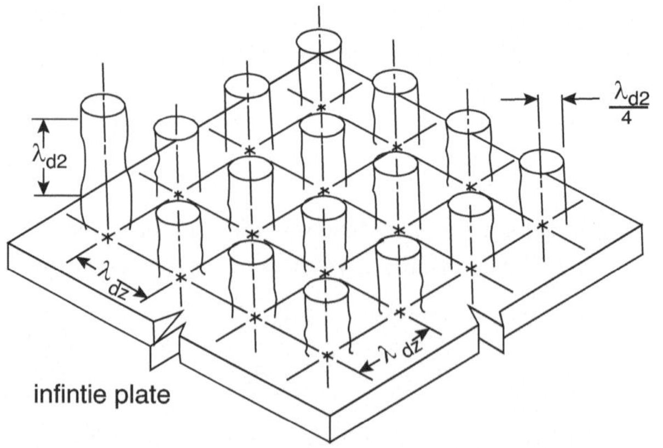

FIGURE 6.4 Vapor removal configuration on an "infinite" flat plate heater. (Obtained from Lienhard and Dhir [8].)

$$U_l = -\frac{\rho_v}{\rho_l} U_v \left[\frac{\frac{\pi}{16}}{1 - \frac{\pi}{16}} \right] \tag{6.55}$$

Upon substitution for U_l from eq. (6.55) in eq, (6.53), an expression of critical velocity of vapor is obtained as

$$U_{vc} = \sqrt{\frac{2\pi\sigma(\rho_l + \rho_v)}{\lambda_{HC}\rho_l\rho_v}} \left[1 + \frac{\rho_v}{\rho_l} \frac{\pi}{16 - \pi} \right]^{-1} \tag{6.56}$$

The next important question is the specification of the dominant wavelength, λ_{HC}. Zuber assumed that the unstable wavelength was the critical Rayleigh unstable wavelength on a liquid jet in vacuum. This wavelength is equal to the circumference of the jet. Upon substitution for A_v/A and U_{vc}, eq. (6.52) becomes

$$q_{\max Z} = \rho_v h_{fg} \frac{\pi}{16} \sqrt{\frac{4\sigma(\rho_l + \rho_v)}{\lambda\rho_l\rho_v}} \left[1 + \frac{\rho_v}{\rho_l} \frac{\pi}{16 - \pi} \right]^{-1} \tag{6.57}$$

Zuber could not decide as to whether the jet spacing λ should be equal to λ_{c2}, the two-dimensional critical Taylor wavelength or, λ_{d2}, the "most dangerous" Taylor wavelength. Substituting in eq. (6.57) for $\lambda = \lambda_{c2}$, or $\lambda = \lambda_{d2}$, an expression for the maximum heat flux is obtained as

$$q_{\max Z} = \frac{\pi}{24} \rho_v h_{fg} \left[\frac{\sigma g(\rho_l - \rho_v)}{\rho_v^2} \right]^{\frac{1}{4}} \left[\frac{\rho_l + \rho_v}{\rho_l} \right]^{\frac{1}{2}} \left[\frac{\rho_l(16 - \pi)}{\rho_l(16 - \pi) + \pi\rho_v} \right] [1.196 \text{ or } 0.909] \tag{6.58}$$

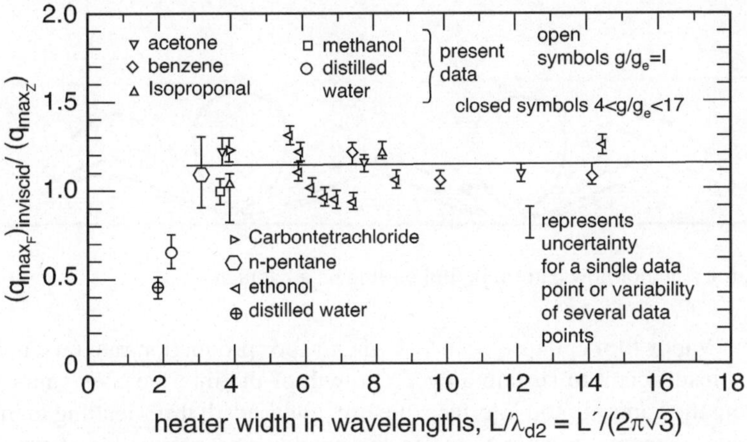

FIGURE 6.5 Maximum heat flux on flat-plate heaters with vertical side walls [9].

where the constant in the last square bracket on the right-hand side of eq. (6.58) is 1.196 when $\lambda = \lambda_{c2}$ and 0.909 when $\lambda = \lambda_{d2}$. Zuber took the constant in the last square bracket to be unity, which is a good mean value. At low pressures, the terms in second and the third square brackets are nearly equal to unity.

Thus, for low pressures, Zuber obtained for a very complex process an elegant expression for maximum heat flux on an infinite flat plate as

$$q_{\max Z} = \frac{\pi}{24}\rho_v h_{fg}\left[\frac{\sigma g(\rho_l - \rho_v)}{\rho_v^2}\right]^{\frac{1}{4}} \tag{6.59}$$

Subsequently, Lienhard et al. [9] argued that it was more probable that the dominant wavelength at which vapor jets became unstable was the "most dangerous" Taylor wavelength, λ_{d2}, rather than the longer Rayleigh unstable wavelength having a length equal to the circumference of the vapor column. The rationale was justified on the basis that on the vapor–liquid interface near the solid surface "most dangerous" Taylor wavelength was dominant. Under this assumption, they suggested that the maximum heat flux on large flat plates should be about 15% higher than that given by eq. (6.59) of Zuber. Figure 6.5 shows correlation of data as a function of width of the plate. The data were obtained for different fluids that wetted the surface well and at gravity levels up to 17 times higher than earth normal gravity. It should be noted that data obtained with water is almost half the predicted value. The reasons for this behavior will be discussed later in the context of a number of variables including wettability and size of the surface that influence the maximum heat flux.

6.4 MINIMUM HEAT FLUX

Zuber was also successful in obtaining an expression for the minimum heat flux on an infinite flat plate purely from hydrodynamic considerations. He rationalized that in transition boiling near the minimum heat flux condition, Taylor wavelength was

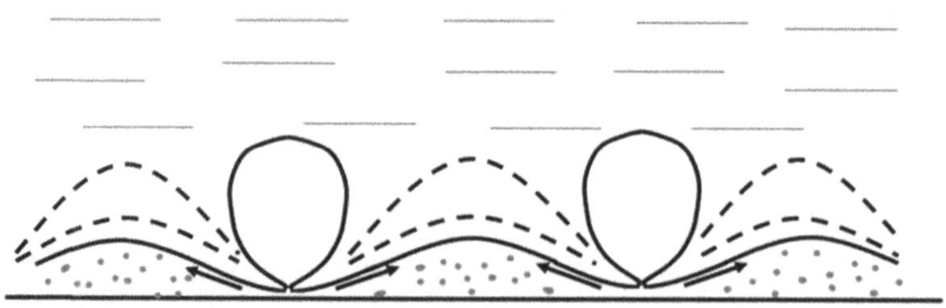

FIGURE 6.6 Vapor flow pattern in film boiling on a flat plate

dominant. Vapor film collapse occurred when vapor production rate on a horizontal plate was insufficient to sustain a cyclic growth of the interface. As vapor bubbles break from the vapor–liquid interface, the film underneath thins leading to increased evaporation. Generated vapor is fed to the neighboring anti-nodes of the wave as shown in Figure 6.6, which in turn become bubble-releasing nodes for the next half cycle and the process repeats. Thus, an expression for the minimum heat flux in a saturated liquid was obtained as

$$q_{\min} = \rho_v h_{fg} N_b V_b f_b \tag{6.60}$$

where N_b is the number of bubbles that are released per unit area of the plate, V_b is the volume of a bubble, and f_b is the bubble release frequency. The bubble release frequency was determined by noting the time taken by the interface to travel a distance equal to Taylor wavelength, λ. This distance is calculated by assuming that bubbles are released alternatively from the nodes and anti-nodes of a Taylor wave. The bubbles grow to a height 0.4λ and time to travel a distance equivalent to 0.1λ is lost during bubble break off at each site. The average interface speed was obtained by averaging the linear growth rate over the distance traveled by the interface. As such, the average growth rate of the interface during its travel was obtained as

$$\frac{d\overline{\eta}}{dt} = \frac{1}{0.4\lambda}\int_0^{0.4\lambda} \frac{d\eta}{dt}\cdot d\eta \tag{6.61}$$

$$= 0.2\lambda(-\omega_i) \tag{6.62}$$

or the expression for the bubble frequency, f_b is written as

$$f_b = \frac{1}{\lambda}\frac{d\overline{\eta}}{dt} = 0.2(-\omega_i) \tag{6.63}$$

For a two-dimensional configuration, Zuber assumed that two bubbles were released per $\lambda_{d,2}^2$ area—one bubble during each half of the cycle. In reality, the vapor removal configuration is three-dimensional. Three-dimensional Taylor instability analysis

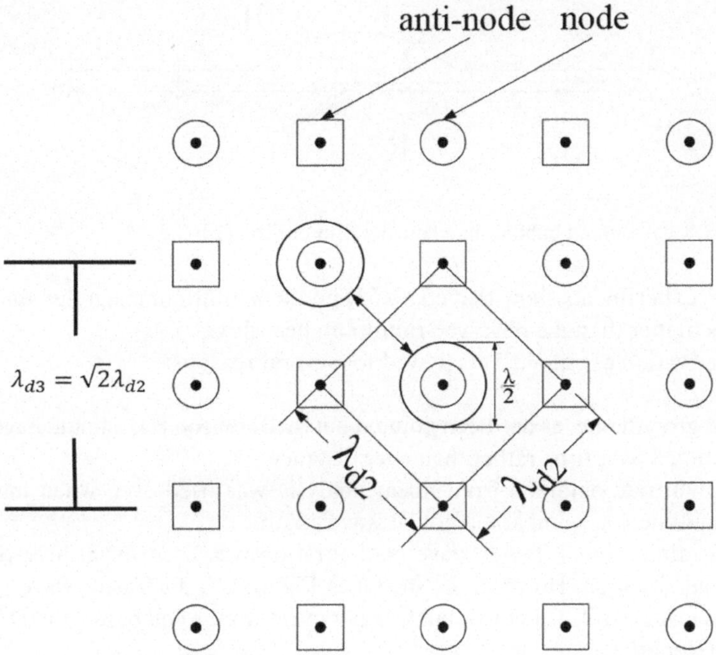

FIGURE 6.7 Three-dimensional Taylor wave pattern on an infinite flat plate [10].

carried out by Sernas et al. [10] in a manner similar to that described in Section 6.2 yields "most dangerous" three-dimensional wavelength as $\lambda_{d,3} = \sqrt{2}\lambda_{d,2}$. Figure 6.7 shows the perceived three-dimensional Taylor wave pattern envisioned during film boiling on a flat plate. Interestingly, this pattern gives release of two bubbles per $\lambda_{d,2}^2$ area per cycle, which is consistent with Zuber's assumption. One bubble from the corners of the square $\frac{1}{4}$ each from the four corners and the second bubble from the center. If bubble diameter is assumed to be half of the "most dangerous" two-dimensional Taylor wavelength, λ_{d2}, eq. (6.60) becomes

$$q_{\min} = \rho_v h_{fg} \cdot 2\left(\frac{4}{3}\pi\left(\frac{\lambda_{d2}}{4}\right)^3\right)0.2(-\omega_i)/\lambda_{d2}^2$$
$$= \frac{\pi}{24} \cdot \rho_v h_{fg} \cdot (0.2)(-\omega_i)\lambda_{d2} \tag{6.64}$$

If $(-\omega_i)$ is taken to be the frequency corresponding to the fastest growing two-dimensional wave, eq. (6.64) becomes, after substitution for $(-\omega_i)_{max}$ and λ_{d2} from eqs. (6.27) and (6.26) respectively

$$q_{\min} = 0.177\rho_v h_{fg}\sqrt[4]{\frac{\sigma g(\rho_l - \rho_v)}{(\rho_l + \rho_v)^2}} \tag{6.65}$$

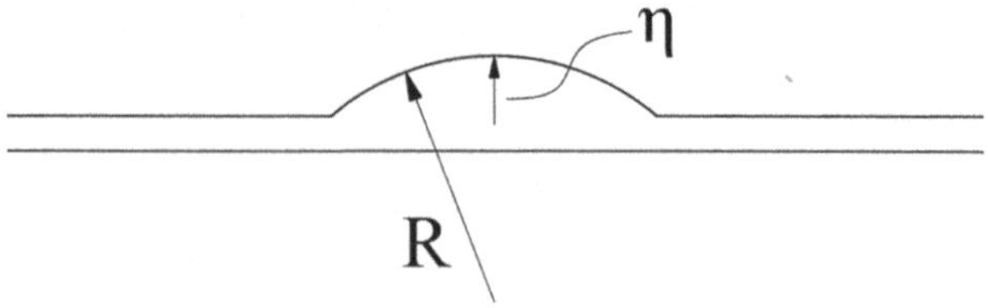

FIGURE 6.8 Assumed bubble shape during film boiling [12]

Data from experiments show that eq. (6.65) predicts minimum heat flux that is about two times higher than the observed minimum heat flux.

The derivation of eq. (6.65) is flawed for several reasons:

1. The growth rate, as has been pointed out by Berenson [11], should have been averaged over time rather than over distance.
2. Growth rate obtained from linear analysis was used even when interface amplitude was comparable to the wavelength.
3. Two-dimensional Taylor wave configuration was used for a three-dimensional situation. However, as shown in Figure 6.7, 3D Taylor wave pattern also leads to release of two bubbles per cycle from a cell based on 2D Taylor wavelength.

Berenson [11] suggested that because of these deficiencies in the model and unavailability of data on growth rate, it was more appropriate to determine the lead constant by comparing predictions from eq. (6.65) with the data. As such, he suggested a value of 0.09 for the lead constant.

Subsequently, Lienhard and Dhir [12] used the interface growth behavior to develop an expression for the minimum heat flux. They assumed that at any interface position, the bubble as shown in Figure 6.8 could be assumed as a segment of a sphere. The volume of such a segment can be written as

$$V = \pi R \eta^2 \left(1 - \frac{\eta}{3R}\right) \tag{6.66}$$

where R is the radius of curvature, and η is the displacement of the interface above the mean position of the film. During the growth of the interface, both R and η change; thus, the rate of change of bubble volume with time can be written as

$$\frac{dV}{dt} = \pi\left(2R\eta - \eta^2\right)\frac{d\eta}{dt} + \pi\eta^2 \frac{dR}{dt} \tag{6.67}$$

Visual observations show that at the time of detachment, the bubble is larger than a hemisphere. Since vapor is continuously added to the bubble, the radius of the bubble cannot decrease after it becomes a hemisphere.

As such with $\frac{dR}{dt}$ set equal to zero, eq. (6.67) becomes

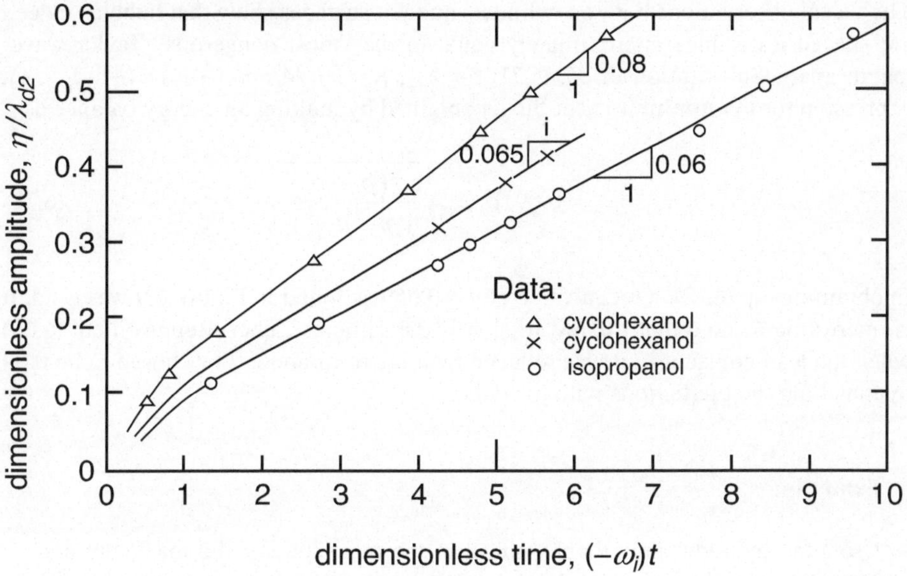

FIGURE 6.9 Typical observations of late bubble growth during film boiling on a horizontal cylinder [12].

$$\frac{dV}{dt} = \pi\left(2R\eta - \eta^2\right)\frac{d\eta}{dt} \tag{6.68}$$

Bubble growth rate during the late stages of growth was obtained from the data such as those plotted in Figure 6.9. These data were obtained from the cine movies of film boiling on a horizontal cylinder. From Figure 6.9, it is seen that bubble growth rate can be written as

$$\frac{d\eta}{dt} = (-w_i)(c\lambda_{d2}) \tag{6.69}$$

where c is an empirical constant. The substitution of $\frac{d\eta}{dt}$ from eq. (6.69) in eq. (6.68) yields

$$\frac{dV}{dt} = \pi\left(2R\eta - \eta^2\right)(-w_i)(c\lambda_{d2}) \tag{6.70}$$

The rate of change of bubble volume as given by eq. (6.70) is maximum when $\eta = R$. Lienhard and Dhir [12] argued that since bubbles grow larger than a hemisphere in saturated film boiling, the vapor production rate per cell should be at least equal to the maximum rate demanded by the growth process. With $\eta = R$,

$$\frac{dV}{dt} = \pi c\lambda_{d2}(-w_i)R^2 \tag{6.71}$$

The visual observations both on cylinders and flat surfaces show that bubbles indeed are spaced a distance approximately equal to the "most dangerous" Taylor wavelength apart. Substitution in eq. (6.71) for λ_{d2}, $R = \lambda_{d2}/4$ and $(-w_i) = (-w_i)_{max}$, an expression for the minimum heat flux is obtained by making an energy balance as

$$q_{min} = 0.091 \rho_v h_{fg} \sqrt[4]{\frac{g\sigma(\rho_l - \rho_v)}{(\rho_l + \rho_v)^2}} \tag{6.72}$$

In obtaining eq. (6.72), a mean value of $c = 0.069$ as found in Figure 6.9 was used. It is interesting to note that this expression is the same as Zuber–Berenson eq. (6.65) when the lead constant in it was reduced by a factor of about 2 or chosen to be 0.09 by matching the predictions with the data.

Example:

Using the hydrodynamic theory, calculate the magnitude of the maximum and minimum heat fluxes for boiling of saturated water on a large flat plate at 1 atm pressure.

Solution:

Since at 1 atm pressure $\rho_v \ll \rho_l$, we use appropriate form of eq. (6.58) for maximum heat flux as

$$q_{max} = \frac{\pi}{24} \rho_v h_{fg} \left[\frac{\sigma g(\rho_l - \rho_v)}{\rho_v^2} \right]^{\frac{1}{4}}$$

$$= \frac{\pi}{24} \times 0.596 \times 2.257 \times 10^6 \left[\frac{58.9 \times 10^3 \times 9.81 \times (957.9 - 0.596)}{(0.596)^2} \right]^{\frac{1}{4}}$$

$$= 1.76 \times 10^5 \times 6.28$$

$$= 1.1 \times 10^6 \text{ W/m}^2$$

$$q_{min} = 0.177 \rho_v h_{fg} \left[\frac{\sigma g(\rho_l - \rho_v)}{(\rho_l + \rho_v)^2} \right]^{\frac{1}{4}}$$

$$= 2.38 \times 10^5 \times 0.1566$$

$$= 3.7 \times 10^4 \text{ W/m}^2$$

As noted in the text, constant 0.177 should be reduced to 0.091. Also, we have used properties at the saturation temperature. Since wall temperature in film boiling can be much higher than that at q_{max}, vapor density should be evaluated at the mean temperature. This in turn will require knowledge of the wall temperature, which can be only obtained if heat transfer coefficient in film boiling is evaluated, which will require an iterative procedure. Furthermore, it should be recognized that minimum heat flux is 30–60 times smaller than the maximum heat flux.

6.5 ALTERNATIVE CONSIDERATIONS OF MAXIMUM HEAT FLUX

Since heat is mostly removed from the surface by evaporation, limit on the maximum heat flux can be imposed either by the maximum rate at which vapor can be removed from the surface under the forces of buoyancy and surface tension in inviscid fluids or by maximum vapor production rate, which ever is less. Limit set by vapor removal rate can lead to accumulation of vapor at and near the surface resulting in degradation of rate of heat transfer from the surfaces. This was the premise of the hydrodynamic theory of Zuber described earlier. On the other hand, maximum heat flux can be limited by availability of liquid contacting the surface to accommodate the imposed evaporation rate. A number of investigators have hypothesized the latter as the possible mechanism for occurrence maximum heat flux condition. However, such a limit will give maximum heat fluxes that are lower than those predicted from vapor removal limit.

Early proponents of the latter approach were Haramura and coworkers, e.g., Haramura and Katto [13]. They envisioned that in fully developed nucleate boiling vapor bubbles merge while sitting at the heater surface to form mushroom type of bubbles. Smaller bubbles trap liquid film between them (macro layer). If the heat flux is such that the macrolayer dries out before the large vapor bubble leaves the heater surface, the surface will dry out. This corresponding heat flux then will be the maximum heat flux. Figure 6.10 shows their conceptual model. The trapped liquid film thickness was assumed to be equal to ¼ of the Helmholtz wavelengths at which vapor velocity in the vapor stems representing small bubble reaches its critical velocity for the interface to become unstable. The growth period or hovering period of a bubble was determined by knowing the bubble departure diameter and vapor production rate, which was assumed to remain constant during the growth period. They eventually obtained an expression similar to that obtained by Zuber (eq. 6.59) solely from hydrodynamic considerations. The empirical constant that depended on liquid to vapor density ratio and fractional area of the heater occupied by vapor was adjusted to be the same as in Zuber's expression, i.e., $\frac{\pi}{24}$.

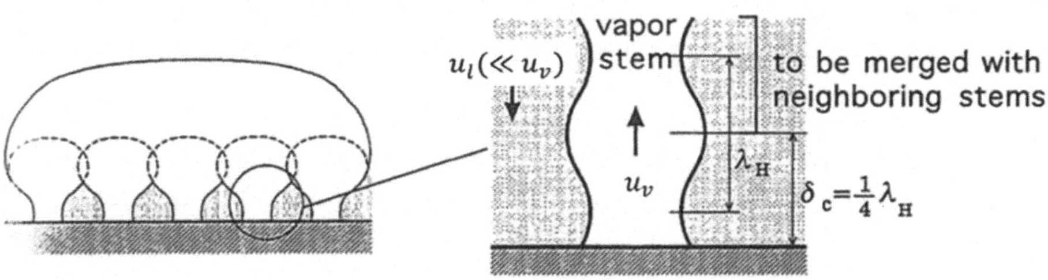

FIGURE 6.10 Hydrodynamic instability of a vapor stem and vapor mass formation [13].

Kandlikar [14] has developed a model for critical heat flux on surfaces inclined at 0°–90° to the horizontal. The model considers expansion of vapor bubble–liquid interface parallel to heater surface. Forces due to momentum flux resulting from evaporation, hydrostatic head, and surface tension are considered. The model uses receding contact angle as a correlating parameter and also includes the effect of liquid subcooling on critical heat flux.

Dhir and Liaw [15] have argued that maximum heat flux is set by the lower of the vapor removal limit or the vapor production limit. The former occurs on well-wetted surfaces with contact angles $\phi \simeq 0$ while latter occurs when surface is partially wetted $\phi > 0$. We will discuss this further in Chapter 8.

More recently, a number of investigators have experimentally studies maximum heat flux on microstructured surfaces and have found that the maximum heat flux can be more than two times higher than that on plane surfaces as given by Zuber's model. It is claimed that it is the result of better liquid supply to the surface with microstructures. However, they have not paid attention to the effect of the size of the surface on maximum heat flux. As has been shown by Lienhard et al. [9], maximum heat flux on well-wetted surfaces can exceed Zuber's prediction for an infinite plate when heater width is less than the most dangerous two-dimensional wavelength. We will elaborate on this further in Chapter 8.

6.6 SUMMARY

In this chapter, we have developed hydrodynamic basis of maximum and minimum heat fluxes during pool boiling as originally proposed by Zuber. Alternative approaches suggested in the literature have been briefly described.

EXERCISES

1. Calculate maximum heat flux on a large flat plate for saturated pool boiling at 1 atm pressure for water, ethanol, methanol, and FC-72. Discuss your results.
2. Calculate and plot for saturated water maximum heat flux using Zuber's model on an infinite flat plate for pressures of 1, 61, 100, and 221.2 bars. Discuss your result.
3. Using Rohsenow's correlation for nucleate pool boiling and maximum heat flux obtained in Prob. 1, calculate the wall superheat at maximum heat flux for the four fluids.
4. A kettle is designed to boil water at 1 atm pressure. The kettle surface is made of nickel, and the highest temperature is limited to 108°C. Calculate the power needed to boil water at this temperature if kettle diameter is 20 cm. If all of the energy goes into vapor generation, calculate the mass rate of steam production. What is the maximum possible heat transfer rate from the kettle surface in nucleate boiling and what will be the corresponding temperature?
5. Calculate the critical velocity at which a vapor column in saturated liquid at 1 atm pressure will become unstable when the dominant wavelength on the

interface is equal to the circumference of the column or the "most dangerous" two-dimensional Taylor wavelength, λ_{d2}. Assume the column diameter to be half of λ_{d2}.

6. For two-dimensional Taylor wave, calculate the critical and the "most dangerous" wavelength for saturated water at 1 atm pressure. How will the wavelengths change for gravity levels of 10^{-2} and $10^{-6}\, g_e$.
7. Calculate the magnitude of frequency corresponding to the "most dangerous" two-dimensional Taylor wavelength. Assume the fluid to be saturated water at a pressure of 1 atm.
8. Extend the two-dimensional Taylor instability analysis to three dimensions and obtain expressions for critical and most dangerous wavelength and corresponding growth rates.
9. Calculate minimum heat flux for water on a flat plate at pressures of 1 and 61 bars. Assume a wall superheat of 100°C.
10. Using your earlier results for $q_{\max}$, plot the ratio of $q_{\max}/q_{\min}$ as a function of P/P_{critical}. Comment on your result.

NOMENCLATURE

A Constant
c Wave velocity
g Gravitational acceleration
k Wave number
p Pressure
R Radius of curvature
U Free stream velocity
t Time
u Velocity in the x-direction
v Velocity in the y-direction
x Horizontal coordinate
y Vertical coordinate
V Volume

SUBSCRIPTS

b Bubble
c Critical
$c2$ Two-dimensional critical
$d2$ Two-dimensional most dangerous
H Helmholtz
i Imaginary
o At undisturbed interface
l Liquid
r Real
ol On the liquid side of unperturbed interface
ov On the vapor side of unperturbed interface

v	Vapor
max	Maximum
min	Minimum

GREEK LETTERS

η	Interface position
Φ	Potential function
ω	Wave frequency
ρ	Density
σ	Surface tension
λ	Wavelength

REFERENCES

1. Bonilla, C. and Perry, C.W. (1941), "Heat Transmission to Boiling Binary Liquid Mixtures." *Trans. Am. Inst. Chem. Eng.*, Vol. 37, p. 685.
2. Kutateladze, S.S. (1948), "On the Transition to Film Boiling Under Natural Convection." *Kotloturbostroenie*, No. 3, p. 10.
3. Kutateladze, S.S. (1951), "A Hydrodynamic Theory of Changes in the Boiling Process under Free Conduction." *Izv. Akad. Nauk. S.S.S.R., Otd. Tekh. Nauk.*, No. 4, p.529.
4. Zuber, N. (1959), "Hydrodynamic Aspect of Boiling Heat Transfer." Ph.D. Dissertation, UCLA.
5. Lamb, H. (1957), *Hydrodynamics*, Dover Publications, New York.
6. Taylor, G.I. (1950), "The Instability of Liquid Surfaces when Accelerated in a Direction Perpendicular to their Plane." *Proc. R. Soc. Lond.*, Vol. 201A, p. 192.
7. Bellman, R. and Pennington, R.H. (1954), "Effect of Surface Tension and Viscosity on Taylor Instability." *Q. Appl. Math.*, Vol. 12, pp. 151–162.
8. Lienhard, J.H. and Dhir, V.K. (1973), "Extended Hydrodynamic Theory of Maximum and Minimum Heat Fluxes", NASA CR-2270.
9. Lienhard, J.H., Dhir, V.K., and Riherd, D.M. (1973), "Peak Pool Boiling Heat Flux Measurements on Finite Horizontal Plates." *J. Heat Transfer*, Vol. 95, No. 4, pp. 447–482.
10. Sernas, V., Lienhard, J.H., and Dhir, V.K. (1972), "The Taylor Wave Configuration during Boiling from a Flat Plate." *Intl. J. Heat Mass Transfer*, Vol. 16, pp. 1820–1822.
11. Berenson, P.J. (1961), "Transition Boiling Heat Transfer from a Horizontal Surface." *ASME J. Heat Transfer*, Vol. 83C, pp. 351–358.
12. Lienhard, J. H. and Dhir, V.K. (1980), "On the Prediction of Minimum Film Boiling Heat Flux." *J. Heat Transfer*, Vol. 102, pp. 380–382.
13. Haramura, Y. and Kato, Y. (1983), "A New Hydrodynamic Model of Critical heat Flux Applicable to Both Pool Boiling and Forced Convection Boiling on Submerged Bodies in Saturated Liquids." *Intl. J. Heat Mass Transfer*, Vol. 26, pp. 389–399.
14. Kandlikar, S.G. (2001), "A Theoretical Model to Predict Pool Boiling CHF Incorporating Effects of Contact Angle and Orientation." *ASME J. Heat Transfer*, Vol. 123, pp. 1071–1079.
15. Dhir, V.K. and Liaw, S.P. (1989), "Framework for a Unified Model for Nucleate and Transition Pool Boiling." *ASME J. Heat Transfer*, Vol. 111, pp. 739–746.

7 Film and Transition Boiling Heat Transfer

INTRODUCTION

In this chapter, we focus on the development of models and correlations for film and transition boiling. As we discussed in the previous chapter, maximum and minimum heat fluxes are the two extremes of transition boiling. Transition boiling is a regime that lies between film and nucleate boiling regimes of the boiling curve. Transition boiling has features of both nucleate and film boiling and is characterized by existence of nucleate boiling where liquid contacts the solid surface and large vapor patches that cover the heater surface and behave like a vapor film with distorted interface. The fractional area of the heater covered by vapor increases with superheat, and the rate of heat transfer in vapor and liquid occupied regions varies both spatially and temporally. As such, modeling of transition boiling is very challenging. Film boiling, although associated with low heat fluxes, is generally amenable to simple analysis. We begin with film boiling followed by transition boiling on a horizontal surface and subsequently include the effect of other system variables such as heater geometry and liquid subcooling on the processes. Combining expressions for the minimum heat flux and film boiling heat transfer coefficient, we are able to determine the superheat, ΔT_{min}, at which minimum heat flux occurs. Figure 7.1 shows a conceptualization of transition boiling, whereas Figure 7.2 shows a photograph of film boiling on a horizontal plate [1] and on a cylinder oriented horizontally [2]. As noted in Chapter 6, the observed pattern is dictated by the Taylor instability of the vapor–liquid interface. Visual observations of the process show, for saturated film boiling, a cyclic release of vapor bubbles alternatively from nodes and anti-nodes of the Taylor wave.

7.1 FILM BOILING HEAT TRANSFER COEFFICIENT ON A HORIZONTAL PLATE

Chang [3] was the first to propose that film boiling and natural convection were analogous processes and that time-averaged heat transfer coefficient could be predicted

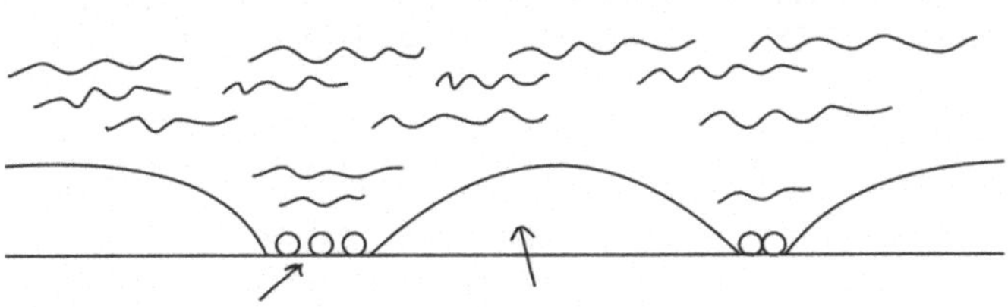

FIGURE 7.1 Conceptualization of transition boiling.

DOI: 10.1201/9781032668437-7

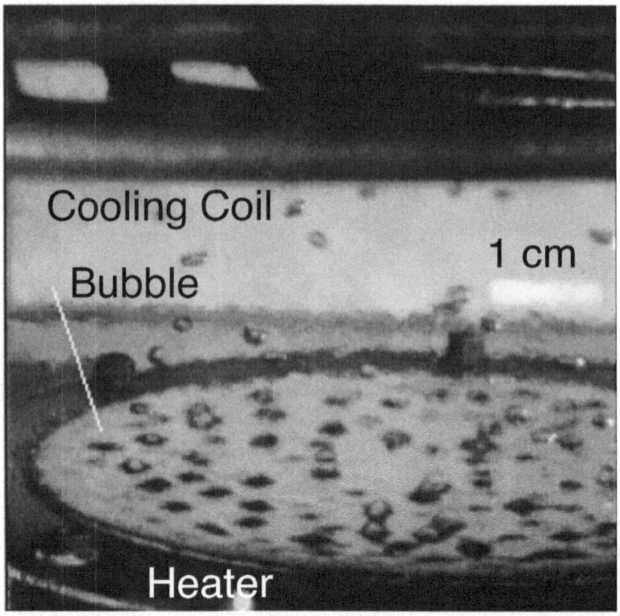

FIGURE 7.2a Digital image of non-departing bubbles arranged on concentric rings on a horizontal circular plate during subcooled film boiling [1].

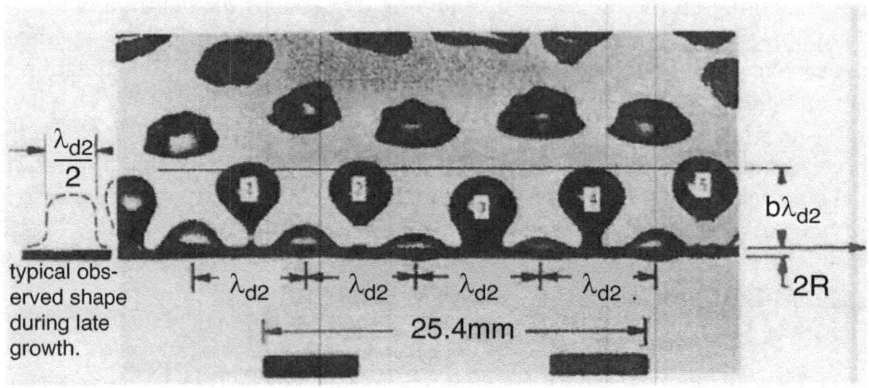

FIGURE 7.2b Typical bubble departure pattern during film boiling on a wire (Radius of wire, $R=0.323$mm, $q=1.3\times10^5$ W/m^2, fluid is acetone at 1atm, $R'=0.0203$, $b=0.65$) [2].

from correlations similar to those for turbulent natural convection. Berenson [4] was probably the first one to develop a mechanistic model for heat transfer coefficient during laminar film boiling on an infinite horizontal plate. Berenson's model was simple, but it reflected a significant insight of the phenomenon. Although the vapor–liquid interface is constantly evolving with time, Berenson's model was a static one. A key assumption made in Berenson's model is that energy transferred across the vapor film is utilized for evaporation at the vapor–liquid interface. Vapor flows

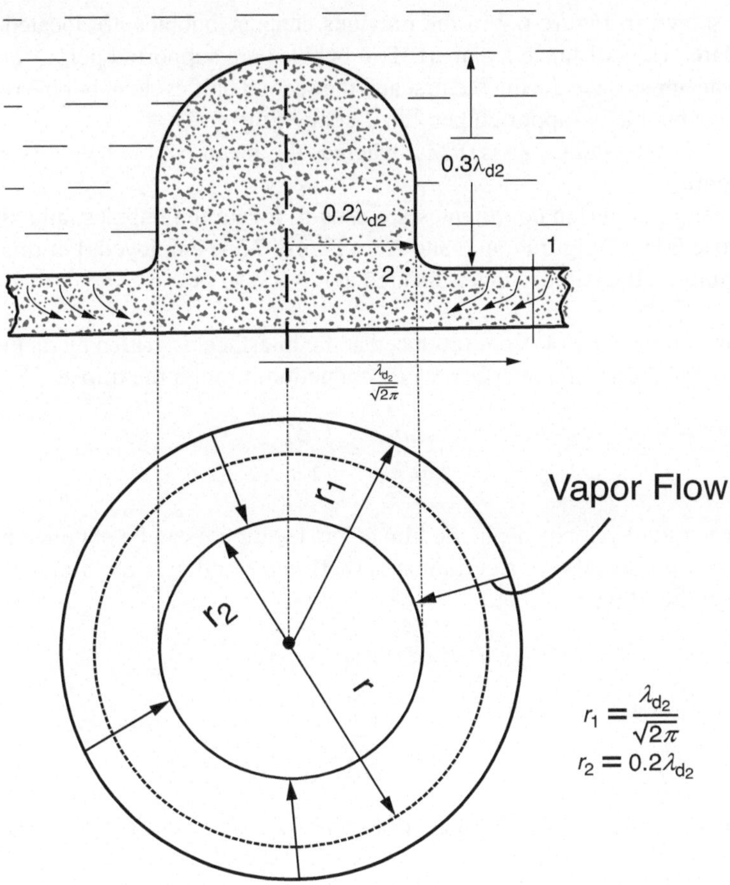

FIGURE 7.3 Bubble shape assumed in Berenson's [4] analysis and vapor flow in the film.

through the film and into the protruding interface, which acts as a vapor reservoir. The friction pressure drop encountered by vapor as it flows through the film is balanced by the hydrostatic head difference between the thin film region and the protruding interface. Figure 7.3 shows the assumed vapor flow configuration for liquid at its saturation temperature corresponding to the system pressure. Various other assumptions made in Berenson's analysis are:

i. Vapor film thickness between adjacent bubbles is constant and vapor flow through the film is modeled as one-dimensional flow.
ii. Fluids are incompressible.
iii. Inertia and convective terms in the momentum and energy equation can be ignored.
iv. Interface between vapor and liquid is free of any ripples.
v. Heat transfer by radiation across the vapor film is neglected.
vi. Properties are evaluated at the mean temperature between wall and the saturation temperature.

vii. As shown in Figure 6.7 in the previous chapter, bubbles are located on a square grid a distance λ_{d_2} apart. Two bubbles are supported per $\lambda_{d_2}^2$ area of the heater surface during the first and the second half cycle of bubble release or one bubble is supported per $\lambda_{d_2}^2/2$ at any time.
viii. Mean bubble radius, r_2, is $0.2\lambda_{d_2}$ whereas the mean bubble height is $0.3\lambda_{d_2}$ (Figure 7.3).
ix. Although cartesian coordinates are used to locate the bubbles, an axisymmetric flow configuration as shown in Figure 7.3 is employed. For this configuration, the outer radius is located at $r_1 = \lambda_{d_2}/\sqrt{2\pi}$.

The downward velocity of vapor produced at the interface is written by making mass and energy balance at the interface with conduction through the film as

$$v = \frac{q}{\rho_v h_{fg}} = \frac{k_v \Delta T}{\delta \rho_v h_{fg}} \tag{7.1}$$

The average radial velocity, $\bar{u}$, in the film at any location r, can be obtained by equating the vapor produced over a certain area (between r_1 and r) of the film to that flowing through the film as

$$\bar{u} \cdot 2\pi r \delta = \pi\left(r_1^2 - r^2\right) v \tag{7.2}$$

$$\bar{u} = \frac{r_1^2 - r^2}{2r\delta^2} v \tag{7.3}$$

Substituting for v in eq. (7.3) from eq. (7.1), an expression for $\bar{u}$ is obtained as

$$\bar{u} = -\frac{\left(r_1^2 - r^2\right)}{2r\delta} \frac{k_v \Delta T}{\rho_v h_{fg}} \tag{7.4}$$

In eq. (7.4), negative sign indicates vapor flows in a direction opposite to increasing r. In the absence of inertia, vapor momentum equation in the radical direction can be written as

$$\frac{dp}{dr} = \mu_v \frac{\partial^2 u}{\partial y^2} \tag{7.5}$$

It should be noted that eq. (7.5) is only approximate as the effect of downward velocity of the vapor generated at the interface is not included. The solution of eq. (7.5) requires two boundary conditions: the first boundary condition is no slip at the wall, i.e., $u=0$ at $y=0$, and the second condition is at the interface. The latter is more difficult to specify since no attempt is being made here to solve for the complex liquid motion adjacent to the interface. As such, the extreme boundary conditions can be a rigid interface, i.e., $u = 0$ at $y = \delta$ or a slip condition (dynamic interface), i.e., $\frac{\partial u}{\partial y} = 0$ at $y = \delta$. Integration of eq. (7.5) yields

$$u = \frac{1}{\mu_v}\frac{dp}{dr}\left(\frac{y^2}{2} - \frac{\delta y}{2}\right) \quad \text{or} \quad \frac{1}{\mu_v}\frac{dp}{dr}\left(\frac{y^2}{2} - \delta y\right) \tag{7.6}$$

The first expression in eq. (7.6) is for a rigid interface, whereas the second expression results when the interface is assumed to be dynamic. The average vapor velocity in the film is obtained as

$$\bar{u} = \frac{1}{\delta}\int_0^\delta u dy = -\frac{1}{\mu_v}\frac{dp}{dr}\cdot\frac{\delta^2}{\beta} \tag{7.7}$$

where $\beta = 12$ for rigid interface and $\beta = 3$ for a dynamic interface.

Substitution of $\bar{u}$ from eq. (7.7) in eq. (7.4) yields

$$\frac{dp}{dr} = \frac{\beta \cdot \mu_v k_v \Delta T}{\rho_v h_{fg}\delta^4}\frac{\left(r_1^2 - r^2\right)}{2r} \tag{7.8}$$

Integration of eq. (7.8) between locations 1 and 2 shown in Figure 7.3 gives

$$\left(p_1 - p_2\right) = \frac{\beta \cdot \mu_v k_v \Delta T}{2\rho_v h_{fg}\delta^4}\left[r_1^2 \ln\frac{r_1}{r_2} - \frac{r_1^2}{2} + \frac{r_2^2}{2}\right] \tag{7.9}$$

Since $r_1 = \dfrac{\lambda_{d2}}{\sqrt{2\pi}}$ and $r_2 = 0.2\lambda_{d2}$, eq. (7.9) becomes

$$p_1 - p_2 = \frac{0.079\beta \cdot \mu_v k_v \Delta T}{\rho_v h_{fg}\delta^4}\frac{\lambda_{d2}^2}{\pi} \tag{7.10}$$

The hydrostatic head difference between location 1 and 2 is obtained for a bubble with hemispherical dome as

$$p_1 - p_2 = (\rho_l - \rho_v)g\left(0.3\lambda_{d_2}\right) - 2\sigma/0.2\lambda_{d_2} \tag{7.11}$$

Eliminating $\left(p_1 - p_2\right)$ between eqs. (7.10) and (7.11) and substituting for λ_{d2}, the two-dimensional "most dangerous" Taylor wavelength, we obtain

$$\sqrt{\sigma g\left(\rho_l - \rho_v\right)} = \frac{1.27\beta \cdot \mu_v k_v \Delta T \sigma}{\rho_v h_{fg}\delta^4 g\left(\rho_l - \rho_v\right)} \tag{7.12}$$

$$\delta^4 = \frac{1.27\beta \cdot \mu_v k_v \Delta T}{g\left(\rho_l - \rho_v\right)\rho_v h_{fg}}\sqrt{\frac{\sigma}{g\left(\rho_l - \rho_v\right)}} \tag{7.13}$$

If a mean value of β between 12 and 3 (= 7.5) is used in eq. (7.13), an expression for the vapor film thickness becomes

$$\delta = 1.76\left[\frac{\mu_v k_v \Delta T}{g(\rho_l-\rho_v)\rho_v h_{fg}}\sqrt{\frac{\sigma}{g(\rho_l-\rho_v)}}\right]^{1/4} \tag{7.14}$$

One should note that the use of a mean value has no physical meaning. Since the bubble height is much larger than the film thickness and thermal conductivity of vapor is small, heat transfer over the bubble occupied region can be neglected. Thus, the total heat loss from one cell shown in Figure 7.3 will be

$$Q = \frac{k_v \Delta T}{\delta}\left(\frac{\lambda_{d2}^2}{2} - \pi\left(0.2\lambda_{d_2}\right)^2\right) \tag{7.15}$$

Thus, if $\bar{\delta}$ is the effective film thickness over total area of the cell, the heat loss from the heater per unit cell is

$$Q = \frac{k_v \Delta T}{\bar{\delta}}\frac{\lambda_{d2}^2}{2} \tag{7.16}$$

Elimination of Q between eqs. (7.15) and (7.16) yields a relation between the effective film thickness and actual film thickness as

$$\bar{\delta} = 1.34\delta \tag{7.17}$$

Substitution for δ from eq. (7.14) in eq. (7.17) gives

$$\bar{\delta} = 2.36\left[\frac{\mu_v k_v \Delta T}{g(\rho_l-\rho_v)\rho_v h_{fg}}\sqrt{\frac{\sigma}{g(\rho_l-\rho_v)}}\right]^{1/4} \tag{7.18}$$

or the average film boiling heat transfer coefficient on a large horizontal plate is given by

$$\bar{h} = 0.42\left[\frac{k_v^3 \rho_v g(\rho_l-\rho_v) h_{fg}}{\mu_v \Delta T\sqrt{\frac{\sigma}{g(\rho_l-\rho_v)}}}\right]^{1/4} \tag{7.19}$$

In eq. (7.19), $\sqrt{\frac{\sigma}{g(\rho_l-\rho_v)}}$ represents the characteristic length. Although vapor is produced at the interface at saturation temperature, vapor becomes superheated as it flows along the plate surface before entering the bubble dome. To account for energy utilized in superheating of vapor, Berenson suggested that latent heat of vaporization used in eq. (7.19) should be corrected as

$$h'_{fg} = h_{fg} + 0.5c_{pv}\Delta T$$

where c_{pv} is specific heat of vapor evaluated at the mean temperature. With heat transfer coefficient in laminar film boiling varying the ΔT as $\Delta T^{-1/4}$. The heat flux will vary as $\Delta T^{3/4}$ when radiation contribution is neglected. In laminar natural convection, however, heat flux varies with ΔT as $\Delta T^{5/4}$.

Example:

Using Berenson's expression, eq. (7.19), calculate the film boiling heat transfer coefficient for water on a large horizontal plate. Water is at saturation temperature corresponding to 1 atm pressure. Wall superheat is 100°C. Neglect radiation contribution.

Solution:

According to eq. (7.19),

$$\bar{h} = 0.42\left[\frac{k_v^3\rho_v g(\rho_l - \rho_v)h'_{fg}}{\mu_v\Delta T\sqrt{\dfrac{\sigma}{g(\rho_l - \rho_v)}}}\right]^{1/4}$$

We evaluate the thermophysical properties at a mean temperature of 150°C (423K). Surface tension and latent heat are evaluated at 100°C. Latent heat of vaporization corrected for sensible heating is found to be:

$$h'_{fg} = h_{fg}\left(1 + 0.5\frac{c_{pv}\Delta T}{h_{fg}}\right)$$

$$= 2.257\times10^6\left(1 + \frac{0.5\times2000\times100}{2.257\times10^6}\right)$$

$$= 2.356\times10^6 \text{ J/kg}$$

$$\bar{h} = 0.42\left[\frac{(27.8\times10^{-3})^3\times0.525\times9.81\times(957.38)\times2.356\times10^6}{142.7\times10^{-7}\times100\times\left(\dfrac{58.9\times10^{-3}}{9.81\times957.38}\right)^{1/2}}\right]^{1/4}$$

$$= 216 \text{ W/m}^2\text{K}$$

As we would expect, heat transfer coefficient in film boiling is much smaller than that in nucleate boiling.

Vapor properties in eq. (7.19) are evaluated at mean temperature $\left(0.5(T_w + T_{sat})\right)$. It should also be noted that for film boiling at higher temperatures, several of the assumptions made in arriving at eq. (7.19) are violated. These include not accounting for the contribution of radiative heat transfer, not including the effect of vapor inertia and convection contributions; ripples can also appear on the interface; and spacing between neighboring bubbles can become irregular. The consequence of these factors is generally to lead to higher heat transfer coefficients than that predicted from eq. (7.19). For inclusion of heat transfer by radiation, one would have to modify eq. (7.1) and redo the complete derivation. This is left as an independent exercise. However, if one assumes that radiation simply represents an additional mode of heat transfer and does not affect the effective film thickness, eq. (7.19) can be modified to add the radiation contribution as:

$$h_r = \frac{\sigma_r}{\left(\dfrac{1}{\varepsilon_w} + \dfrac{1}{\varepsilon_l} - 1\right)} \frac{\left(T_w^4 - T_{sat}^4\right)}{\left(T_w - T_{sat}\right)} \tag{7.20}$$

In eq. (7.20), σ_r is the Stefan Boltzmann's constant, and ε_w and ε_l are the emissivities of the heater surface and liquid respectively. Equation (7.20) is obtained by dividing the radiative heat flux with wall superheat. It should be noted that in writing eq. (7.20), vapor is not assumed to be a participating medium. In simply adding radiative contribution from eqs. (7.20) to (7.19) to obtain overall heat transfer coefficient in film boiling, a correction for increased film thickness should be made. For example, Hamill and Baumeister [5] have suggested that radiative heat transfer coefficient from eq. (7.20) should be multiplied by 0.88 before it is added to eq. (7.19).

Combining the expressions for the minimum heat flux developed in Chapter 6 and eq. (7.19) for film boiling heat transfer coefficient excluding contribution of radiation, an expression for the wall superheat at the occurrence of minimum heat flux or at film collapse is obtained as

$$\Delta T_{min} = \frac{q_{min}}{\bar{h}} = 0.13 \frac{\rho_v h_{fg}}{k_v} \left[\frac{g(\rho_l - \rho_v)}{(\rho_l + \rho_v)}\right]^{2/3} \left[\frac{\mu_v}{g(\rho_l - \rho_v)}\right]^{1/3} \left[\frac{\sigma}{g(\rho_l - \rho_v)}\right]^{1/2} \tag{7.21}$$

In evaluating ΔT_{min}, the vapor properties are to be evaluated at the mean temperature. As such, an iterative procedure will be required in obtaining ΔT_{min}. Also, it should be noted that in obtaining eq. (7.21) the numerical constant in the expression of minimum heat flux was taken to have a value of 0.09. Data from experiments are generally consistent with prediction from eq. (7.21), although significant scatter in the data is observed.

7.2 LEIDENFROST TEMPERATURE AND MINIMUM FILM BOILING TEMPERATURE

Often, minimum film boiling temperature is considered qualitatively to be synonymous to Leidenfrost temperature [6]. In the eighteenth century, Leidenfrost observed that water droplets placed on a hot surface danced on the surface if the temperature of the surface was greater than a certain value, which has been subsequently called the Leidenfrost temperature. The dancing behavior is caused by a cushion of vapor that forms nearly instantaneously between the hot surface and the droplet. Vaporization of the droplet as a result of heat transfer through the vapor causes cushion pressure to build up under the droplet, which pushes it upward. Subsequently, the droplet falls back on the hot surface due to gravity and the process repeats. At temperatures below the Leidenfrost temperature, liquid droplets momentarily contact the surface and shatter as a result of very rapid nucleation of vapor bubbles upon contact of liquid droplets with the surface. Minimum film boiling temperature on the other hand represents the temperature that is just sufficient to sustain a cyclic Taylor wave pattern on the vapor film present between liquid and the solid surface. Breakdown of the pattern leads to liquid–solid contacts in the valleys of the Taylor wave and formation of vapor bubbles as a result of nucleation. These liquid contacts are of very short duration before the interface is pushed back as a result of rapid vapor production. Nucleation leads to improvement in heat transfer and turnaround of the boiling curve. The turnaround depends on the extent of the liquid contacts and their duration as well as the ability of the surface temperature to recover after liquid contacts occur and heat transfer from the surface improves. Surface roughness, wettability, thickness of the substrate, and its thermophysical properties influence the experimentally observed minimum film boiling temperature. Depending on the experimental conditions, large variations in minimum film boiling temperature and consequently minimum heat flux can occur. At film collapse, significant spatial and temporal variations can occur in the surface temperature. In the regions where liquid contacts occur, the surface will cool down, whereas the surface covered with vapor will be at a higher temperature. The magnitude of spatial and temporal variations will depend on the extent and duration of the liquid–solid contacts and thermophysical properties of the solid. Generally, the surface temperature at which turnaround in the boiling curve occurs is measured with thermocouples embedded in the solid at some distance away from the surface. As such, measured temperature will be different than the local surface temperature at which vapor film collapses. Assuming that ($\Delta T_{\text{min, iso}}$), obtained by combining q_{min} and the heat transfer coefficient in stable film boiling, as given by eq. (7.21) is representative of an isothermal surface, Henry [7] obtained a correlation for ΔT_{min}, which accounted for the thermophysical properties of the substrate and liquid as

$$\frac{\Delta T_{\text{min}} - \Delta T_{\text{min, iso}}}{\Delta T_{\text{min, iso}} - \left(T_l - T_{\text{sat}}\right)} = 0.42\left(\sqrt{\frac{k_l \rho_l c_{pl}}{k_s \rho_s c_{ps}}}\,\frac{h_{fg}}{c_{ps}\Delta T_{\text{min, iso}}}\right) \tag{7.22}$$

Equation (7.22) is applicable for both subcooled and saturated liquids and also includes the effect of ratio of latent heat of vaporization to thermal energy stored in the solid. Thickness of the substrate also plays a role, although its effect has not been included in eq. (7.22). A more generalized correlation for ΔT_{min} that includes data for liquid metals and cryogenic liquids as well has been developed by Baumeister and Simon [8].

7.3 TRANSITION BOILING

As noted earlier, transition boiling is a mixed-mode boiling that combines features of both film and nucleate boiling. In transition boiling at any instance, a large fraction of the heater area is covered with vapor. Discrete spots exist on the surface that may experience short periods of liquid contacts followed by long periods during which vapor is present. During short contacts of liquid with the solid, rapid nucleation occurs and liquid is pushed away from the surface while vapor continues to persist on the surface. Transition boiling is characterized by a reduction in heat flux with increase in wall superheat as liquid contact areas and duration of liquid contacts decrease, i.e., current decreases as potential is increased. Thus, it is an unstable process, and it is difficult to obtain steady-state data in transition boiling when heat flux to the surface is controlled. This fact has led to relatively fewer studies on transition boiling. It is not surprising, therefore, that transition boiling is one of the least understood and modeled modes of boiling.

Two types of steady-state experiments have generally been performed in obtaining transition boiling data. In temperature-controlled experiments, vapor of a high boiling point liquid is condensed on the back side of the surface over which transition boiling is to be studied. Temperature of the surface is controlled by changing the pressure of the condensing vapor and thereby the saturation temperature of the vapor. Such experiments have, for example, been carried out by Berenson [9], Hessse [10], and Ramilison and Lienhard [11]. The difficulty with these experiments is that only few data points are accessible in transition boiling because of the limitations imposed by the sum of resistances between the heat source and the heat sink. In the heat flux-controlled experiments such as those performed by Bui and Dhir [12] on a vertical surface, it is the mismatch between the heat input and heat flux at the heated surface that causes the process to become time-dependent. This problem, however, can be circumvented if a feedback loop to control heat input is implemented. As such, only for a limited range of heat fluxes, steady-state transition boiling data have been obtained. Transient transition boiling data can, of course, be obtained either in the heat flux or the temperature-controlled experiments. A convenient way of obtaining transient transition boiling data is through experiments in which starting at a temperature in film boiling heater is quenched.

Berenson's [9] experiments are probably the first well-documented and carefully conducted experiments of transition boiling. In his experiments, Berenson boiled n-pentane and carbon tetrachloride on a horizontal disc heated by condensation of steam underneath the disc. The data points accessed by Berenson when going from the nucleate boiling side were different from those obtained when accessing from the film boiling side. Using maximum and minimum heat flux points as a guide, Berenson faired a single curve through a limited dataset available in transition boiling. Witte and Lienhard [13] have questioned the uniqueness of the transition boiling curve obtained by Berenson. They have argued that two different sets of transition boiling data are obtained when accessing from nucleate and film boiling sides of the boiling curve. Subsequent experiments of Bui and Dhir [12], Liaw and Dhir [14], and Maracy and Winterton [15] show unequivocally the existence of two transition boiling curves. The data of Liaw and Dhir obtained on a vertical surface and plotted in Figure 7.4 show that the difference in the transition boiling curves accessed from nucleate and film

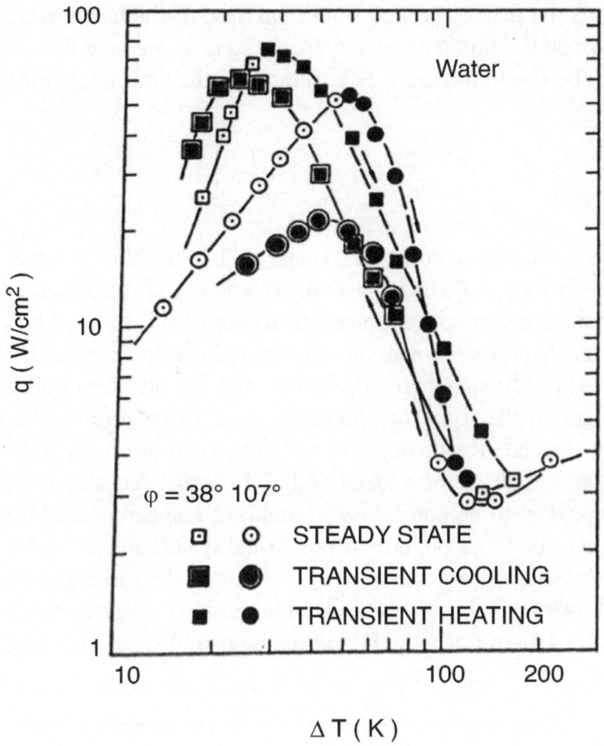

FIGURE 7.4 Transition boiling curves obtained on a vertical surface at different contact angles [14].

boiling sides is dependent on the wettability of the surface. The difference decreases as surface became more wettable. For a well-wetted surface, there is hardly any difference between transition boiling curves accessed from film or nucleate boiling sides of the boiling curve. It has been argued by Auracher [16], for example, that the differences in the two transition boiling curves obtained by Bui and Dhir, Liaw and Dhir, and Maracy and Winterton [15] are due to the transient nature of the data. Had the data been taken under strict steady-state conditions, a unique transition boiling curve would have been obtained. We will discuss this further in Chapter 8.

At present, no credible mechanistic model for dependence of transition boiling heat flux on wall superheat exists in the literature. Most of the empirical correlations rely on the fact that the upper limit of the transition boiling heat flux is the maximum heat flux, whereas the lower limit corresponds to the minimum heat flux. Nucleate boiling is assumed to prevail over the liquid occupied (wet) region, whereas conditions similar to film boiling exist on the dry area. In one of the approaches, an expression for the transition boiling heat flux by summing the time-averaged heat fluxes over wet and dry areas of the heater is written as

$$q = F_l q_l + (1 - F_l) q_v \tag{7.23}$$

In eq. (7.23), F_l is the time-averaged fractional area of the heater occupied by liquid, q_l is the heat flux on the liquid occupied area, and q_v is the heat flux on the area occupied by vapor. The fractional area, F_l, is related to the wall temperature as

$$F_l = \left[\frac{T_{\min} - T}{T_{\min} - T_{\max}}\right]^n \tag{7.24}$$

In eq. (7.24), n is an empirical constant. Equation (7.24) implies that the heater is totally covered with liquid at $q_{\max}$ ($T_{\max}$) and is totally dry at $q_{\min}$ ($T_{\min}$). This is not really correct as in nucleate boiling a certain fraction of the heater area is covered by vapor and some liquid–solid contacts may exist neat the minimum heat flux condition. The heat fluxes q_l and q_v are evaluated using correlations for nucleate and film boiling, respectively. There is a danger in this type of approach since while forcing the correlations to match the data, dependence of dry and wet fractional areas on wall superheat may be artificially skewed. Bjonard and Griffith [17] somewhat alleviated this difficulty by assuming q_l and q_v to be equal to $q_{\max}$ and $q_{\min}$ respectively. More recently, Garg and Dhir [18] have shown through their numerical simulation of the boiling curve that transition boiling is simply an extension of nucleate boiling to high superheats.

Figure 7.5 shows the wet area fractions in transition boiling as measured by several investigators. The data of Lee et al. [19] and Dhuga and Winterton [20] were obtained

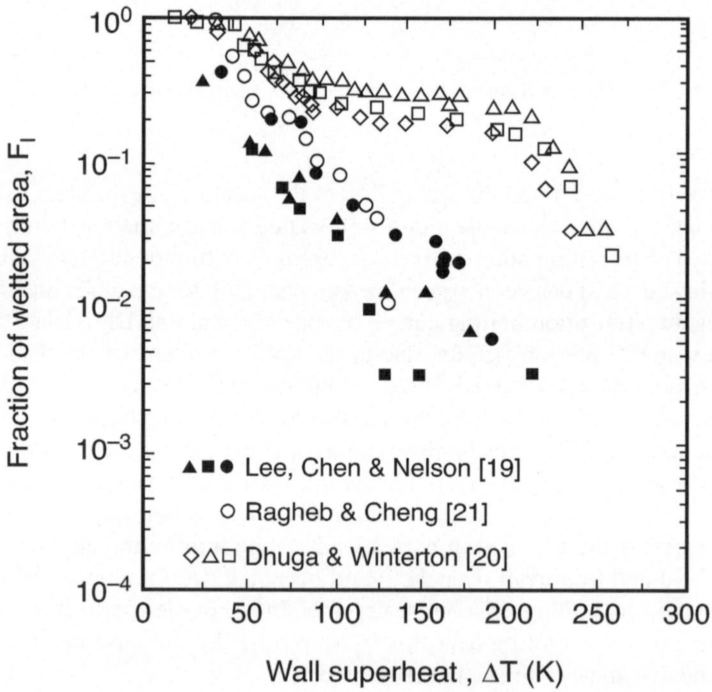

FIGURE 7.5 Wet area fractions during transition boiling of water [19–21].

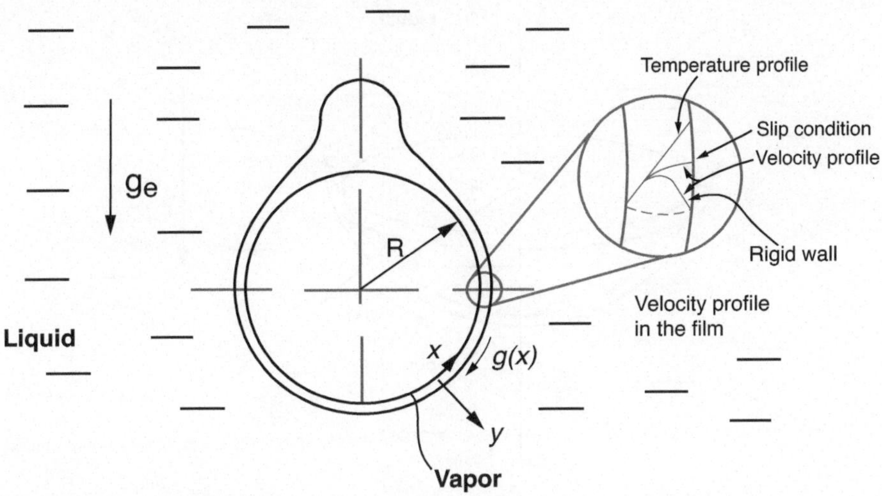

FIGURE 7.6 Film boiling on a horizontal cylinder.

when experiments were conducted in quenching mode. Significant differences exist between the data of Dhuga and Winterton and other investigators such as Lee et al. [19] and Ragheb and Cheng [21]. Data such as that shown in Figure 7.5 can be used to check the consistency of the wetted area fractions used in eq. (7.23). A number of physical variables including the manner in which data are taken (steady state or transient), surface wettability, whether transition boiling curve is accessed from nucleate or transition boiling, etc., affect the transition boiling curve. This aspect is elaborated on further in Chapter 8.

7.4 FILM BOILING ON A HORIZONTAL CYLINDER AND A SPHERE

Photograph of film boiling on a horizontal cylinder is shown in Figure 7.2b. A vapor film covers the solid surface and vapor bubbles formed on the protruding interface are released cyclically. In a given cycle, bubbles are released alternatively from nodes and anti-nodes.

Bromley [22] was the first one to provide a model for film boiling heat transfer coefficient on horizontal cylinders. He employed Nusselt's analysis for laminar film condensation on horizontal tubes, to obtain an expression for the film boiling heat transfer coefficient. Figure 7.6 shows the vapor flow configuration used by Bromley to carry out the analysis. Here following the approach of Dhir and Lienhard [23] for laminar film condensation, we will carry out a generalized analysis, which will be valid for laminar film boiling on plane (e.g., a vertical wall) and for axisymmetric bodies (e.g., cylinders and spheres). Figure 7.7 shows a film boiling on an axisymmetric surface. Various assumptions made in carrying out the steady-state analysis are:

i. Vapor film thickness is much smaller than the radius of curvature of the axisymmetric body.
ii. Fluids are incompressible.

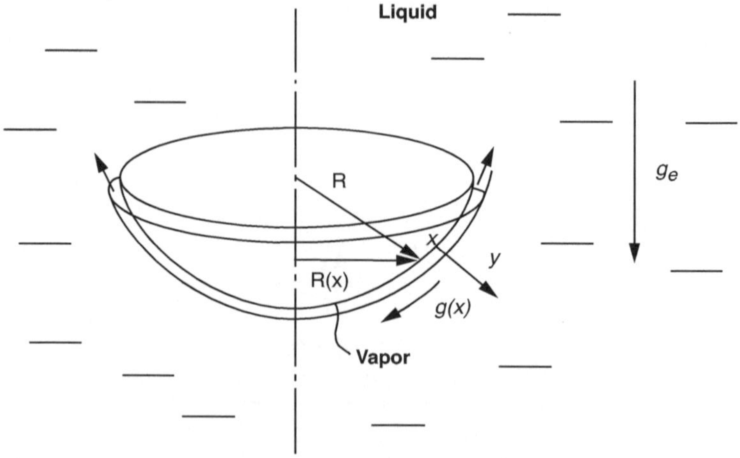

FIGURE 7.7 Film bubbling on general axisymmetric bodies.

iii. Flow in the film is laminar.
iv. Inertia and convective terms can be neglected in the momentum and energy equations respectively.
v. Vapor properties are evaluated at the mean film temperature and radiation contribution across the film is neglected.
vi. Interface between vapor and liquid is free of any ripples.
vii. Flow along the periphery is dominant and any cross-flow effect is neglected.

Since large velocity gradients occur normal to the surface, and inertia terms are neglected, the momentum equation for the vapor can be written as

$$0 = -\frac{\partial p}{\partial x} - \rho_v g(x) + \mu_v \frac{\partial^2 u}{\partial y^2} \tag{7.25}$$

In writing eq. (7.25), coordinate, x, is along the surface and y is normal to the surface. Also, $\frac{\partial p}{\partial x}$ represents the pressure gradient in the liquid and is equal to $-g(x)\rho_l$ with $g(x)$ being component of g_e along x. As such, the momentum equation becomes

$$0 = g(x)(\rho_l - \rho_v) + \mu_v \frac{\partial^2 u}{\partial y^2}$$

or

$$\frac{\partial^2 u}{\partial y^2} = -\frac{g(x)(\rho_l - \rho_v)}{\mu_v} \tag{7.26}$$

We require two boundary conditions to solve for velocity, u, from eq. (7.26). One boundary condition for eq. (7.26) is zero velocity at the wall $(y = 0)$. The second boundary condition is specified at the interface. Since we are not solving for the flow field in the liquid, it is difficult to specify an appropriate boundary condition. As such, two limiting possibilities, as was the case for the horizontal plate are: slip or rigid wall boundary condition at the interface. For slip boundary condition, $\frac{\partial u}{\partial y} = 0$ at $y = \delta$, whereas for the rigid wall boundary condition, $u = 0$ at $y = \delta$. If we solved for the flow field in the liquid as well, appropriate conditions at the interface will be matching of velocities and shear stress in the two phases. With integration of eq. (7.26) and use of the above described boundary conditions, an expression of the vapor velocity is obtained by

$$u = \frac{g(x)(\rho_l - \rho_v)}{2\mu_v}\left(-y^2 + c\delta y\right) \tag{7.27}$$

where $c = 1$ for a rigid interface and is equal to 2 for a dynamic interface (slip condition).

Since convective terms in the steady-state energy equation are neglected and vapor temperature variation normal to wall is dominant and conduction in the x direction is neglected, the energy equation can be written as

$$\frac{\partial^2 T}{\partial y^2} = 0 \tag{7.28}$$

The boundary conditions for eq. (7.28) are: a specified wall temperature, T_w and saturation temperature at the interface when evaporative resistance at the interface is neglected. As such, an expression for the temperature distribution in the vapor film is obtained as

$$T - T_{\text{sat}} = \left(1 - \frac{y}{\delta}\right)(T_w - T_{\text{sat}})$$

$$= \left(1 - \frac{y}{\delta}\right)\Delta T \tag{7.29}$$

where $\Delta T = T_w - T_{\text{sat}}$

An energy balance at any location x along the surface (Figure 7.7) yields

$$\frac{k_v (T_w - T_{\text{sat}})}{\delta(x)} 2\pi R(x) = \frac{\partial}{\partial x}\int_0^\delta \rho_v u 2\pi R(x) h_{fg} dy$$

$$+ \frac{\partial}{\partial x}\int_0^\delta \rho_v u 2\pi R(x) c_{pv}(T - T_{\text{sat}}) dy \tag{7.30}$$

The first term on the right-hand side results from the rate at which energy is used to produce vapor that is added at the interface of the film along the surface. The second term represents the increase with distance of the sensible heat content of the vapor flowing in the film because of the vapor temperature being above the saturation temperature. Substitution for u from eq. (7.27) and $(T - T_{sat})$ from eq. (7.29) in eq. (7.30) yields:

$$\frac{k_v (T_w - T_{sat}) R(x)}{\delta(x)} = \frac{h_{fg} \rho_v (\rho_l - \rho_v)}{c_1 \mu_v} \frac{\partial}{\partial x} \left(R(x) g(x) \delta^3 \left(1 + c_2 \frac{c_{pv} \Delta T}{h_{fg}} \right) \right) \tag{7.31}$$

where $c_1 = 12$ and $c_2 = 1/2$ for a rigid interface and $c_1 = 3$ and $c_2 = 3/8$ for a dynamic interface.

Defining

$$h'_{fg} = h_{fg} \left(1 + c_2 \frac{c_{pv} \Delta T}{h_{fg}} \right)$$

and $\Delta = \delta R^{1/3} g^{1/3}$, eq. (7.31) becomes

$$\frac{k_v \Delta T R^{4/3} g^{1/3}}{\Delta} = \frac{h'_{fg} \rho_v (\rho_l - \rho_v)}{c_1 \mu_v} \frac{d}{dx} \Delta^3$$

or

$$k_v \Delta T R^{4/3} g^{1/3} = \frac{3 h'_{fg} \rho_v (\rho_l - \rho_v)}{4 c_1 \mu_v} \frac{d}{dx} \Delta^4 \tag{7.32}$$

Integrating eq. (7.32) with respect to x gives

$$\Delta^4 = \frac{4 c_1 \mu_v k_v \Delta T}{3 h'_{fg} \rho_v (\rho_l - \rho_v)} \int_0^x R^{4/3} g^{1/3} dx \tag{7.33}$$

In writing eq. (7.33) we have used the condition that $\Delta = 0$ at $x = 0$. Since Δ is defined as a product of δ, $R^{1/3}$ and $g^{1/3}$, it will be zero when at least one of them is zero. For a vertical wall, it is δ that is zero at $x = 0$, for a horizontal cylinder R is constant, and $g(x) = 0$ at $x = 0$ and for a sphere $g(x)$ and $R(x)$ both are zero at $x = 0$.

For a vertical wall, with $g(x) = g_e$ and $R(x) \to \infty$, eq. (7.33) yields

$$\delta^4 = \frac{4 c_1 \mu_v k_v \Delta T x}{3 \rho_v (\rho_l - \rho_v) g_e h'_{fg}}$$

or

$$\delta = \left[\frac{4 c_1 \mu_v k_v \Delta T x}{3 \rho_v (\rho_l - \rho_v) g_e h'_{fg}} \right]^{1/4} \tag{7.34}$$

or local heat transfer coefficient is

$$h = \frac{k_v}{\delta} = \left[\frac{3k_v^3 h'_{fg} \rho_v (\rho_l - \rho_v) g_e}{4c_1 \mu_v \Delta T x} \right]^{1/4} \tag{7.35}$$

and the average heat transfer coefficient is

$$\bar{h} = \frac{1}{x} \int_0^x h \, dx = \frac{4}{3} \left(\frac{3}{4c_1} \right)^{1/4} \left[\frac{k_v^3 h'_{fg} \rho_v (\rho_l - \rho_v) g_e}{\mu_v \Delta T x} \right]^{1/4} \tag{7.36}$$

or

$$\bar{h} = 0.943 \text{ or } 0.67 \left[\frac{k_v^3 h'_{fg} \rho_v (\rho_l - \rho_v) g_e}{\mu_v \Delta T x} \right]^{1/4} \tag{7.36a}$$

With substitution for c_1, the overall numerical constant in eq. (7.36a) is 0.943 for a dynamic interface and is 0.67 for a rigid interface. Generally, the film boiling heat transfer coefficient measured on a vertical surface is higher than that given by eq. (7.36a) with a constant of 0.943. We will discuss this more in detail in a subsequent section. It should be noted that according to eq. (7.35), local heat transfer coefficient decreases with distance from the leading edge, at $x^{-1/4}$ and with wall superheat as $\Delta T^{-1/4}$.

For a cylinder, $g(x) = g_e \sin\left(\frac{x}{R}\right)$ and radius, R, along the surface is constant and normal to the plane of the paper is infinite. Substituting for $g(x)$ and R in Δ, eq. (7.33) gives

$$\delta^4 = \frac{4c_1 \mu_v k_v \Delta T}{3\rho_v (\rho_l - \rho_v) g_e h'_{fg}} \frac{\int_0^x \left[\sin\left(\frac{x}{R}\right) \right]^{1/3} dx}{\left[\sin\left(\frac{x}{R}\right) \right]^{4/3}} \tag{7.37}$$

with local heat transfer coefficient being equal to k_v / δ, eq. (7.37) leads to

$$h = \left[\frac{3k_v^3 \rho_v (\rho_l - \rho_v) g_e h'_{fg}}{4c_1 \mu_v \Delta T} \right]^{1/4} \left[\frac{\left(\sin\left(\frac{x}{R}\right) \right)^{4/3}}{\int_0^x \left(\sin\left(\frac{x}{R}\right) \right)^{1/3} dx} \right]^{1/4}$$

or the average heat transfer coefficient is obtained as

$$\bar{h} = \frac{1}{\pi R}\left[\frac{3k_v^3\rho_v(\rho_l-\rho_v)g_e h'_{fg}}{4c_1\mu_v\Delta T}\right]^{1/4}\left\{\int_0^{\pi R}\left[\frac{\left(\sin\left(\frac{x}{R}\right)\right)^{4/3}}{\int_0^x\left(\sin\left(\frac{x}{R}\right)\right)^{1/3}dx}\right]^{\frac{1}{4}}dx\right\}$$

Numerical value of the term in the curly parenthesis is found to be $0.868\pi R^{3/4}$. Or, the expression for the average heat transfer is

$$\bar{h} = 0.73\left(\frac{3}{c_1}\right)^{1/4}\left[\frac{k_v^3\rho_v(\rho_l-\rho_v)g_e h'_{fg}}{\mu_v\Delta TD}\right]^{1/4} \tag{7.38}$$

where D is the diameter of the cylinder. For a dynamic interface, $c_1 = 3$ and numerical constant in front of the expression in eq. (7.38) will be simply 0.73. For a rigid interface, the constant will be 0.52. For a sphere, $R(x) = R\sin\frac{x}{R}$ and $g(x) = g_e\sin\frac{x}{R}$. By following a development similar to that for a cylinder, an expression for the average heat transfer coefficient on a sphere is found as

$$\bar{h} = 0.81\left(\frac{3}{c_1}\right)^{1/4}\left[\frac{k_v^3\rho_v(\rho_l-\rho_v)g_e h'_{fg}}{\mu_v\Delta TD}\right]^{1/4} \tag{7.39}$$

For a dynamic interface, the numerical constant in eq. (7.39) will reduce to 0.81, whereas for a rigid interface, it will be 0.57. Vapor flowing through the film accumulates in a bubble formed at the downstream stagnation point. This bubble is released cyclically from the heater surface. Since bubble height is much larger than the film thickness, the heater area under the bubble is effectively insulated. To account for the presence of a bubble, Bromley [22] reduced the average heat transfer coefficient on a cylinder by 15%. As a result, numerical constant in eq. (7.38) for $c_1 = 3$ (slip condition) was reduced from 0.73 to 0.62. A similar argument will reduce for a sphere the numerical constant from 0.81 to 0.68.

Equation (7.39) with a numerical constant of 0.62 was found by Bromley to predict film boiling heat transfer coefficient from horizontal cylinders varying in diameter from 6.35mm to 19mm. Subsequently, Breen and Westwater [24] conducted film boiling heat transfer experiments on horizontal cylinders with wide range of diameters—both larger and smaller than those used by Bromley. They argued that for cylinders with a diameter much larger than the two-dimensional critical Tayler wavelength, λ_{c2}, the surface would behave like a horizontal surface, whereas for very small cylinders, the vapor flow was more axial rather than along the circumference. As such, they identified a parameter λ_{c2}/D to correlate film boiling heat transfer coefficient on horizontal cylinders. Here λ_{c2} is the critical two-dimensional Taylor wavelength on a horizontal surface and was obtained in Chapter 6. Their correlation is

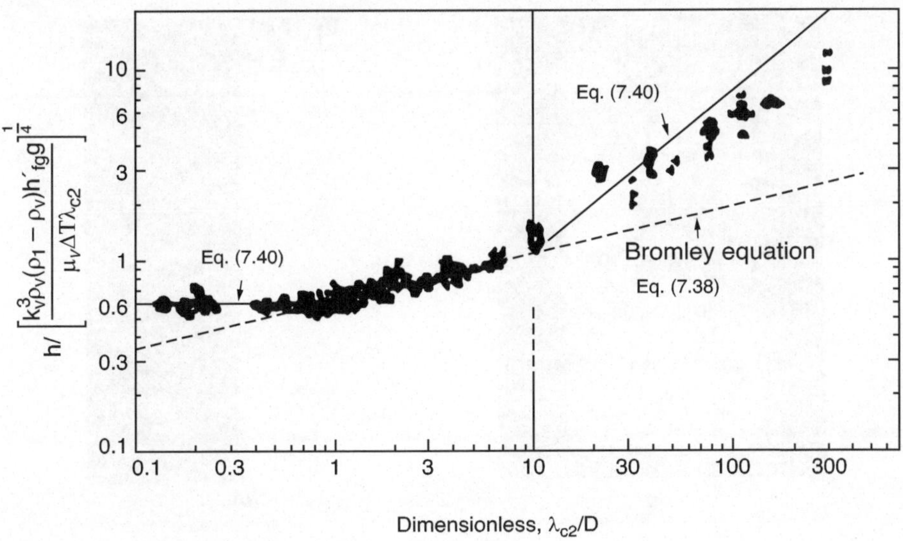

FIGURE 7.8 General correlation for film boiling on horizontal cylinders. Range of diameters is from 0.0056 to 48.13 mm. (From Breen and Westwater [24].)

$$\bar{h} = \left[\frac{k_v^3 \rho_v (\rho_l - \rho_v) g_e h'_{fg}}{\mu_v \Delta T \lambda_{c2}}\right]^{\frac{1}{4}} \left(0.59 + 0.069 \frac{\lambda_{c2}}{D}\right) \tag{7.40}$$

A comparison of the data against correlation eq. (7.40) of Breen and Westwater and eq. (7.38) with numerical constant of 0.62 is shown in Figure 7.8. Data were obtained with hydrocarbon fluids.

Data obtained by Merte and Clark [25] for film boiling of liquid nitrogen on a 25.4 mm diameter sphere have been found to be consistent with that given by eq. (7.39) with numerical constant of 0.68. Dhir and Purohit [26] found that during film boiling of saturated water on spheres, the data were higher than prediction from eq. (7.39) with the numerical constant 0.68 when contribution of radiation was accounted for as well. They found the data to be correlated well when the numerical constant was increased to 0.8. The increase was justified on the basis of presence of capillary waves on the vapor–liquid interface. Since heater surface temperatures in film boiling can be quite high, radiation contribution similar to that for a flat plate should be included in eqs. (7.38) and (7.39).

7.5 FILM BOILING ON A VERTICAL WALL

The laminar flow-based analysis with a smooth interface such as that described in the last section is valid only for a short distance from the leading edge, as waves eventually develop on the interface. Because of presence of these waves, the observed heat transfer coefficient is larger than that predicted from eq. (7.36) and is nearly

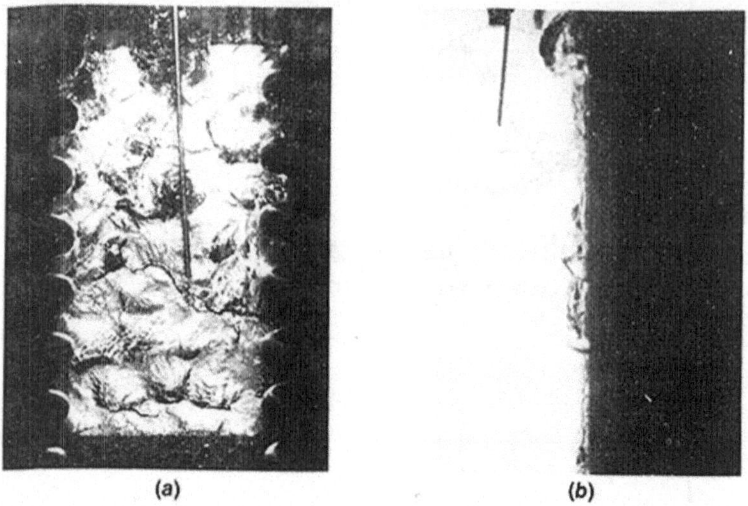

FIGURE 7.9 Natural convection film boiling of saturated water on a plane vertical surface ($\Delta T = 386\text{K}$): (a) front view; (b) side view [27].

independent of the distance from the leading edge. Frontal and side views shown in Figure 7.9 of photographs obtained by Bui and Dhir [27] during their studies of film boiling on a vertical surface clearly show large bulges and short waves formed on the vapor–liquid interface. In the past, several investigators have speculated that observed behavior of heat transfer coefficient (independent of distance and magnitude higher than predicted from laminar analysis) was a result of vapor flow in the film becoming turbulent. Hsu and Westwater [28] proposed that vapor film becomes turbulent when Reynold's number, *Re*, based on vapor film thickness, δ, and vapor velocity at the interface, U_i is, i.e., $\mathrm{Re} \cong \frac{U_i \delta}{\nu_v} = 100$. Hsu and Westwater also developed a model for turbulent vapor film. In the model, velocity in the laminar sublayer was assumed to increase linearly with distance away from the wall, thereafter the velocity remained constant in the film. An expression for the heat transfer coefficient for turbulent film boiling was obtained by employing analysis similar to that used in deriving eq. (7.36) but with a turbulent velocity profile as described above.

The assumption of a turbulent film is a conjecture in the absence of any experimental validation. Photographs of Bui and Dhir [27] as shown in Figure 7.9 clearly demonstrate the formation of large vapor bulges on the film with intervening thin film regions. Ripples or capillary waves are also present on the interface. Based on these observations, Bui and Dhir [27] proposed that

i. Bulges resulting from interfacial instability interrupt the vapor flow in the film and thereby reduce the vapor flow path length to the distance between each bulge instead of the distance from the leading edge of the plate. After each bulge, thin film region begins again. The structure continuously traverses the solid surface; thus, all locations beyond the initial wave free film

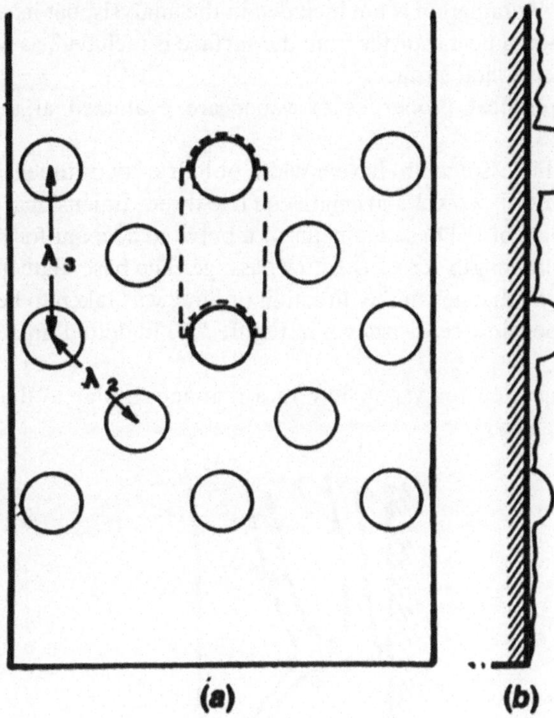

FIGURE 7.10 Configuration of the vapor–liquid interface for the heat transfer model: (a) front view; (b) side view [27].

region experience alternatively the presence of bulges and the thin film region between the bulges as the interface moves upwards with a velocity U_i.

ii. Bulges (long waves) provide additional buoyancy force, which increases the vapor velocity at the interface.
iii. Capillary waves (short wavelength) or ripples on the thin film improve heat transfer.

Figure 7.10 shows the conceptualized large wave pattern, with three-dimensional wave aligned along the height of the plate. Envisioned flow pattern in the liquid relative to the bulges as large vapor bulges move upward is shown in Figure 7.11. In carrying out the analysis for the heat transfer in the thin film region, Bui and Dhir [27] made several assumptions:

i. The vertical wall is held at a constant temperature.
ii. Vapor and liquid are incompressible.
iii. Flow in the thin film is laminar.
iv. Inertia terms in the momentum equation and convection terms in the energy equation can be neglected. However, correction for the sensible heat carried by vapor is made.

v. Heat transfer by radiation is not included in the analysis, but its contribution to the total rate of heat transfer from the surface is included, as was done for the case of horizontal surface.
vi. The thermophysical properties of vapor are evaluated at a mean film temperature.
vii. One flow cell is assumed to have a width of half of two-dimensional wavelength, λ_2, which is $\lambda_3/\sqrt{2}$ and length equal to three-dimensional wavelength reduced by half of the base diameter of a bulge to account for reduction in vapor flow path length across the flow passage. The base diameter, *d*, of the bulges, though, changes in the direction of flow, was taken to be half of λ_2. A typical vapor flow cell is shown in Figure 7.10 in dotted lines.

The momentum equation for vapor flow in a manner similar to that described in Section 7.4 is written as

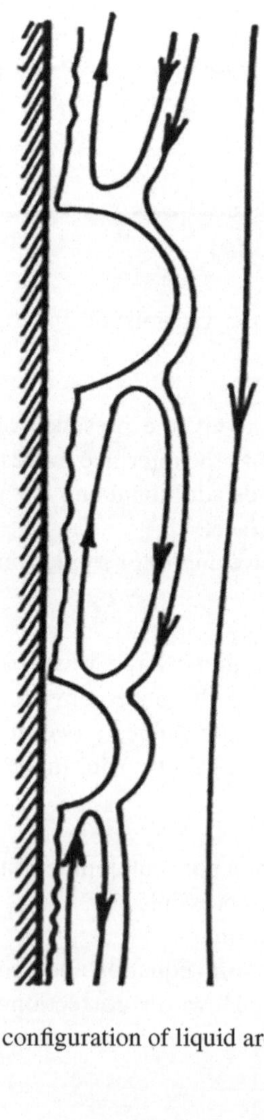

FIGURE 7.11 The relative flow configuration of liquid around bulges [27].

$$\mu_v \frac{\partial^2 u}{\partial y^2} = -g(\rho_l - \rho_v) \tag{7.41}$$

The velocity profile satisfying the boundary conditions:

$$u = 0 \text{ at } y = 0$$

$$u = U_i \text{ at } y = \delta$$

can be written as

$$u = \frac{g(\rho_l - \rho_v)}{2\mu_v}(y\delta - y^2) + \frac{U_i y}{\delta} \tag{7.42}$$

Mass and energy balance applied to a section of thin film gives

$$\frac{d}{dx}\left[h_{fg}\int_0^\delta \rho_v u dy\right] + \frac{d}{dx}\left[c_{pv}\int_0^\delta (T - T_{\text{sat}})\rho_v u dy\right] = \frac{k_v \Delta T}{\delta} \tag{7.43}$$

where ΔT is the wall superheat. The first term on the left-hand side of the above equation represents the rate at which energy is utilized in conversion of liquid into vapor at the film interface, whereas the second term accounts for superheating of vapor above the saturation temperature. For a linear temperature profile in the film, eq. (7.43) can be written as

$$\frac{d}{dx}\left[h'_{fg}\int_0^\delta \rho_v u dy\right] = \frac{\kappa_v \Delta T}{\delta} \tag{7.44}$$

where

$$h'_{fg} = h_{fg} + 1/2 c_{pv}\Delta T\left[1 - \frac{1}{\left(3 + \frac{g(\rho_l - \rho_v)\delta^2}{2\mu_v U_i}\right)}\right] \tag{7.45}$$

Substitution of velocity profile given by eq. (7.42) in eq. (7.44) yields

$$\frac{d}{dx}(\delta^4) + \frac{4\mu_v U_i}{g(\rho_l - \rho_v)}\frac{d}{dx}(\delta^2) = \frac{16\mu_v \kappa_v \Delta T}{\rho_{v(\rho_l - \rho_v)} g h'_{fg}} \tag{7.46}$$

Integration of eq. (7.46) with x while assuming $\delta = 0$ at $x = 0$ yields

$$\delta^4 + \frac{4\mu_v U_i}{g(\rho_l - \rho_v)}\delta^2 - \frac{16\mu_v \kappa_v \Delta T x}{\rho_{v(\rho_l - \rho_v)} g h'_{fg}} = 0 \tag{7.47}$$

The assumption of $\delta = 0$ at $x = 0$ is strictly valid only at the leading edge of the vertical plate. At subsequent portions of the film between each bulge, the film thickness is nonzero at the beginning of each cell. However, the error due to assumption of zero thickness leads to a change of only a few percent in the average heat transfer coefficient. In a dimensionless form, eq. (7.47) can be written as

$$\delta^{*4} + 4U_i^* \delta^{*2} - 16x^* = 0 \tag{7.48}$$

where

$$\delta^* = \frac{\delta}{\left(\frac{\nu_v^2}{g}\right)^{1/3}} \tag{7.49}$$

$$U_i^* = \frac{U_i}{\frac{g}{\mu_v}(\rho_l - \rho_v)\left(\frac{\nu_v^2}{g}\right)^{1/3}} \tag{7.50}$$

$$x^* = \frac{x}{\left(\frac{\nu_v^2}{g}\right)^{1/3}}\left[\frac{(\rho_l - \rho_v)}{\rho_v}\right]^3 \frac{\kappa_v \Delta T}{\mu_v h_{fg}} \tag{7.51}$$

A non-negative solution of quadratic eq. (7.48) in δ^{*2} is

$$\delta^* = \sqrt{2}\left[\left(U_i^{*2} + 4x^*\right)^{1/2} - U_i^*\right]^{1/2} \tag{7.52}$$

From eq. (7.52), we see that for $U_i^* = 0$, film thickness will increase as $x^{*\,1/4}$ as was the case for analysis carried out earlier for a laminar film with rigid or a dynamic interference. Local heat transfer coefficient along the thin film is

$$h(x) = \frac{\kappa_v}{\delta} = \frac{\kappa_v}{\delta^*}\left(\frac{\nu_v^2}{g}\right)^{-1/3} \tag{7.53}$$

The average heat transfer coefficient for the thin film region is

$$\bar{h}_f = \frac{1}{l}\int_0^l h(x)\,dx \tag{7.54}$$

In eq. (7.54), l is the effective flow path length, which as described earlier is approximately, $\lambda_2 - \frac{d}{2}$. The contribution of heat transfer rate from the thin film region to the total heat transfer rate over a cell area as marked by dotted lines in Figure 7.10 is

$$q_f = \bar{h}_f \frac{A_f}{A_c} \Delta T = \bar{h}_f \frac{\left(\lambda_2^2 - \frac{\pi}{4} d^2\right)}{\lambda_2^2} \Delta T$$
$$= \bar{h}_f \left(1 - \frac{\pi}{4} \frac{d^2}{\lambda_2^2}\right) \Delta T \tag{7.55}$$

Three issues still remain to be addressed in calculation of q_f from eq. (7.55), i.e., the evaluation of the effect of presence of short waves and ripples on the vapor–liquid interface, calculation of interface velocity as a result of presence of large waves or bulges on the interface and the two-dimensional wavelength, λ_2, of the large waves. The presence of ripples improves heat transfer over plane interface. This is due to the fact that heat transfer rate through the film varies inversely with the film thickness. Coury and Dukler [29] have shown that for sinusoidal waves, right-hand side of eq. (7.55) should be multiplied by a constant, c_3, defined as

$$c_3 = \frac{1}{\sqrt{1 - \eta^2}} \tag{7.56}$$

where η is the ratio of amplitude of ripples to the mean thickness of the film. For a guessed value of $\eta = 0.5$, constant c_3, representing enhancement in heat transfer coefficient due to ripples in the thin film region would have a value of 1.15. Thus, right-hand side of eq. (7.55) should have a multiplier of 1.15 in front of $\bar{h}_f$.

For calculation of the velocity of interface, Bui and Dhir [27] assumed that the interface moved at the same speed as the vapor bulge. Figure 7.11 shows the envisioned flow field. Bulges not only experience buoyancy and viscous drag, but the liquid flow also introduces drag on the thin film region. Because in the momentum eq. (7.25), inertia terms in the vapor are ignored and the flow field of the liquid is not solved, velocity and shear stress on both vapor and liquid side of the interface are not matched. To account for the liquid drag, form drag was included in the force balance for the bulge and was arbitrarily taken to be equal to the viscous drag, which was assumed to be equal to that on a disc of bulge base diameter, d. Buoyancy force was assumed to act on a volume of half of a sphere of diameter d. By balancing the drag and buoyancy forces, an expression for U_i was obtained as

$$U_i = 0.37 \left[\frac{g(\rho_l - \rho_v)}{\rho_l \nu_l^{1/2}} \right]^{2/3} d \tag{7.57}$$

Prediction of the two-dimensional wavelength, λ_2, of the long wave that evolves into vapor bulges is much more intricate. Stability analysis carried out by Sherman [30] of film boiling on a vertical surface shows that dominant wavelength, λ_d, has a functional form

$$\lambda_d = f(\text{Re}, \text{Pr}, \text{Sh}) \tag{7.58}$$

where

$$\text{Re} = \frac{\bar{u}_v x}{\nu_v}$$

$$\text{Pr} = \frac{\mu_v c_{pv}}{\kappa_v}$$

and

$$\text{Sh} = \frac{c_{pv} \Delta T_w}{h_{fg}}$$

Thus, dominant wavelength will not only depend on the thermophysical properties of vapor but also on vapor superheat and distance from the leading edge of the plate. Noting from the work of Dhir and Taghavi-Tafreshi [31] that the peaks of the instability wave always locate about half a wavelength away from the leading edge, for the wave pattern to develop, the minimum distance from the leading edge should be about $1\frac{1}{2}\lambda_d$. Thus, in calculating the dominant two-dimensional wavelength, *Re* in eq. (7.58) should be calculated at $x = 1\frac{1}{2}\lambda_d$. This requires an iterative procedure. Reynolds number is evaluated using laminar film analysis as developed in Section 7.4.

To determine the total heat transfer from one cell, we must also account for rate of heat transfer from underneath the bulge. In Berenson's analysis for a horizontal plate, rate of heat transfer under the bulges or vapor bubbles was assumed to be negligible. On a vertical wall, the size and shape of the bulges change as the vapor film travels upward. Bui and Dhir [27] found that rate of heat transfer underneath the bulges was non-negligible. However, they found that the ratio of rate of heat transfer underneath the bulges to the average heat transfer over thin film region varied only from 0.34 to 0.25 as height-to-base diameter (~0.01m) ratio is varied from 0.1 to 1. Thus, although their analysis showed that heat transfer underneath the bulges was non-negligible, little error would be made if a mean value of 0.3 was used for ratio of heat transfer over the bulges to that over the thin film region.

Generally, surface temperatures are high during film boiling. As such, contribution of heat transfer by radiation across the film must be included in the overall rate of heat transfer. Because radiation represents an additional mode of heat transfer to conduction, in principle thin film analysis should be carried out by including contribution of heat transfer by radiation, with the net effect being thickening of the film. However, this correction is generally not significant and can be ignored while just including the contribution of radiation as a parallel mode of heat transfer.

Since the vapor bulges and film alternatively move over a given location on the surface, time and area-averaged heat transfer coefficient over one cell can be written as

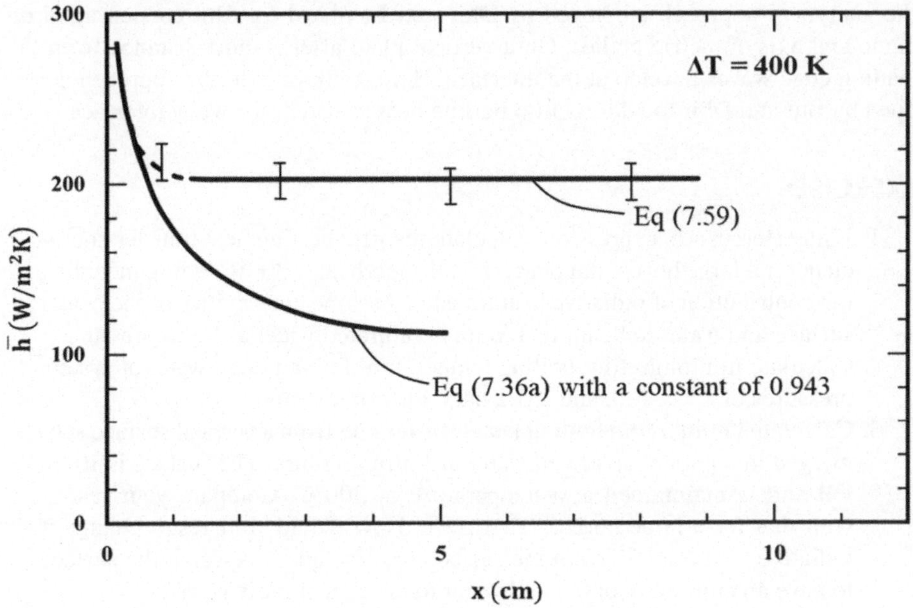

FIGURE 7.12 Comparison of predicted and observed dependence of heat transfer coefficient on distance from leading edge [27].

$$\bar{h} = c_3 \bar{h}_f \left(1 - \frac{\pi}{4}\frac{d^2}{\lambda_2^2}\right) + c_4 \bar{h}_f \frac{\pi}{4}\frac{d^2}{\lambda_2^2} + \frac{1}{\left[\frac{1}{\varepsilon_w} + \frac{1}{\varepsilon_l} - 1\right]} \frac{T_w^{\,4} - T_{sat}^{\,4}}{T_w - T_{sat}} \tag{7.59}$$

where second term represents rate of heat transfer under a bulge and third term contribution of radiation.

As noted earlier value of constant c_3 is 1.15 and c_4 is 0.3. In eq. (7.59) ε_w and ε_l are emissivities of the wall and liquid respectively.

Figure 7.12 shows a comparison of the experimental data with the prediction from eq. (7.59) for film boiling of water at 1 atm pressure and a wall superheat of 400K. It is seen that when $\varepsilon_w = \varepsilon_l = 1$, eq. (7.59) does a good job in describing the data, which shows heat transfer coefficient except near the leading edge to be independent of distance from the leading edge. In comparison, laminar film analysis with plane interface as described earlier would have predicted heat transfer coefficient to decrease as $x^{-1/4}$.

7.6 SUMMARY

In this chapter, we have described film boiling on a horizontal surface, a horizontal cylinder, a sphere, and a vertical plate. In developing expression for film boiling heat transfer coefficient on a horizontal surface, we have used Berenson's model, whereas for laminar film boiling on a vertical surface, a cylinder, and a sphere, we have used the analytical approach employed by Dhir and Lienhard for film condensation on plane and axisymmetric bodies. On a vertical plate after a short distance from the leading edge waves develop at the interface. Here we have used the approach developed by Bui and Dhir to address film boiling heat transfer on a wavy interface.

EXERCISES

1. Using Berenson's expression, calculate the film boiling heat transfer coefficient for a large horizontal plate, if wall superheat is 300°C, while including the contribution of radiative heat transfer. Assume emissivities of the heater surface and water to be unity. Liquid is saturated water at 1 atm pressure.
2. Calculate minimum film boiling temperature for saturated water at system pressures of 1, 61, 100, and 221.2 bars. Plot your results.
3. Calculate laminar film boiling heat transfer rate from a vertical surface submerged in a pool of saturated water at 1 atm pressure. The surface is 10 cm toll and is maintained at a temperature of 300°C. Compare your results with that for a large horizontal surface. How would your result change if radiative heat transfer from the surface was included? Assume the surface to have an emissivity of 0.5 and water to have an emissivity of 1.
4. Film boiling occurs on a flat plate in a pool of water maintained at 1 atm pressure and 80°C. Using Henry's correlation, obtain the minimum film boiling temperature. The plate is made of stainless steel.
5. A nichrome wire 5mm in diameter is held horizontally in saturated water at 1 atm pressure. If the wire temperature is 200°C, calculate the heat flux or power input needed to sustain stable film boiling on the wire.

NOMENCLATURE

A_c, A_v	Cell and vapor film occupied areas respectively
c	Constant
c_{pl}	Specific heat of liquid
c_{pv}	Specific heat of vapor
d	Diameter
h	Heat transfer coefficient
h_{fg}	Latent heat of vaporization
k_v	Thermal conductivity of vapor
p	Pressure
r	Radial Coordinate
U_i	Interface velocity
u	Radial velocity of vapor
$\bar{u}$	Average radial velocity of vapor
v	Vertical velocity of vapor
R	Radius of curvature of a surface
Q	Heat transfer rate
q	Heat flux
T	Temperature

x	Axial coordinate
y	Coordinate normal to the surface

SUBSCRIPTS

l	Liquid
max	Maximum
min	Minimum
v	Vapor
r	Radiative
2	Two-dimensional
3	Three-dimensional

GREEK LETTERS

η	Interface position
μ	Molecular viscosity
ρ	Density
σ	Surface tension
λ	Wavelength

REFERENCES

1. Banerjee, D. and Dhir, V.K. (2005), "Study of Subcooled Film Boiling on a Horizontal Disc: Parts I & 2." *J. Heat Transfer*, Vol. 47, No. 4, pp. 271–293.
2. Lienhard, J.H. and Dhir V.K. (1980), "On the Prediction of the minimum Film Boiling Heat Flux." *J. Heat Transfer*, Vol. 102, pp. 380–382.
3. Chang, Y.P. (1959), "Wave Theory of Heat Transfer in Film Boiling." *ASME J. Heat Transfer*, Vol. 81, pp. 1–12.
4. Berenson, P.J. (1961), "Film Boiling Heat Transfer from a Horizontal Surface." *ASME J. Heat Transfer*, Vol. 83, pp. 351–362.
5. Hamill, J.D. and Baumeister, K.J. (1967), "Effect of Subcooling and Radiation on Film Boiling Heat Transfer from a Flat Plate." NASA TND-3925.
6. Leidenfrost, J.G. (1756), "De Aquae Communis Norinullus Qualitatibus Tractatus (A tract about some qualities of common water)." University of Duisburg, Also published in *Int. J. Heat Mass Transfer*, Vol. 9, p. 1153 (1966).
7. Henry, R.E. (1974), "A Correlation for Minimum Film Boiling Temperature." *AIChE Symp. Ser.* 70, Vol. 138, p. 8.
8. Baumeister, K.J. and Simon, F.F. (1973), "Leidenfrost Temperature—Its Correlation for Liquid Metals, Cryogenics, Hydrocarbons and Water." *J. Heat Transfer*, Vol. 96, No. 2, pp. 166–173.
9. Berenson, P.J. (1960), "Transition Boiling Heat Transfer for a Horizontal Surface." *MIT Tech. Rept.* No. 17. pp. 1–119.
10. Hesse, G. (1973), "Heat Transfer in Nucleate Boiling, Maximum Heat Flux and Transition Boiling." *Int. J. Heat Mass Transfer*, Vol. 16, pp. 1611–1627.
11. Ramilison, J.M. and Lienhard, J.H. (1987), "Transition Boiling Heat Transfer and the Film Transition Regime." *ASME J. Heat Transfer*, Vol. 109, pp. 746–752.
12. Bui, T.D. and Dhir, V.K. (1985a), "Transition Boiling Heat Transfer on a Vertical Surface." *ASME J. Heat Transfer*, Vol. 107, pp. 756–763.

13. Witte, L.C. and Lienhard, J.H. (1982), "On the Existence of Two Transition Boiling Curves." *Int. J. Heat Mass Transfer*, Vol. 25, pp. 771–779.
14. Liaw, S.P. and Dhir, V.K. (1986), "Effect of Surface Wettability on Transition Boiling Heat Transfer from a Vertical Surface." *Proceedings of International Heat Transfer Conference*, Vol. 4, pp. 2031–2036, San Francisco.
15. Maracy, M. and Winterforn, R.H.S. (1988), "Hysteresis and Contact Angle Effects in Transition Boiling of Water." *Int. J. Heat Mass Transfer*, Vol. 31, pp. 1443–1449.
16. Aurucher, H. (1992), "Transition Boiling in Natural Convection Systems." *Pool and External Flow Boiling*, Eds V. K. Dhir and A. E. Bergles, Engineering Foundation conference, pp. 219–236. Santa Barbara, California.
17. Bjornard, T.A. and Griffith, P. (1977), "PWR Blowdown Heat Transfer." *Symposium on the Thermal and Hydrodynamic Aspects of Nuclear Reactor Safety*, Vol. 1, pp. 17–41. ASME conference Atlanta, Georgia, USA.
18. Garg, D. and Dhir, V.K. (2019), "A Unified Three-dimensional Numerical Model for Boiling Curve in a Temperature Controlled Mode." *J. Heat Transfer*, Vol. 141, No. 1. pp. 1–13.
19. Lee, L.Y.W., Chen, J.C., and Nelson, R.A. (1985), "Liquid-Solid Contact Measurement Using a Surface Thermocouple Temperature Probe in Atmospheric Pool Boiling Water." *Int. J. Heat Mass Transfer*, Vol. 28, pp. 1415–1423.
20. Dhuga, D.S. and Winterton, R.H.S. (1985), "Measurement of Surface Contact Transition Boiling." *Int. J. Heat Mass Transfer*, Vol. 28, pp. 1869–1880.
21. Ragheb, H.S. and Cheng, S.C. (1979), "Surface Wetted Area during Transition Boiling in Fixed Convective Flow." *ASME J. Heat Transfer*, Vol. 101, pp. 381–383.
22. Bromley, L.A. (1950), "Heat Transfer in Stable Film Boiling." *Chem. Eng. Prog. Ser.*, Vol. 46, pp. 221–227.
23. Dhir, V.K. and Lienhard, J.H. (1971), "Laminar Film Condensation on Plane and Axi-Symmetric Bodies in Non-Uniform Gravity." *ASME J. Heat Transfer*, Vol. 93, pp. 97–100.
24. Breen, B.P. and Westwater, J.W. (1962), "Effect of Diameter of Horizontal Tubes on Film Boiling Heat Transfer." *Chem. Eng. Prog. Symp. Ser.*, Vol. 58, pp. 67–72.
25. Merte, H. and Clark, J.A. (1961), "Boiling Heat Transfer Data for Liquid Nitrogen at Standard and Non-Zero Gravity." *Cryogenic Engineering Conference*, Ann Arbor, MI.
26. Dhir, V.K. and Purohit, G.P. (1978), "Subcooled Film Boiling Heat Transfer from Spheres." *Nucl. Eng. Design*, Vol. 47, pp. 49–66.
27. Bui, T.D. and Dhir, V.K. (1985b), "Film Boiling Heat Transfer on an Isothermal Vertical Surface." *J. Heat Transfer*, Vol. 107, pp. 764–771.
28. Hsu, Y.Y. and Westwater, J.W. (1960), "Approximate Theory for Film Boiling on Vertical Surfaces." *Chem. Eng. Prog. Symp. Ser.*, Vol. 56, No. 30, pp. 15–24.
29. Coury, G.E. and Dukler, A.E. (1970), "Turbulent Film Boiling on Vertical Surfaces. A Study Including the Influence of Interfacial Waves." *Proceedings of the International Heat Transfer Conference*, Paris, Paper No. B.3.6.
30. Sherman, D.C. (1980), "Natural Convection Film Boiling on a Vertical Surface." Ph.D. Dissertation, California Institute of Technology, Pasadena, CA.
31. Dhir, V.K. and Taghavi-Tafreshi, K. (1981), "Hydrodynamic Transitions during Dripping of a Liquid from Underside of a Horizontal Tube." ASME Paper No. 81-WA/HT-12, presented at the ASME Winter Annual Meeting, November 15–20, 1981, Washington, DC.

8 Variables Affecting Nucleate, Transition, and Film Boiling

INTRODUCTION

Several experimental variables including heater surface characteristics can affect the boiling process. The influence on the boiling curve of a few of these variables was briefly discussed earlier. A description of the role played by some of the more important variables is given in this chapter.

8.1 SURFACE FINISH

The early data of Berenson [1] on a horizontal surface using pentane as the test liquid showed (Figure 5.7) that the effect of increased surface roughness was to shift the nucleate boiling curve to the left. This in turn corresponded to a higher heat flux at a given wall superheat for a rougher surface. The test liquids used in Berenson's experiments wetted the surface well, and it was noted that rougher surfaces also yielded slightly higher maximum heat fluxes. More recently, Bui and Dhir [2] obtained nucleate and transition boiling data using water as the test liquid for vertical copper surfaces of different finishes. Figure 8.1 shows their data for a clean mirror finish and an emery 600 finish surface. The behavior observed in nucleate boiling is similar to that reported by Berenson. However, during transient cooling, the rough surface yields transition boiling heat fluxes that are lower than those observed on a smooth surface.

Generally, it is argued that on rough surfaces, nucleation of larger cavities especially on poorly wetted surfaces with lower inception superheat leads to early onset of nucleate boiling. The rate at which new nucleation sites are added with increasing wall superheat is also higher for rough surfaces. Singh et al. [3] found that for nucleate boiling of water on a copper surface, the exponent on ΔT increased slightly with roughness, indicating a stronger dependence of wall heat flux on wall superheat. Shoukri and Judd [4], however, observed a trend opposite to that reported by Singh et al. [3]. Bier et al. [5] from their Freon data obtained on four surfaces prepared by following different procedures concluded that a relation of the type

$$h \sim R_p^n \tag{8.1}$$

did not exist between heat transfer coefficient and roughness. In eq. (8.1), R_p is the surface roughness parameter, and n is an empirical constant. It should be mentioned that three of the four surfaces used in their experiments had almost the same mean roughness but probably a different cavity size distribution because of the differences

DOI: 10.1201/9781032668437-8

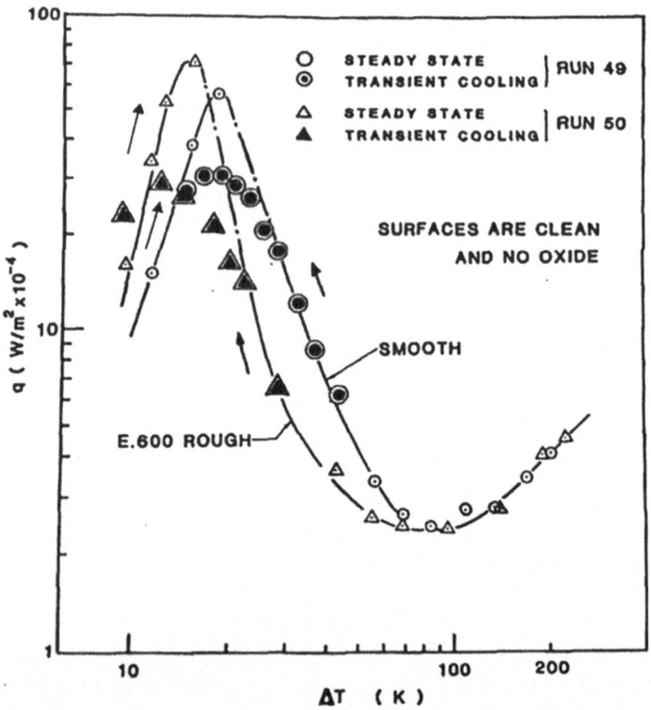

FIGURE 8.1 Boiling curves accessed from nucleate and film boiling sides [2].

in procedures followed to prepare the surfaces. The work of Bier et al. [5] suggests that it is important to distinguish between roughness and cavity size distribution, and any heat transfer model should account for both. It should also be noted that increased roughness can improve the wettability of the surface and in turn reduce the number of cavities that trap gas or vapor, which can nucleate at a lower superheat. Although significant increase in nucleate boiling heat flux at a given superheat is observed as surface roughness increases, the increase in maximum heat flux is minimal when the surface is well wetted. Berenson's data plotted in Figure 5.7 for pentane on copper, which is expected to wet the surface well, show a small increase in maximum heat flux with increased roughness but it is within the uncertainty of measurements. Data of Bui and Dhir [2] with water plotted in Figure 8.1 show an increase in maximum heat flux for the rough surface. This is attributed to the improvement in wettability of the surface with increase in roughness. The explanation for this behavior will be presented in the next section on wettability.

In transition boiling obtained in a cooling mode, rougher surfaces give lower heat fluxes at a given wall superheat. The reason for this is that for rougher surfaces, a larger fraction of the surface is covered with vapor. Heat transfer in film boiling is not affected by roughness, as stable and continuous vapor film covers the heater surface. Minimum film boiling temperature and minimum heat flux, however, can be affected by the manner in which transition boiling is accessed and whether the data are taken in steady state or in transient mode. The key reason for this behavior is the occurrence of liquid–solid contacts near the minimum heat flux condition. These contacts are altered also by surface roughness and wettability, which we will discuss next.

8.2 SURFACE WETTABILITY

With increase in surface wettability, the gas/vapor trapping/retention capability of the cavities on a surface diminishes. This in turn delays the onset of nucleate boiling and shifts the boiling curve to the right. In liquids that wet the surface very well, inception of boiling at relatively high superheats can lead to hysteresis in the nucleate boiling curve, as is the case for liquid metals discussed in Chapters 5 and 10. Figure 8.2 shows the nucleate boiling data of Liaw and Dhir [6] for different contact angles with water as the test liquid. In this figure, fully developed nucleate boiling data of Nishikawa et al. [7] are also plotted. Two sets of data show a similar dependence of heat flux on wall superheat. However, data of Nishikawa et al. [7] appear to correspond to a contact angle of about 69°. It is seen that with increase in the wettability of the surface, the nucleate boiling curve shifts to the right, the functional dependence of heat flux on wall superheat remains nearly unchanged. As discussed earlier, the maximum heat flux on a partially wetted surface $(\varphi \lesssim 90°)$ is lower than that given by the hydrodynamic theory. It appears that for partially wetted surfaces, the maximum heat flux condition is determined by the maximum rate at which vapor can be generated at the surface rather than by the rate at which vapor

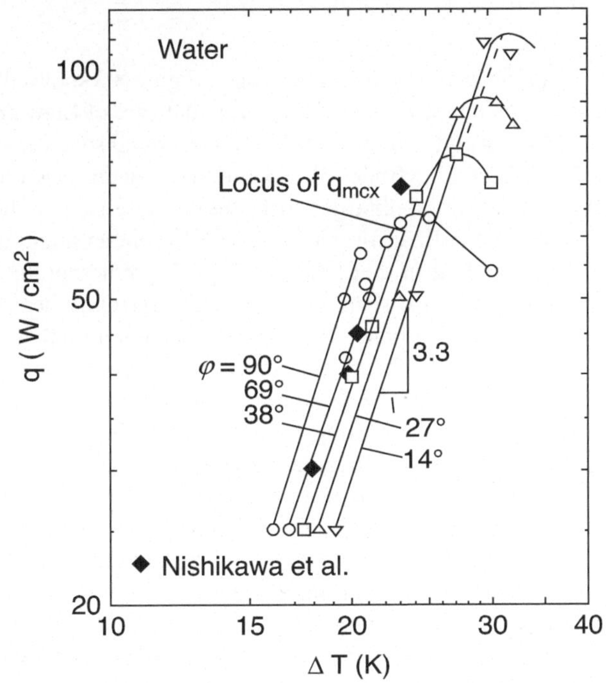

FIGURE 8.2 Effect of surface wettability on nucleate boiling heat transfer on a vertical surface [6].

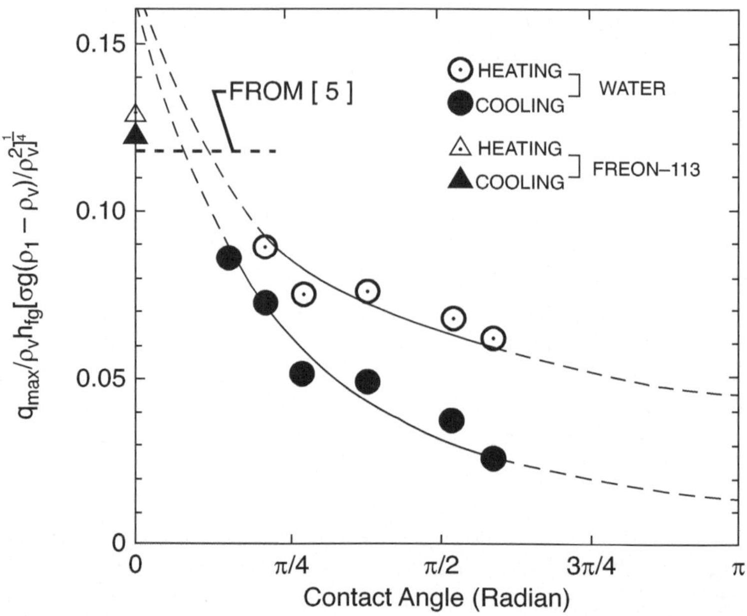

FIGURE 8.3 Effect of contact angle on maximum heat flux obtained on a vertical surface while accessing the boiling curve from nucleate boiling (heating) and film boiling (cooling sides) [6].

can be removed under the available buoyancy force. Figure 8.3 shows the dependence of maximum heat flux on contact angle. The plotted data are of Liaw and Dhir [6] for a vertical surface. Maracy and Winterton [8] for a horizontal surface, and Hahne and Diesselhorst [9] for horizontal cylinders 1–7.5 mm in diameter have reported similar effect of contact angle. The data obtained at 1 atm pressure clearly show that maximum heat flux decreases nonlinearly with increase in contact angle, and at contact angle of 90°, can be about half of that given by Zuber's prediction for an infinite flat plate. Smaller size plates are also seen to give higher maximum heat fluxes that are higher than those for a large flat plate. This will be discussed further in Section 8.4. As such, we have to be very careful in specifying in the experiments the wettability of the surface and size of the surface. It should be further noted that the critical heat flux obtained while accessing the boiling curve from film boiling side is lower. The hysteresis disappears when surface becomes fully wetted.

In transition boiling, the observed level of hysteresis between transient heating and cooling (traversing the transition boiling from nucleate or film boiling side) is affected by the surface wettability. For less wetting surfaces (high contact angle), the wet area fraction while accessing the transition boiling curve from nucleate boiling side is more than that when transition boiling data are obtained from the film boiling side. However, the wet area fractions for small contact angles (well-wetted surfaces) are about the same when transition boiling is accessed from either side. The difference increases with increase in contact angle. Since the cavity size, shape, and the contact angle affect the vapor/gas retention capability, any predictive model for density of active nucleation sites must rely on the information about the size distribution and the shape of the cavities that are present on the surface. Transition boiling curve on a fully wetted surface shows little or no hysteresis (Hohl et al. [10]), while the level of hysteresis increases with contact angle. Within the uncertainty of data, minimum heat flux is not found to be affected much by wettability, but the corresponding wall superheat (minimum film boiling temperature) is affected by wettability and whether data are taken in steady state or transient heating or cooling mode. Wettability has no effect on heat transfer coefficient in stable film boiling above the minimum film boiling temperature as no liquid–solid contacts occur.

8.3 SURFACE CONTAMINATION

Physicochemical changes of a boiling surface can take place in several ways: deposition of inert matter contained in the host liquid; slow chemical reaction of the surface with the gasses dissolved in the liquid or with the vapor; and strong chemical reaction of the metal with the concentrated solutions of electrolytes. The latter two processes continued over a long period of time can lead to a loss of material or corrosion of the surface. Erosion of the surface can also result from repeated collapse of vapor bubbles in a subcooled liquid and collisions of bubbles with the heated surface during flow boiling.

The presence of foreign material on the surfaces can affect the nucleation behavior of the surface by deactivating some cavities, or by providing new nucleation sites, and by changing the wettability of the surface. As yet no basis exists for quantification of the effect of surface cleanliness on nucleate and transition boiling. Bui and Dhir [2] obtained water data on clean and dirty surfaces. The dirty surfaces were obtained by prolonged nucleate boiling of water that had been exposed to the laboratory environment. Over time, the surface became contaminated with dirt particles accumulated in water. Various degrees of dirt deposits were obtained by varying the boiling time and the elapsed time between consecutive experiments. Dirt deposits were found to have a small effect on the nucleate boiling curve. However, the maximum heat flux on the surface subjected to longest boiling period was about 25% higher than that on a clean surface. The increase is attributed to the improved wettability of the surface in the presence of dirt deposits. The difference between transition boiling data obtained during heating and cooling of the surface decreased as the surface became dirtier. This behavior is similar to that observed with improvement in wettability of the surface. For the surface subjected to the longest boiling time, the two transition boiling curves obtained during heating and cooling nearly overlapped.

The presence of oxygen in the host liquid leads to contamination of the surface via slow chemical reaction. The process, if continued over a long period, can lead to aging of the surface or a permanent change in the boiling surface characteristics. Metal ions from the pits or cavities on the heated surface or from the fittings placed in the hardware can be present in the host liquid. The metal ions combine with oxygen to form metal oxides that precipitate at the heated surface. Joudi and James [11] identified the precipitate in the form of small brown spots around the cavities or at the base of the bubbles where high heat and mass transfer gradients exist. The effect of contamination of the surface was to push the boiling curve to the right, i.e., to increase the wall superheat for a given heat flux. This effect is similar to that of improved wettability of the surface. Joudi and James found that after one or two runs with water, the stainless steel surface could be rejuvenated by boiling refrigerant 113. Some of the variability in the data reported in the literature can be attributed to the aging of the surface.

Ross [12], in his review article on corrosion and heat transfer, reports results of long-term boiling (hundreds of hours) of water on iron, copper, and aluminum surfaces. It was found that mass loss rate increased with concentration of oxygen in water. At low heat fluxes, the mass loss rate was independent of heat flux, but at high heat fluxes, a reduction in mass loss rate was observed as heat flux was increased. When sodium chloride (electrolyte) was dissolved in water, the mass loss rate curves showed a local minima with respect to nucleate boiling heat flux. At low heat fluxes, the mass loss rate decreased with heat flux, but at high heat fluxes, an opposite trend was observed. It was suggested that minima in mass loss rate may be due to competing effects of oxygen and chloride ions at the interface. Although, no data on the effect of aging in transition boiling are available, the effect of aging will depend on the extent of change in contact angle. No effect of aging on the film boiling heat transfer coefficient is expected, except thicker contaminant layer can add additional resistance to flow of heat.

8.4 HEATER SIZE, GEOMETRY, AND MATERIAL THICKNESS

The boiling inception superheat and partial nucleate boiling heat fluxes can be strongly affected by convective flow patterns and bubble movement on and around the surface. The convective flow itself is influenced by the heater size, geometry, and to some extent by the size and shape of the container surrounding the test surface.

Investigations of different modes of boiling have been conducted over the last half of the century for a variety of heater sizes and geometries. These include horizontal and inclined surfaces, horizontal cylinders, and spheres. In the following, we describe the boiling heat transfer data obtained on these surfaces and any models that have been developed to explain the data.

8.4.1 Heater Size

Very recently, She [13] has systematically investigated the effect of heater size on nucleate boiling heat flux and maximum heat flux on horizontal surfaces. In his experiments, the size of horizontal disks, contact angle, and size of the vessels surrounding the heater were varied parametrically. Saturated water at 1 atm pressure was used as the test liquid. Copper disks of diameter 1, 1.5, and 2 cm were used as the test surfaces. For a given contact angle, the size of the disk and the container size had only a small effect in nucleate boiling. The effect was limited only to low heat fluxes when convection contribution is significant. However, the heater size and contact angle had a significant effect on the maximum heat flux. Data plotted in Figure 8.4 show maximum heat flux as a function of contact angle for three different size surfaces when the size of the surrounding vessel was equal to the size of the heater. For a contact angle of about 12°, the maximum heat flux on 1 cm diameter disk is found to be about two times that predicted from Zuber's model for an infinite flat plate as given in Chapter 6. Figure 8.5 shows She's data for maximum heat flux on difference size heaters for a contact angle of about 12°. She's data are the same as the

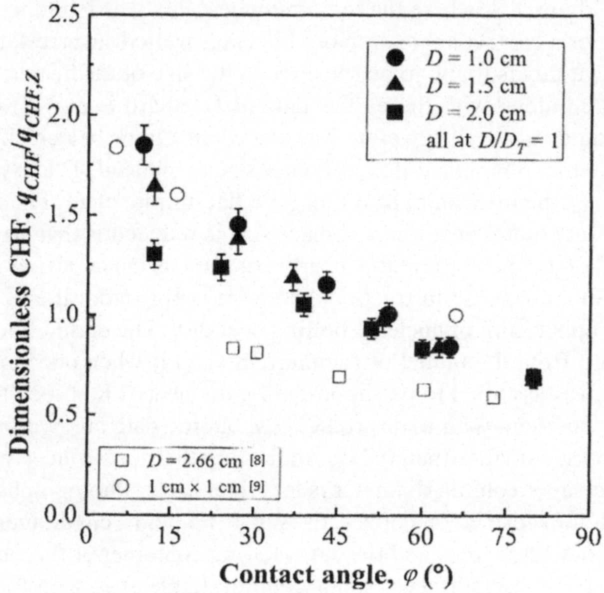

FIGURE 8.4 Dimensionless CHF as a function of contact angle φ for the disks of diameters D = 1.0, 1.5, and 2.0 cm. [13].

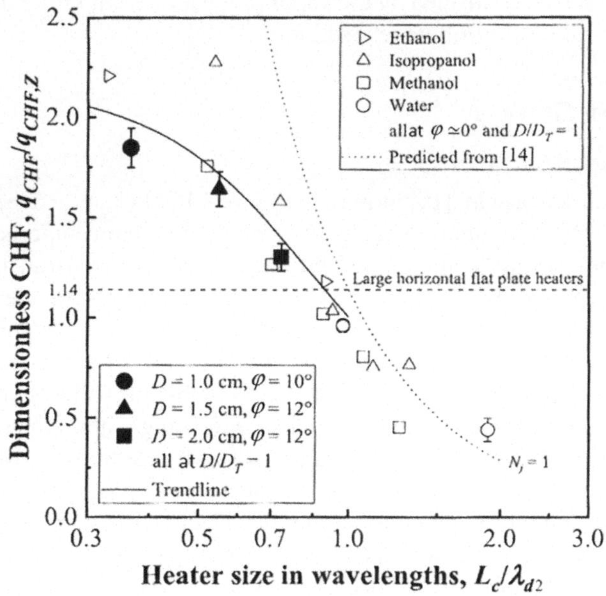

FIGURE 8.5 Dimensionless CHF as a function of heater size in wavelengths L_c/λ_{d2} for the well-wetted disks [13].

data plotted in Figure 8.4, where the maximum heat flux was found to decrease with contact angle. For a contact angle of about 12° (well-wetted surfaces), the magnitude of maximum heat flux is found to be sensitive to the size of the heater. In Figure 8.5, these data plotted along with the earlier data of Lienhard et al. [14] obtained with well-wetting liquids when an egg crate was placed on a large heater, show an asymptotic limit to the maximum heat flux as heater size is reduced. The asymptotic value is about two times the maximum heat flux for a flat infinite plate. The key reason for the increase in maximum heat flux as heater size is reduced is that size of the vapor column formed on the heater remains nearly constant as heater size is reduced. This in turn results in an increase in fractional area for vapor removal and a corresponding increase in upper limit of nucleate boiling heat flux. The dotted line in Figure 8.5 shows prediction from the model of Lienhard et al. [14] when one vapor column of diameter $\lambda_{d_2}/2$ was assumed to be supported by the heater. It is seen that data show a trend that is consistent with their predictions, but the data are overpredicted when heater size becomes smaller than $0.7\lambda_{d_2}$, indicating that for surfaces with width less than $0.7\lambda_{d_2}$, the vapor column diameter is less than $\lambda_{d_2}/2$. Photographs in Figure 8.6 show vapor columns formed on copper disks of 1, 1.5, and 2 cm diameter with a contact angle of about 12 degrees and the surrounding container of the same size as the heater surface. For these surfaces, vapor column diameter is seen to be influenced by the heater diameter. Increasing the size of the container did not have a significant effect in increasing the magnitude of the maximum heat flux. For partially wetted surfaces as seen in Figure 8.4, the heater size effect is small. This observation further corroborates the hypothesis made earlier that on partially wetted surfaces, the maximum heat flux is controlled by the vapor production limit rather than the hydrodynamic vapor removal limit, which is higher.

8.4.2 Heater Geometry

8.4.2.1 Inclined Surfaces

The data of Nishikawa et al. [15] plotted in Figure 5.10 of Chapter 5 clearly show the effect of the angle of inclination of the plate on the inception superheat and on the partial nucleate boiling heat fluxes. The angle is measured from the horizontal surface facing upward. Surprisingly, the nucleate boiling heat transfer coefficient is higher when the surface is facing downward than when it is horizontal or vertical. The enhancement in heat transfer on downward-facing surfaces probably results from the disruption of the thermal layer caused by bubbles sliding along the wall. In fully developed nucleate boiling, no such effect of angle of inclination is evident, as the contribution of convective heat transfer to total heat transfer diminishes. The transition from partial to fully developed nucleate boiling is nearly independent of angle of inclination and is given by eq. (5.25). Data of Liaw and Dhir [6], as discussed in Chapter 6, show that maximum heat flux on a well-wetted vertical plate is about 20% less than that for a horizontal plate.

Lienhard and Dhir [16] have extended the hydrodynamic theory for maximum and minimum heat fluxes on an infinite horizontal plate to other heater geometries.

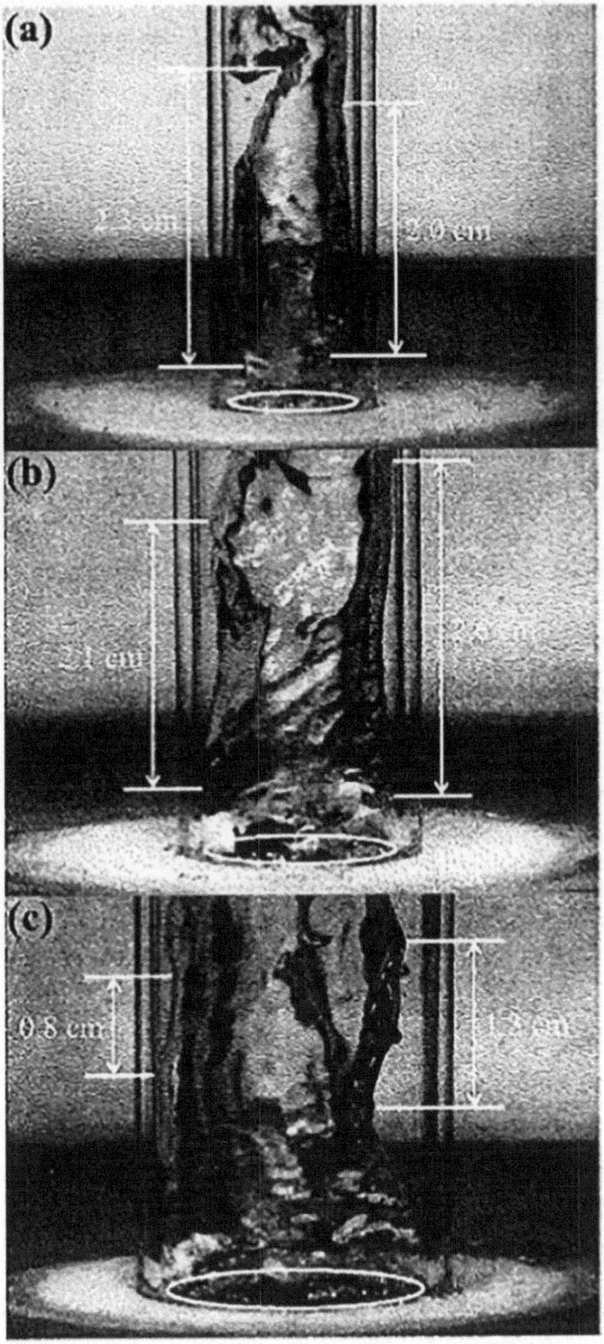

FIGURE 8.6 Vapor columns and interfacial wavelengths on well-wetted disks of diameters (a) $D = 1.0\,\text{cm}$ and (b) $D = 1.5\,\text{cm}$. (c) $D = 2.0\,\text{cm}$ [13].

Making an energy balance at the heater surface, an expression for maximum heat flux can be written as

$$q_{\max} = \rho_v h_{fg} U_{vc} \frac{A_v}{A_h} \tag{8.2}$$

Their extension is based on the premise that for the other geometries, as was the case for a flat plate, vapor leaves the heater in the form of jets, and it is the instability of the jets that determines the maximum vapor removal rate or the maximum heat flux condition. Thus, in prediction of q_{max} from eq. (8.2), one needs to determine U_{vc} from an assessment of the dominant unstable wavelength on the vapor–liquid interface. The ratio $\frac{A_v}{A_h}$ depends on the diameter of a vapor column and number of vapor jets that are supported by an area A_h of the heater. Lienhard and Dhir [16] used experimental measurements along with hypothesis based on physical arguments to obtain U_{vc} and $\frac{A_v}{A_h}$. Throughout this section, we will develop expressions for different heater geometries using Lienhard and Dhir's approach. Figure 8.7 shows the various geometries considered by them. For a ribbon held horizontally in a pool of saturated liquid with the boiling surface oriented in the vertical direction, an expression for $q_{\max}$ with boiling either on one side or both sides of the ribbon is obtained as

$$\frac{q_{\max}}{q_{\max,z}} = 0.9 \tag{8.3}$$

for $H' \geq 2.7$; where H' is defined as

$$H' = \frac{H}{\sqrt{\frac{\sigma}{g(\rho_l - \rho_r)}}} \tag{8.4}$$

for $H' < 2.7$,

$$\frac{q_{\max}}{q_{\max,z}} = \frac{1.18}{\sqrt[4]{H'}} \tag{8.5}$$

8.4.2.2 Horizontal Cylinders

The earliest study of nucleate pool boiling heat transfer on horizontal cylinders is that of Lance and Myers [17], whereas the most recent study is that of Cornwell and Einarsson [18]. Both pool and low flow velocity conditions during nucleate boiling heat transfer on a horizontal tube have been considered by Cornwell and Einarsson. Special care was taken in their experiments to minimize the effect of conduction in the circumferential direction. Figure 8.8 shows their data obtained with saturated Freon 133 at 1 atm. Under pool boiling conditions, the heat transfer coefficient is highest at the lower stagnation point and is lowest at the top. However, the heat

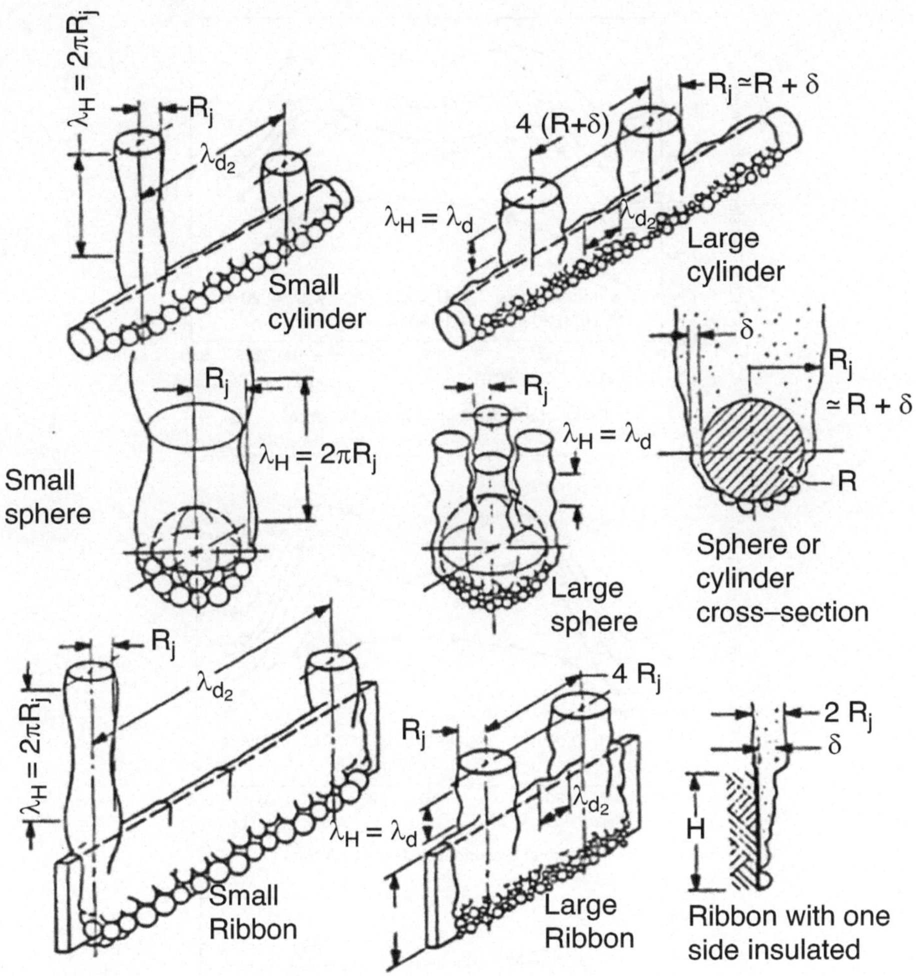

FIGURE 8.7 Vapor removal configurations near maximum heat flux, on a variety of finite heaters [16].

transfer coefficient remains fairly constant over most of the periphery of the cylinder. The heat transfer coefficient profile observed during pool boiling is much flatter than would be expected under single-phase natural convection conditions. At small imposed flow velocities, the location of maximum heat transfer coefficient shifts to the sides of the tube. It was suggested that sliding of the bubbles probably led to a higher heat transfer at the sides. Cornwell and Einarsson concluded that because of the sensitivity of the flow field to chamber geometry and the tube diameter and the dependence of the nucleate boiling process on surface conditions, it is extremely difficult to predict with any reasonable accuracy the nucleate boiling heat fluxes at low velocities and low superheats.

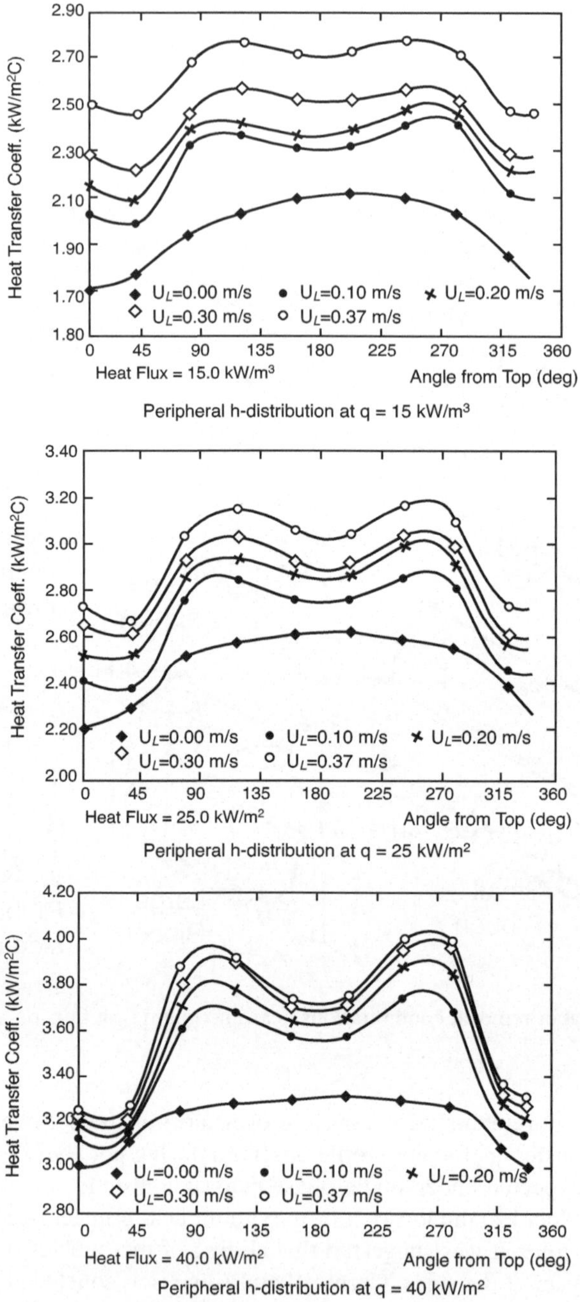

FIGURE 8.8 Circumferential variation of partial nucleate boiling heat fluxes on a horizontal cylinder [18].

Single-phase natural convection boundary layer thickness on a cylinder is a function of position along the circumference. As such, the size range of cavities that nucleate will be a function of position. On the lower portion of the cylinder, the bubbles slide along the cylinder. This leads to cyclic disruption of the thermal layer all along the bubble path. Enhanced evaporation may also occur at the base of the bubbles. The combined effect of these two processes is to increase the heat transfer above that for a flat plate. On the upper portion of the cylinder, however, the mechanisms similar to those on an upward facing flat plate will prevail. Thus, in partial nucleate boiling, the rate of heat transfer averaged over the cylinder surface area at a given superheat will be higher than that on an upward facing flat plate.

In fully developed nucleate boiling, lateral merger of bubbles occurs, and as a result, mushroom type of bubbles supported by vapor stems form. With these bubbles, evaporation at the vapor stems is the main mode of heat transfer. Contributions of other modes of heat transfer (e.g., natural convection, transient conduction) become secondary while geometry plays little role in determining the dependence of heat flux on wall superheat. (Support for the above argument is derived from the observations of Nishikawa et al. (Figure 5.8) on flat plates inclined at different angles with the horizontal).

The hydrodynamic theory of Zuber as described earlier and extended by Sun and Lienhard [19] and subsequently by Lienhard and Dhir [16] can be used to describe the maximum heat flux condition on horizontal cylinders. Again, the premise of the hydrodynamic theory is that there is an upper limit of the rate at which available buoyancy force can remove vapor from the heater. Because the vapor jets cannot lie closer than twice the jet diameter (a tacit assumption supported by the bubble spacing observed in film boiling) and jet diameter depends on the cylinder diameter, a distinction must be made between small and large cylinders (similar to what was done earlier for a vertical surface). Vapor removal configurations for small and large cylinders are shown in Figure 8.7.

By assuming that on a small cylinder (quantitative definition will be given later), the vapor jets are located a distance Taylor wavelength, λ_{d_2} apart and the jet diameter is equal to cylinder diameter plus twice the thickness of the vapor blanket around the cylinder, the ratio A_v/A_h can be written with one jet supported by length λ_{d_2} of the cylinder.

$$\frac{A_v}{A_h} = \frac{\pi(R+\delta)^2}{2\pi R\lambda_{d2}} \tag{8.6}$$

In eq. (8.6), λ_{d_2} is the "most dangerous" two-dimensional Taylor wavelength on a plane interface. This assumption is justified on the basis that uneven placing of bubbles on the heater surface smears any geometrical features. In eq. (8.6), δ is the vapor blanket thickness. The critical vapor velocity, U_{vc}, in vapor jets is obtained in the same manner as that for jets on a flat plate. With the assumption that at low pressures liquid velocity is very small in comparison to the vapor velocity, an expression for U_{vc} is written as

$$U_{vc}=\sqrt{\frac{2\pi\sigma}{\rho_v 2\pi(R+\delta)}}=\sqrt{\frac{\sigma}{\rho_v(R+\delta)}} \tag{8.7}$$

where jet diameter is assumed to be $2(R+\delta)$.

Substituting for A_v/A_h and U_{vc} from eqs. (8.6) and (8.7) respectively in eq. (8.2), we obtain

$$q_{\max}=\rho_v h_{fg}\sqrt{\frac{\sigma}{\rho_v(R+\delta)}}\cdot\frac{\pi(R+\delta)^2}{2\pi R\cdot 2\pi\sqrt{3}\left(\sqrt{\frac{\sigma}{g(\rho_l-\rho_v)}}\right)} \tag{8.8}$$

$$=\frac{\rho_v h_{fg}}{4\pi\sqrt{3}}\frac{(R'+\delta')^{3/2}}{R'}\sqrt[4]{\frac{\sigma g(\rho_l-\rho_v)}{\rho_v^2}} \tag{8.9}$$

where

$$R'=\frac{R}{\sqrt{\frac{\sigma}{g(\rho_l-\rho_v)}}} \tag{8.10}$$

and

$$\delta'=\frac{\delta}{\sqrt{\frac{\sigma}{g(\rho_l-\rho_v)}}} \tag{8.11}$$

The dimensionless radius R' is the square root of Bond number, which is defined as the ratio of the buoyancy to surface tension forces. Sun and Lienhard [19] used photographs of boiling on horizontal cylinders to determine dependence of the thickness of the vapor blanket on horizontal cylinders. They correlated the vapor blanket thickness data on horizontal cylinders as

$$\delta'=R'\left(\sqrt{\frac{3.72}{R'}}-1\right) \tag{8.12}$$

Substitution of δ' from eq. (8.12) in (8.9) yields

$$q_{\max}=(3.72)^{3/4}\frac{\rho_v h_{fg}}{4\pi\sqrt{3}}\sqrt{\frac{\sigma g(\rho_l-\rho_v)}{\rho_v^2}}\cdot\frac{1}{\sqrt[4]{R'}} \tag{8.13}$$

Normalizing the maximum heat flux by Zuber's expression, the dimensionless heat flux on small horizontal cylinders is given by

$$\frac{q_{\max}}{q_{\max,z}} = \frac{0.94}{\sqrt[4]{R'}} \tag{8.14}$$

Equation (8.14) demonstrates that maximum heat flux on small cylinders increases as cylinder radius decreases.

As the cylinder diameter or the jet diameter becomes large, the jet spacing becomes larger than the Taylor wavelength. Since Taylor wavelength will also be shorter than the Helmholtz critical wavelength, and observed vapor jets are too short to accommodate a wavelength equal to the circumference of the jet, it was argued that the two-dimensional Taylor wavelength will be the dominant wavelength at the vapor jet–liquid interface. With these assumptions, expressions for vapor jet to heater area ratio and critical vapor velocity can be written as

$$\frac{A_v}{A} = \frac{\pi(R+\delta)^2}{2\pi R \cdot 4(R+\delta)} = \frac{R+\delta}{8R} \tag{8.15}$$

$$U_{vc} = \sqrt{\frac{2\pi\sigma}{\rho_v \lambda_{d_2}}} = \frac{1}{\sqrt[4]{3}}\sqrt[4]{\frac{\sigma g(\rho_l - \rho_v)}{\rho_v^2}} \tag{8.16}$$

Substitution for A_v/A and U_{vc} from eqs. (8.15) to (8.16) respectively in eq. (8.2) yields the maximum heat flux on large cylinders as

$$q_{\max} = \frac{\rho_v h_{fg}}{8} \cdot \frac{1}{\sqrt[4]{3}}\sqrt[4]{\frac{\sigma g(\rho_l - \rho_v)}{\rho_v^2}}\,\frac{R+\delta}{R} \tag{8.17}$$

Sun and Lienhard found that the vapor blanket thickness on large cylinders varied linearly with radius and correlated the data as

$$\delta = 0.24R \tag{8.18}$$

Substituting for δ from eqs. (8.18) to (8.17) and normalizing the maximum heat flux with Zuber's expression for maximum heat flux yields

$$\frac{q_{\max}}{q_{\max,z}} = 0.9 \tag{8.19}$$

Figure 8.9 shows a comparison of predictions from eqs. (8.14) and (8.19) with the data obtained from large number of fluid–cylinder radius combinations. The transition from small to large cylinders can be approximately determined by equating the most dangerous wavelengths with twice the jet diameter or

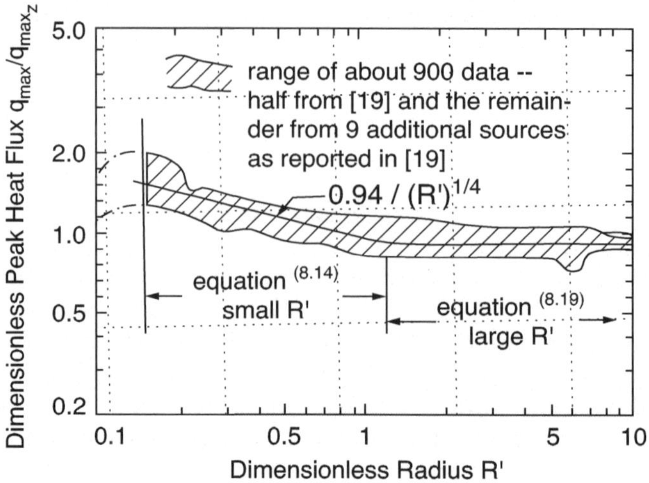

FIGURE 8.9 The maximum heat flux on horizontal cylinders [19].

$$\lambda_{d_2} = 4(R+\delta) \tag{8.20}$$

or

$$2\pi\sqrt{3}\sqrt{\frac{\sigma}{g(\rho_l-\rho_v)}} = 4R(1+0.24) \tag{8.21}$$

or

$$R' \cong 2.2 \tag{8.22}$$

It can be seen from Figure 8.9 that this value of R' approximately coincides with the cylinder radius at which the maximum heat flux attains a constant value.

Example:

A horizontal wire 3 cm long of radius 2 mm is held in saturated water at 1 atm pressure. Calculate the maximum heat flux for the wire. First assess if the wire falls in small or large cylinder case.

Solution:

$$R' = \frac{R}{\sqrt{\frac{\sigma}{g(\rho_l-\rho_v)}}}$$

$$= \frac{2 \times 10^{-3}}{2.5 \times 10^{-3}}$$
$$= 0.8$$

So the wire is a small cylinder ($R' < 2.2$). For this case

$$\frac{q_{max}}{q_{max\,Z}} = \frac{0.94}{\sqrt[4]{0.8}} \approx 1$$

or

$$q_{max} = q_{max\,Z}$$

From Chapter 6,

$$q_{max\,Z} = 1.1 \times 10^6 \text{ W/m}^2$$

or

$$q_{max} = 1.1 \times 10^6 \text{ W/m}^2$$

Most of the data in the study of Sun and Lienhard [19] and Lienhard and Dhir [16] were obtained under well-wetted surface conditions. For partially wetted surfaces, only a few maximum heat flux data on heaters of different sizes and geometries are available to quantify the effect of wettability. From the data plotted in Figure 8.3, it can be inferred that partially wetted cylinders [9] will behave somewhat differently from a horizontal or a vertical plate (Figure 8.3). Nevertheless, it should be mentioned that the data of Hahne and Diesselhorst [9] were obtained on cylinders of different materials and therefore some uncertainty may exist with respect to the measured contact angles.

Heater geometry and size probably influence transition boiling as well. However, not enough data under well-defined heater surface conditions are available in the literature to make any conclusive statement.

8.4.2.3 Minimum Heat Flux on a Horizontal Cylinder

The derivation of minimum heat flux on horizontal cylinders can be obtained by following a procedure similar to that used by Lienhard and Dhir [20] for horizontal plates. On a horizontal cylinder, one bubble is supported on the surface area of a cylinder one Taylor wavelength long. It should be noted here that the Taylor wavelength on a horizontal cylinder depends on the radius and decreases as the cylinder radius becomes small. Since the diameter of the cylinder is much smaller than the width of the bubble at the top, the bubble is assumed to be a spherical wedge rather than the bubble with a hemisphere cap as was the case for a flat plate. Figure 8.10 shows a sketch of the bubble from [21]. For a spherical wedge whose dihedral angle is assumed to be one radian, and radius of curvature and arc length are equal to one-fourth of a Taylor wavelength, λ_{d2}, on a cylinder, the rate of change of volume with time is written as

$$\dot{V} = \pi \left(\frac{\lambda_{d2}}{4} \right)^2 (-w_i) c \lambda_{d2} \tag{8.23}$$

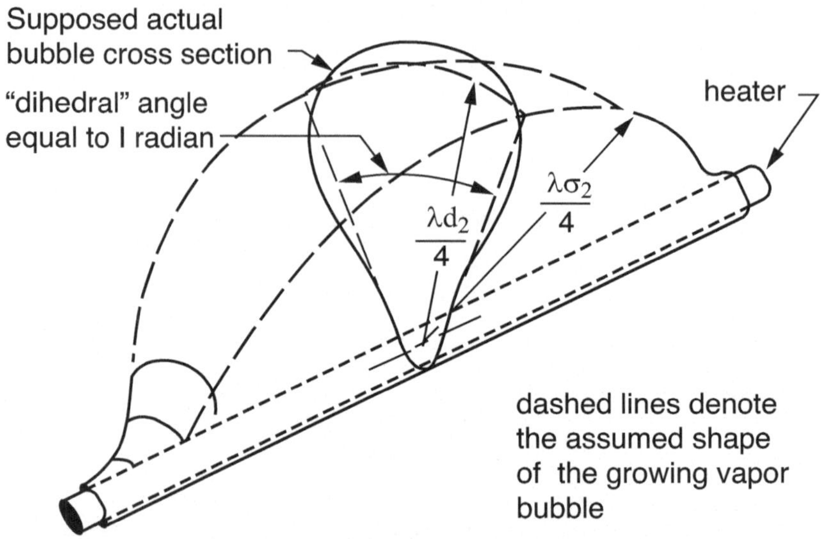

FIGURE 8.10 Idealization of interface growing on a horizontal cylinder during film boiling. (Adapted from Lienhard and Dhir [21].)

where $(-w_i)$ is the interface growth rate per unit heater area as was described in Chapter 6 for a flat plate and c is an empirical constant. The energy required to sustain this rate of bubble volume change in film boiling is given as

$$Q = \rho_v h_{fg} \dot{V} = \frac{\rho_v h_{fg}}{16} \lambda_{d_2}^3 c(-w_i) \tag{8.24}$$

Division of eq. (8.24) by the surface area of the cylinder that supports one bubble yields an expression for the minimum heat flux as

$$q_{\min} = \frac{\rho_v h_{fg}}{16} \frac{\lambda_{d2}^3 c(-w_i)}{2\pi R \lambda_{d_2}} \tag{8.25}$$

$$= \frac{\rho_v h_{fg}}{32\pi} \frac{\lambda_{d2}^2}{R} c(-w_i) \tag{8.26}$$

The most dominant Taylor wavelength on a cylinder is given by the expression [20]

$$\lambda_{d_2} = 2\pi\sqrt{3}\sqrt{\frac{\sigma}{g(\rho_l - \rho_v)}}\left(1 + \frac{1}{2R'}\right)^{-1/2} \tag{8.27}$$

The corresponding expression for the growth rate of the most dangerous Taylor wavelength was obtained as

$$-w_i = \sqrt[4]{\frac{g(\rho_l - \rho_v)}{3\sigma}}\left(1 + \frac{1}{2R'^2}\right)^{1/4}$$

$$\cdot\left[\frac{g(\rho_l - \rho_v)}{\rho_l + \rho_v} - \frac{g(\rho_l - \rho_v)}{3(\rho_l + \rho_v)}\cdot\left(1 + \frac{1}{2R'^2}\right) + \frac{g(\rho_l - \rho_v)}{g(\rho_l + \rho_v)R'^2}\right]^{1/2} \tag{8.28}$$

In writing eqs. (8.27) and (8.28), the vapor blanket thickness is assumed to be much smaller than the cylinder radius. Substituting for λ_{d_2} and $-w_i$ from eqs. (8.27) and (8.28) respectively in eq. (8.26) and taking c to be 0.069 (as was the case for a flat plate), the expression for the minimum heat flux on cylinders is obtained as:

$$q_{\min} = 0.06\rho_v h_{fg}\sqrt[4]{\frac{\sigma g(\rho_l - \rho_v)}{(\rho_l + \rho_v)^2}}\left[R'^2\left(2R'^2 + 1\right)\right]^{-1/4} \tag{8.29}$$

A comparison of predictions from eq. (8.29) with relatively small number of data reported in the literature [21], [22], [23] and [24] is shown in Figure 8.11. The predictions from eq. (8.29) fall in the middle of the data obtained with several fluids including CO_2. The CO_2 data were obtained at a pressure near critical. The data exhibit a scatter of about 50%. Such a scatter is inherent in minimum heat flux data because of the end effects. In most instances, vapor film collapse occurs prematurely at the end of the cylinder because of the fin effect at the ends, which leads to additional cooling there. At pressures near critical, vapor film thickness can be comparable to the radius of the cylinder. This, in turn, will cause the observed heat flux to be lower than that given by eq. (8.29). This would be expected from the model if cylinder radius was increased to $R + \delta$.

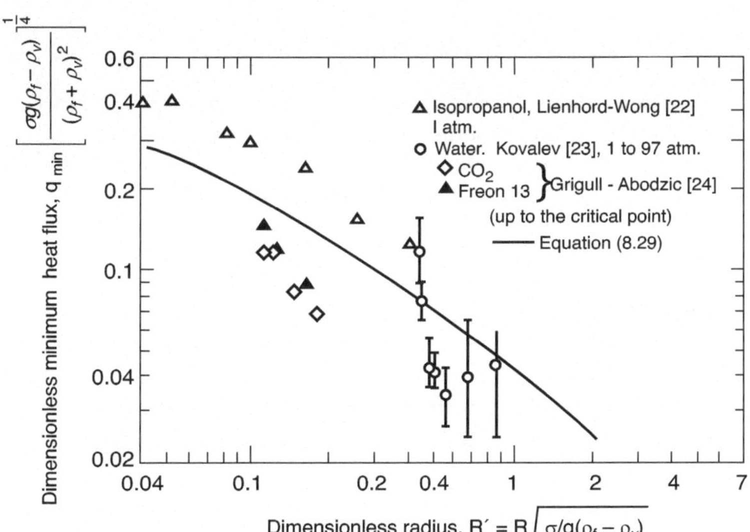

FIGURE 8.11 Comparison of the q_{min} prediction with the highly scattered available data [21].

Example:

A wire 2 mm in radius and a length of 25 mm is placed in saturated water at 1 atm pressure. Calculate the minimum heat flux at which vapor film on the wire is expected to collapse.

Solution:

From eq. (8.29)

$$q_{\min} = 0.06\rho_v h_{fg} \sqrt[4]{\frac{\sigma g(\rho_l - \rho_v)}{(\rho_l + \rho_v)^2}} \left[R'^2 (2R'^2 + 1)\right]^{-1/4}$$

Since vapor density depends on temperature, begin by first guessing a wall temperature at the minimum heat flux. Let $T_{\min} = 180°\text{C}$, $\bar{T} = \frac{100 + 180}{2} = 140°\text{C}$, or $\rho_v = 0.54 \text{ kg/m}^3$.

With this, we obtain $q_{\min}$ as

$$q_{\min} = 0.06(0.54)(2.257 \times 10^6) \sqrt[4]{\frac{58.9 \times 10^{-3} \times 9.81(957.9 - 0.54)}{(957.9 + 0.54)^2}}$$

$$\cdot\left[0.8^2 (2 \times 0.8^2 + 1)\right]^{-1/4}$$

$$= 1.042 \times 10^4 \text{ W/m}^2$$

If we had evaluated ρ_v at T_{sat}

$$q_{\min} = 1.254 \times 10^4 \text{ W/m}^2$$

Thus, we see that with the guessed temperature at $q_{\min}$, the correction to $q_{\min}$ could be about 20% lower if all properties were evaluated at T_{sat}.

An appropriate procedure would be to first guess $T_{\min}$ and evaluate both $q_{\min}$ and $h_{\min}$, calculate $\Delta T_{\min}$ or $T_{\min}$. Compare the guessed value with the calculated value. If the difference is small, we have a converged solution. If not, repeat the process. Usually with one additional iteration, one can obtain near exact values.

8.4.2.4 Spheres

Only limited nucleate boiling heat transfer data on spherical heaters have been reported in the literature. Ded and Lienhard [25] have reported maximum heat flux data obtained during quenching of spheres and have developed a model to explain the data obtained on different size spheres. Their expression for small spheres, $R' \leq 4$ is

$$\frac{q_{\max}}{q_{\max,z}} = \frac{1.734}{\sqrt{R'}} \quad \text{for} \quad R' \leq 4 \tag{8.30}$$

$$\text{where} \quad R' = \frac{R}{\sqrt{\dfrac{\sigma}{g(\rho_l - \rho_v)}}} \tag{8.31}$$

However, they did not report a lower limit of R' at which eq. (8.30) ceases to be applicable.

For large spheres

$$\frac{q_{\max}}{q_{\max,z}} = 0.84 \quad \text{for} \quad R' > 4 \tag{8.32}$$

Transition from small to large spheres occurs at $R' \simeq 4$. The development of the model is based on the number of vapor columns that are supported on the sphere, vapor column diameter, and the dominant wavelength on the vapor column–liquid interface. The latter is taken to be two dimensional "most dangerous" wavelength on a horizontal plate.

Figure 8.12 shows predictions from Lienhard and Dhir [16] for maximum heat fluxes on various heater geometries as obtained by using extended hydrodynamic theory. It is found that maximum heat flux depends on the characteristic length of

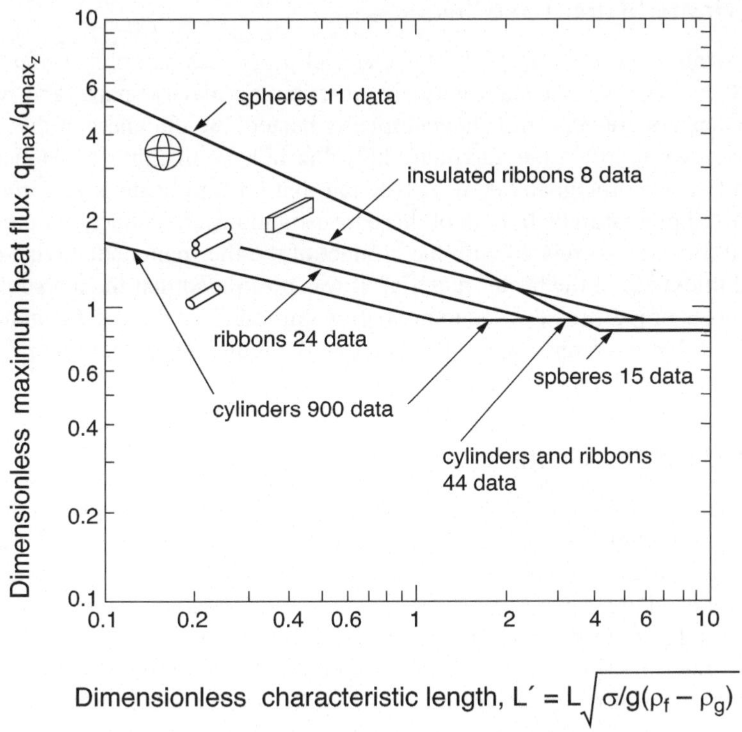

FIGURE 8.12 Prediction of maximum heat flux from the extended hydrodynamic theory [16].

the heater as long as dimensionless characteristic length, L', is less than about 5. Thereafter, maximum heat flux is independent of L'. The asymptotic value in each case is within a few percent of flat plate value as effectiveness of geometry diminishes. It should be further noted the recent data of She [13] obtained on horizontal disks of different sizes show similar dependence of maximum heat flux on L' with transition occurring at a higher value of L'.

Limited data are reported for transition boiling on spheres. Dhir [26] has reported transient transition boiling data obtained during quenching of spheres subjected to upward flow. Data show shift of transition boiling curves upward and to the right. The ratio of q_{max} to q_{min} was also seen to decrease with flow velocity. Dhir and Purohit [27] have correlated minimum film boiling temperatures on spheres of stainless steel, copper, and silver submerged in water at 1 atm pressure as

$$\Delta T_{min} = 101 + 8\Delta T_{sub} \text{ °K} \tag{8.33}$$

Minimum heat flux for saturated water was found to be

$$q_{min} = 2.4 \times 10^4 \frac{\text{W}}{\text{m}^2} \tag{8.34}$$

8.4.3 Heater Material and Thickness

In steady-state experiments, the heater material properties can play a role through such surface characteristics as cavity size distribution, cavity shape, and wettability. If the heaters are very thin and electrically heated, the thickness of the material has been shown to affect the maximum heat flux in a significant way, Houchin and Lienhard [28] and Tachibana et al. [29] found that on thin heaters, maximum heat flux occurred prematurely. In both of these investigations, the reduction in the maximum heat flux was correlated with the product of the thermal conductivity, specific heat, and thickness of the heater material. It was postulated that the hot spots under vapor bubbles or jets probably led to premature burnout. The data of Tachibana et al. also showed that in comparison to DC heating, the reduction in maximum heat flux in thin heaters was larger with AC heating.

8.5 SYSTEM PRESSURE

With an increase in system pressure, the incipient superheat decreases, and the nucleate boiling curve shifts to the left. From the hydrodynamic theory, one can find that the maximum heat flux initially increases with pressure, attains its highest value between reduced pressures (p/p_c) of 0.3 and 0.4, and thereafter decreases with pressure. Fath and Judd [30] showed that Mikic and Rohsenow's [31] mechanistic model for partial nucleate boiling, with the contribution of macro/microlayer evaporation included in it was also applicable at pressures lower than 1 atm. However, several of the parameters used in the model were determined from experiments that also provided the heat transfer data. Gorenflo et al. [32] have correlated the nucleate boiling heat transfer coefficient data for Freon 12, Freon 13, and Propane obtained at reduced pressures, $p^* = p/p_c$ up to 0.93 as

$$\frac{h}{h_0} = \left(\frac{q}{q_0}\right)^{n(p^*)} \cdot F(p^*) \tag{8.35}$$

where

$$F(p^*) = 2.1p^{*0.27} + \left(4.4 + \frac{1.8}{1-p^*}\right)p^* \tag{8.36}$$

and

$$n(p^*) = 0.9 - 0.3p^{*0.3} \tag{8.37}$$

The values of the reference heat transfer coefficient h_0 at the reference heat flux, $q_0 = 2\times10^4$ W/m^2 and the reference pressure, $p^* = 0.03$ for the three liquids are:

	F-12	F-13	C_3H_8
h_0 (W/m^{2}K)	2,300	2,200	2,100

Gorenflo et al. also extended the isolated bubble model of Mikic and Rohsenow to high heat fluxes, with the assumption that despite the overlapping of the areas of influence, the single bubble formulation remained unchanged. They also assumed that the frequency of liquid replacement at dense bubble packings was higher than that for isolated bubbles. Up to reduced pressures of 0.8, the model predictions tended to show only qualitative agreement with the data. For well-wetted surfaces, predictions based on the hydrodynamic theory, yield maximum heat fluxes that are in good agreement with the data obtained over a wide range of pressures.

8.6 GRAVITY

The magnitude and direction of the gravitational acceleration with respect to the heater surface influence the convective hydrodynamic and thermal boundary layers and bubble departure size and trajectory. On a given surface, the thermal layer thickness and the temperature profile in the thermal layer determine the inception superheat. In partial nucleate boiling, heat transfer by convection represents a major fraction of the total heat transfer rate. Thus, gravity plays an important role in this mode of boiling. Figure 8.13 shows the nucleate boiling data obtained at different magnitudes of accelerations normal to the heater. The data are for liquid nitrogen boiling at 1 atm and, as reported by Merte [33], were obtained by Merte and coworkers in the early 1960s. The high gravity data were taken in a centrifuge, the near-zero-gravity data in a drop tower, and the data obtained on a downward facing

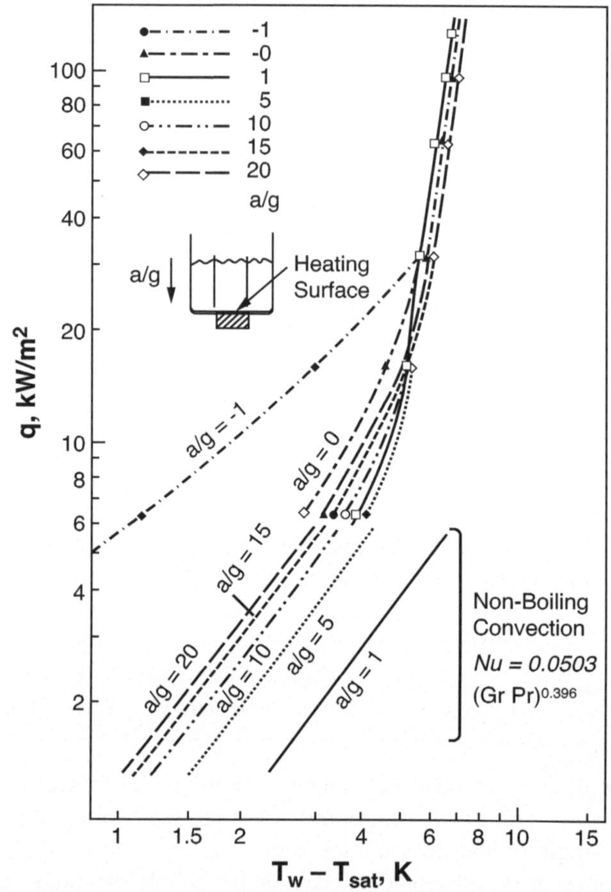

FIGURE 8.13 Effect of level of gravity on nucleate boiling heat transfer [33].

horizontal disk were considered to be −1 *g* data. In the data plotted, *a* is the acceleration normal to the surface, and *g* is the eath normal gravity. All of the data were taken under transient boiling conditions. At low wall superheats, the magnitude of gravity appears to affect the boiling curve. The heat fluxes associated with near zero g and −1g are higher than those obtained when a larger body force was present. The data taken on a downward facing surface were considered to be −1 *g* data. On a downward facing surface, the bubbles move along the surface and thereby enhance the heat removal rate by disrupting the thermal layer over a larger area. Near zero gravity, the bubbles in these transient experiments probably persisted for a longer period and thereby enhanced the evaporative contribution to total heat transfer rate. However, at high nucleate boiling heat fluxes, no effect of magnitude of acceleration is evident.

Reduced gravity nucleate boiling data of Zell et al. [34] are shown in Figure 8.14. These data were obtained with F-12 boiling at different pressures on a platinum wire. Reduced gravity conditions were created in parabolic flight of a KC-135 airplane. The various curves in Figure 8.14 correspond to the best fit through the data obtained at earth normal gravity. The symbols represent reduced gravity data with gravity

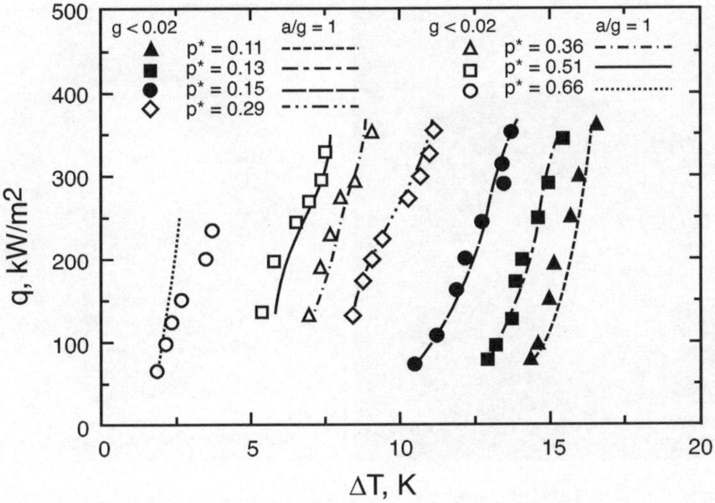

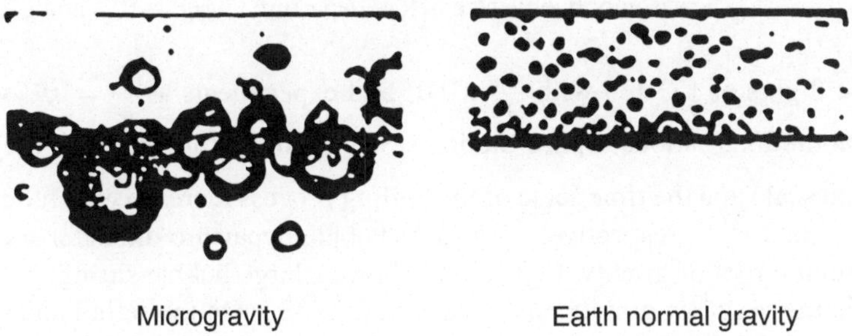

FIGURE 8.14 Effect of gravity level on nucleate boiling heat transfer [34].

level being less than 0.02 g_e. Almost no effect of reduced gravity on heat transfer is noted even though at the microgravity condition bubbles were much larger than those at earth normal gravity. The low-gravity conditions were also found to have little effect on subcooled nucleate boiling.

According to the hydrodynamic theory, the maximum heat flux should vary with gravity as $\sqrt[4]{g}$. The high-gravity centrifuge data do confirm this dependence. The short time drop tower data of Siegel and Usiskin [35] showed that during reduced gravity conditions, a large bubble (≃ 4 cm in diameter) existed near the heated surface while the surrounding enclosure diameter was 7.5 cm. Under reduced gravity conditions, many internal and external factors (e.g. thermocapillary forces, steady or transient nature of experiments, presence of dissolved gases, and vibrations) can influence the process.

Recently, Dhir and coworkers [36,37] have carried out nucleate boiling experiments and numerical simulations under pool boiling conditions at $1 \leq \frac{g}{g_e} \leq 10^{-7}$ or a flat surface. Reduced gravity experiments were conducted [33], [39], [40] and [41] in

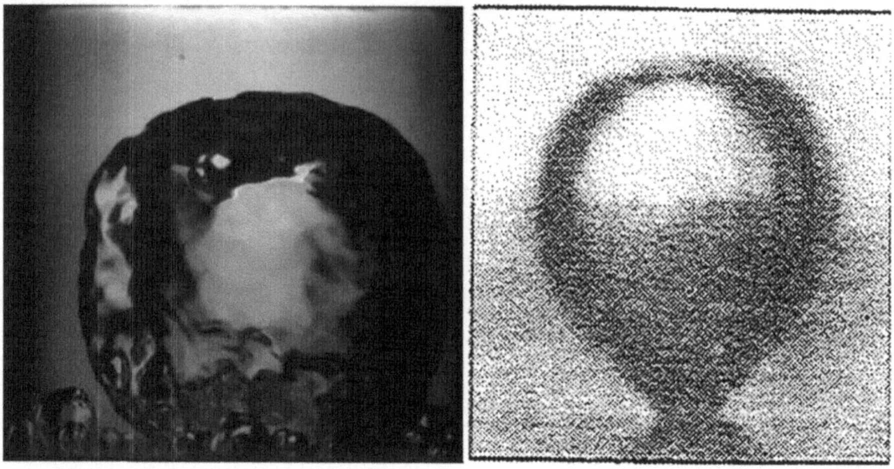

$\frac{g}{g_e} \simeq 10^{-7}$ (PF_5060) $\frac{g}{g_e} = 1$ (Water)

FIGURE 8.15 Comparison of bubble size before departure at $g/g_e \cong 10^{-7}$ and $g/g_e = 1$ [36].

parabolic flights of KC-135 with $\frac{g}{g_e} \simeq 0.01$ and experiments at $\frac{g}{g_e} \simeq 10^{-7}$ were conducted on the International Space Station (ISS). From scaling analysis, they showed that length scale and the time scale of the boiling process increase with reduced gravity as $g^{-1/2}$ and $g^{-3/4}$, respectively. As such, bubble departure diameter increases as inverse square root of gravity. Figure 8.15 shows a large bubble sitting on the heater surface in the very low gravity environment of the ISS. On the right hand side is the bubble size and shape obtained at earth normal gravity though with a different liquid and contact angle. They also obtained nucleate boiling heat transfer data at different levels of gravity. Figure 8.16 shows their data obtained at earth normal gravity and on ISS on similar heaters of the same dimensions. In the experiments, on ISS Perfluoro n-Hexane was used as the test liquid. Whereas in the experiments at earth normal gravity FC-72 was the test liquid. Both liquids have similar thermophysical properties. Although the pressures in the earth normal gravity and on ISS were somewhat different, it is found that at a given superheat, heat transfer coefficient in nucleate boiling is much smaller at $\frac{g}{g_e} \simeq 10^{-7}$ than that at earth normal gravity. Warrier et al. [37] correlated available nucleate boiling heat transfer coefficient data obtained at different gravity levels. Most of the data were for a fixed $\frac{q}{q_{max}}$ ratio of about 0.22. This value corresponds to the lower end of nucleate boiling curve. Figure 8.17 shows these data. In plotting the data, experimental values of q_{max} were used where available. For cases where experimental values of q_{max} were not available, the hydrodynamic prediction from Zuber et al. [38] was used. The heat transfer coefficient for microgravity conditions was normalized with the corresponding value at earth normal gravity. It is noted from Figure 8.17 that $\frac{h}{h_{ge}}$ decreases with decrease in gravity

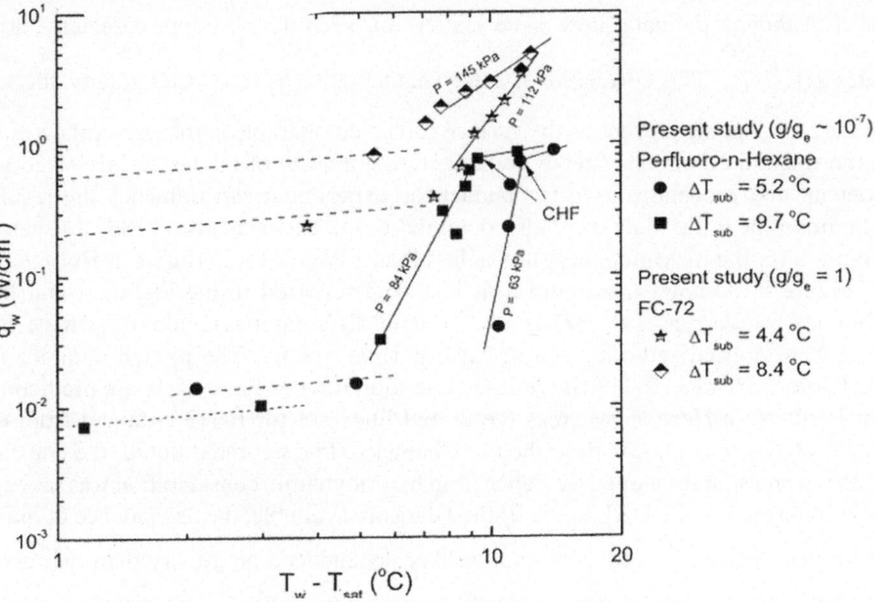

FIGURE 8.16 Comparison of nucleate boiling heat transfer data at $g/g_e = 1$ and $g/g_e = 10^{-7}$ [37].

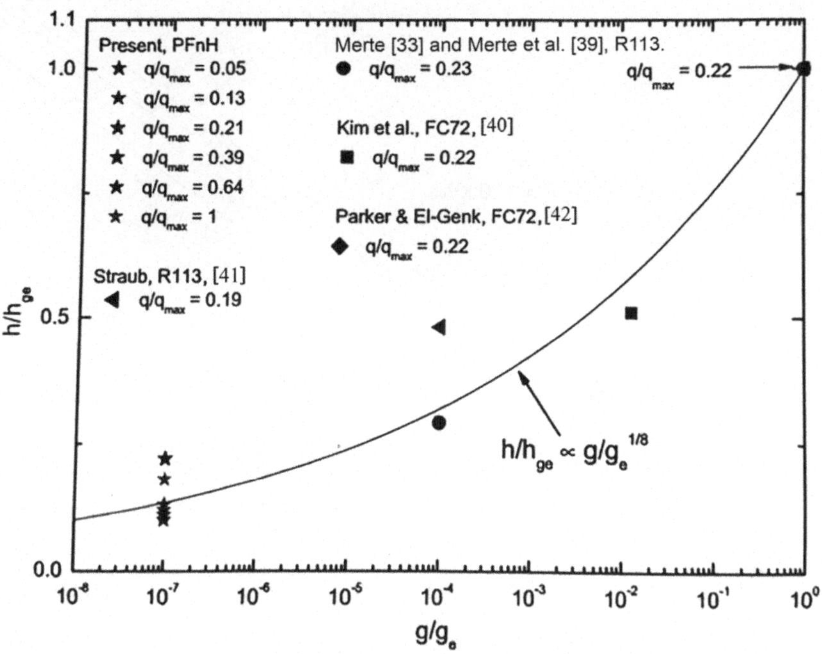

FIGURE 8.17 Correlation of the nucleate boiling heat transfer coefficient at different gravity levels [37].

level. Although the data show some scatter, it is seen that $\frac{h}{h_{ge}}$ approximately correlates as $\left(\frac{g}{g_e}\right)^{1/8}$. This correlation is somewhat tentative as at reduced gravity, the size of the large bubble present on the surface can be comparable to the size of the heater. Experimental conditions (steady state or transient), size of the heater, dissolved gas content, and procedure used to conduct the experiment can influence the results. Maximum heat flux data were also obtained in the experiments on ISS. As seen in Figure 8.16, the maximum heat flux is less than 1 W/cm^2 for Perfluoro-n-Hexane.

Figure 8.18 shows maximum heat flux data reported in the literature and that obtained by Warrier et al. [37] as a function of different magnitudes of gravity. The data were normalized with that at earth normal gravity. The plotted data are for Perfluoro-n-Hexane, R-113, and FC-72. The solid lines in Figure 8.18 are predictions for Perfluoro-n-Hexane, whereas the dotted lines are for R-113 obtained from eq. (6.38) of Zuber et al. [38] described in Chapter 6. For saturated liquid, the constant in the expression developed by Zuber from hydrodynamic consideration was taken to be 0.16 instead of $\pi/24$. Although limited data are available, the dependence of maximum heat flux at $\frac{g}{g_e} < 10^{-4}$ shows a weaker dependence on gravity than that given by the hydrodynamic theory. It should be noted that at reduced gravity, no vapor

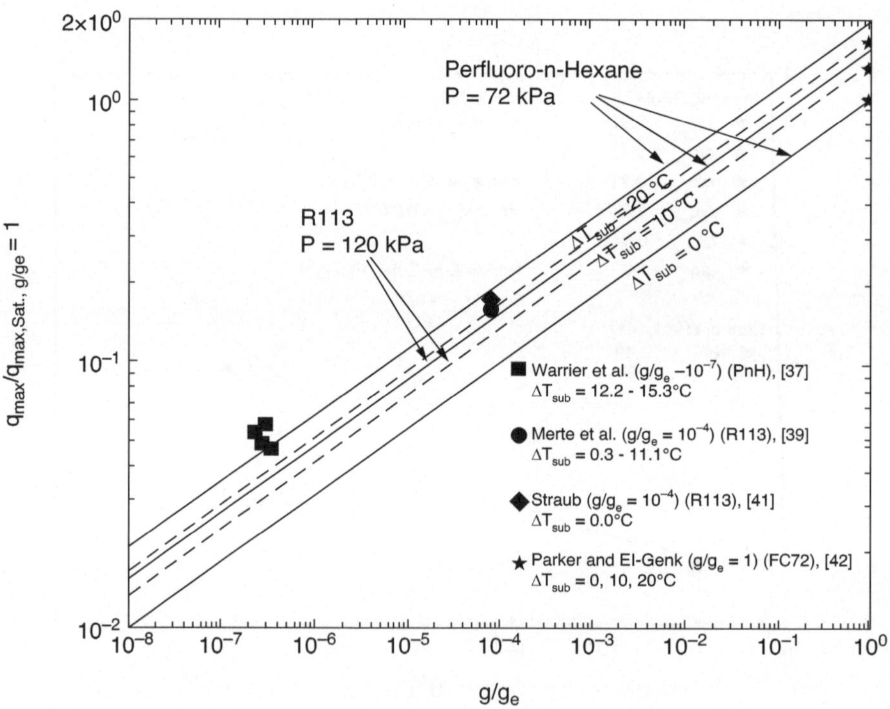

FIGURE 8.18 Maximum heat flux at different magnitudes of gravity [37].

columns are observed to form, which are the cornerstone of the formulation of the hydrodynamic theory. Furthermore at reduced gravity, large bubbles occupy more and more of the heater area. These bubbles continue to persist on the heater surface and serve as a vapor sink for smaller bubbles that form on the heater surface and slide along the heater to merge with the larger bubble. This configuration is much different from that found at high gravity levels. The size of the larger bubbles sitting on the heater surface or hovering over the surface can be comparable to that of the heater or the enclosing vessel. As such, any correlation developed based on the limited data should be used with an abundance of caution. Appropriate approach will be to use numerical simulations based on mechanistic models, while validating them with data from experiments.

8.7 MODE OF THE EXPERIMENTS

The boiling curve can be affected by the manner in which the experiments are conducted, i.e., at steady state or transient. Most of the discussion presented previously was for a steady-state process. One of the earliest studies on transient nucleate boiling heat fluxes is that reported by Johnson [43]. In the experiments, thin metallic ribbons submerged in saturated water were heated electrically. The power to the heaters was increased exponentially. It was concluded that for exponential periods greater than 5 ms, the transient nucleate boiling process including incipience could be described by a steady-state process. However, the transient maximum heat fluxes were found to exceed the steady-state values. Subsequent studies of Sakurai and Shiotsu [44,45] on platinum wires submerged in a pool of water show that for exponential periods varying from 5 ms to 1 s, the incipience heat fluxes increase as the exponential time decreases. In nucleate boiling, the transient heat transfer coefficients were generally found to be lower than those obtained under steady-state conditions. The ratio of the transient to steady-state heat fluxes depended on the magnitude of the heat flux, but values as low as 0.5 have been reported. For periods less than 10 ms, the transient maximum and DNB heat fluxes were found to increase as the exponential time decreased. The DNB heat flux was defined as the highest nucleate boiling heat flux at which a linear relationship between $\ln q$ and $\ln \Delta T$ ceased to exist. Figure 8.19 shows their data for $q_{\max}$ and q_{DNB} obtained at pressures ranging from 1 to 20 atm as a function of transient period. It is seen that the difference in steady state and transient heat fluxes decreases with increase in transient time. Kozawa et al. [46] have also studied transient boiling from a 70-μm thick copper foil submerged in a pool of R-133. Their objective was to determine the magnitude of the wall superheat at the onset of transition boiling rather than the magnitude of the maximum heat flux. Stepwise heating was used, and artificial pinholes were created on the surface to enhance nucleation site density and thereby the rate of heat transfer. An optimum pinhole density was found for enhancement of boiling heat transfer. The authors found that the presence of pinholes limited the temperature overshoot. The onset of boiling was not only affected by the magnitude of the heat flux but also by the wall superheat at boiling incipience.

Models for transient maximum heat fluxes have been proposed by Serizawa [47] and Pasamehmetoglu et al. [48]. These models employ an accepted steady-state

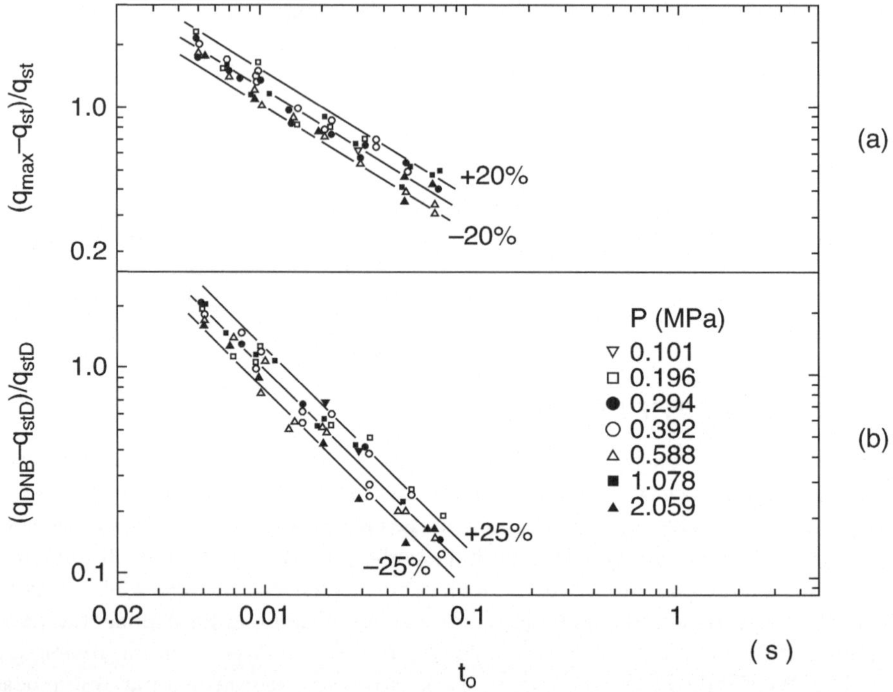

FIGURE 8.19 Effect of transients on DNB and maximum heat flux [45].

critical heat flux model as a starting point. In the work of Pasamehmetoglu et al., the steady-state model of Haramura and Katto [49] was used while including the effect of time-dependent heat flux on the initial liquid layer thickness (hydrodynamic basis) and on the evaporation rate of the liquid layer. The conduction equation in the heater was also solved to relate the surface heat flux to the energy generation rate in the solid. The predictions from the model were reported to be in good agreement with the data. The hovering period of a bubble was an adjustable parameter in the model.

Quenching of a hot body in a saturated or a subcooled liquid has many industrial applications. To circumvent the difficulties and facilitate acquisition of data in transition boiling, quenching (transient cooling) has also been used as an experimental technique for carrying out boiling studies. In the quenching experiments, the time-dependent temperature history in the solid is used to recover the surface heat flux. If quenching is used as an experimental technique, such questions arise as how well the transient boiling curves can be identified with a steady-state process, and what criterion or conditions need to be satisfied before quenching results can be considered to approach steady-state results? Bergles and Thompson [50] and Veres and Florschuetz [51] made the first two attempts to relate the transient boiling curves to the steady-state boiling corves. Bergles and Thompson reported that maximum heat fluxes in their transient and steady-state tests were about the same, while some differences existed in the steady and transient transition boiling data. However, Bergles and Thompson did not use the same heater in the two tests. Veres and Florchuetz, by employing the same heater in the steady state and transient tests, eliminated the issue of the use of different heaters in the experiments of Bergles and Thompson. But they introduced another new variable in terms of induction heating that was used to carry out the steady-state experiments.

Peyavopanakul and Westwater [52] studied the boiling process during quenching by systemically varying the time needed to quench the heater surface. In the experiments, 5.07 cm diameter horizontal copper disks of thickness varying from 0.05 to 5.1 cm were used. Liquid nitrogen at 1 atm was used as the test liquid. Figure 8.20 shows the boiling curves obtained in their experiments. From the data, the authors concluded that disk thickness had no effect on nucleate and transition boiling as long as the thickness was greater than 1.3 cm. For thinner test specimens, the boiling curves were generally shifted to the right. The maximum heat flux decreased with decreasing thickness when the disk was thinner than 2.5 cm. For thickness greater than 2.5 cm, the maximum heat flux was independent of thickness. It was noted that if the time to traverse the top 10% of the boiling curve was greater than 1 s, the boiling process could be called quasi-steady.

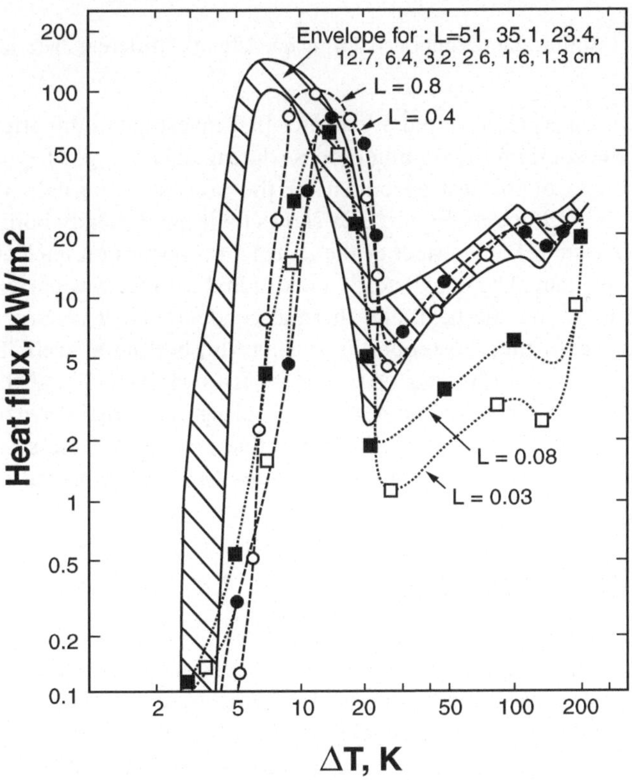

FIGURE 8.20 Boiling curves obtained during quenching in liquid nitrogen of surfaces of different thicknesses [52].

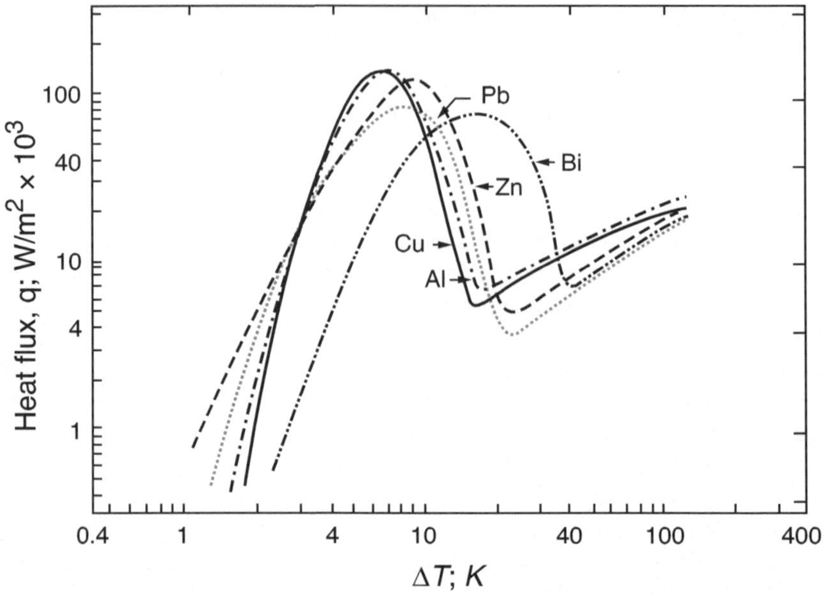

FIGURE 8.21 Boiling curves obtained during quenching of different material surfaces [53].

In a subsequent work, Lin and Westwater [53] investigated the effect of thermal properties of the solid on the boiling process during quenching. They noted that the required thickness of the test specimen for the process to be called quasi-steady was different for each metal. Figure 8.21 shows their quasi-steady boiling curves for the five metals tested. Some effect of the thermal properties on nucleate and transition boiling is evident. The quasi-steady maximum heat flux was correlated with the product of density, specific heat, and thermal conductivity. It is surprising that the authors saw effect of thermal properties during film boiling as well. Transient heat flux data obtained with ramping of power input to the heater and those obtained during quenching demonstrate the need for coupling of the hydrodynamic and thermal processes on the liquid side with the thermal response of the solid. The effect of metal properties during quenching of spheres has been reported by Irving and Westwater [54].

8.8 LIQUID SUBCOOLING

The rate of convective heat transfer increases with liquid subcooling. As a result, liquid subcooling influences the inception and partial nucleate boiling regions of the boiling curve. On a wall flux versus wall superheat plot, convective heat flux and partial nucleate boiling curves lie higher than those for saturated boiling. However, at high nucleate boiling heat fluxes, the subcooled and saturated boiling curves almost overlap. This behavior suggests that at high heat fluxes, the convective contribution becomes small. Also, since the nucleation site density depends only on wall superheat, the number density of active sites must play an important role. At present, no theoretical criterion such as that developed by Moissis and Berenson [55] for saturated boiling exists for transition from partial to fully developed nucleate boiling under subcooled conditions. The maximum heat flux is found to increase with subcooling. Zuber et al. [38] extended the hydrodynamic prediction for saturated liquids to a subcooled liquid by accounting for heat lost to the liquid from the vapor–liquid interface in a transient manner. The net effect of this was to increase the vapor removal rate before vapor columns became unstable. Their expression for maximum heat flux in subcooled liquids is

$$\frac{q_{\max,sub}}{q_{\max,Z}} = \left[1 + C_1 \left\{ g(\rho_l - \rho_v)\rho_v^2 \alpha_l^4 / \sigma^3 \right\}^{1/8} \mathrm{Ja}\right] \tag{8.38}$$

where

$$\mathrm{Ja} = \frac{\rho_l c_{pl} \Delta T_{\mathrm{sub}}}{\rho_v h_{fg}}$$

where $q_{\max,z}$ is the maximum heat flux for saturated liquids. From analysis, the constant C_1 was found to have a value of 5.33. The Jakob number as defined above is in the conventional manner except that ΔT is replaced by liquid subcooling, ΔT_{sub}. According to eq. (8.38), the maximum heat flux increases linearly with subcooling. The prediction from eq. (8.38) was found to be in good agreement with subcooled water data obtained up to 2 atm and subcooled ethanol data obtained up to 10 atm pressure.

Elkassabgi and Lienhard [56] have made an extensive investigation of maximum heat fluxes in subcooled pool boiling on horizontal cylinders. In the experiments, Acetone, Isopropanol. Methanol, and Freon 133 were used as the test liquids and liquid subcoolings as high as 140°C were attained. Figure 8.22 shows their data obtained with isopropanol. Based on the trend of the data, three subcooling regimes for the maximum heat flux were identified. The data that varied linearly with subcooling (low subcoolings) were correlated in a manner similar to eq. (8.38), but the time constant for transient liquid–vapor contacts was obtained from the mechanical energy stability criterion. Their correlation for low subcooling is

$$\frac{q_{\max,\,\mathrm{sub}}}{q_{\max,\,\mathrm{sat}}} = 1 + 4.28\ \mathrm{Ja}/\mathrm{Pe}^{1/4} \tag{8.39}$$

where Peclet number is defined as:

$$\mathrm{Pe} = \frac{\sigma^{3/4}}{\alpha_l \left[g(\rho_l - \rho_v)\right]^{1/4} \rho_v^{\,1/2}}$$

The constant in eq. (8.39) was found to be independent of dimensionless cylinder radius. The maximum heat flux, $q_{\max}$, in saturated boiling was obtained from the

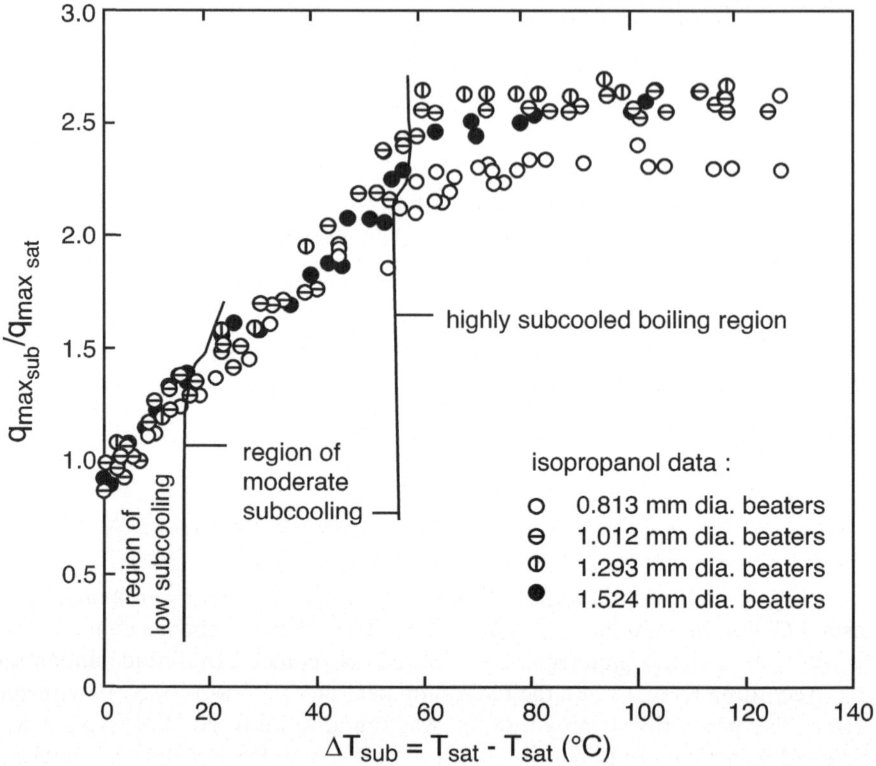

FIGURE 8.22 Effect of liquid subcooling on maximum heat flux on horizontal cylinders [56].

hydrodynamic predictions described earlier in this chapter. Equation (8.39) shows for low subcoolings a weaker dependence on Peclet number than eq. (8.38) of Zuber et al. [38].

At moderate subcoolings, bubbles were observed to surround the heater without detaching. For these subcoolings, the maximum heat flux was considered to be determined by natural convection from the outer edge of the bubble boundary layer. Their correlation for moderate subcoolings is

$$\frac{q_{\text{max, sub}}}{\left(k_l/2R_{\text{eff}}\cdot\Delta T_{\text{sub}}\right)} = 28 + \frac{1.5Ra^{1/4}}{\left(\beta\Delta T_{\text{sub}}\right)^{7/8}} \tag{8.40}$$

$$\text{Ra} \equiv \frac{g\beta\Delta T_{\text{sub}}\left(2R_{\text{eff}}\right)^3}{\alpha_l v_l} \tag{8.41}$$

$$\text{R}_{\text{eff}} = R\left(1 + D_d/R\right) \tag{8.42}$$

where R is cylinder radius, β is coefficient of isobaric expansion, and D_d is the bubble diameter at departure as given by Fritz's expression and is given as

$$D_d = 0.0208\varphi\sqrt{\frac{\sigma}{g(\rho_l - \rho_v)}}$$

and φ is the contact angle in degrees

At very high subcoolings, the maximum heat flux was found to be independent of ΔT_{sub}. Elkassabgi and Lienhard noted that the maximum heat flux was limited by the evaporation rate at the heater surface and not by the rate at which energy could be removed from the bubble boundary by natural convection (the authors called it a molecular effusion limit). The maximum heat flux was correlated as

$$q_{\text{max, sub}} = \rho_v h_{fg}\sqrt{R_G T_{sat}/2\pi}\cdot\left[0.01 + 0.0047\exp\left(1 - 1.11\times10^{-6}x\right)\right] \quad (8.43)$$

where

$$x = R\left(R_G T_{sat}\right)^{1/2}/\alpha_l \quad (8.44)$$

and R_G is the gas constant. Since geometry influences the convective process, and the radius of the cylinder is used as the characteristic length for the molecular effusion limit, eq. (8.43) and eq. (8.40) should be considered specific to the range of cylinder diameters employed in the study.

Transition boiling is influenced strongly by liquid subcooling. Very limited data on subcooled transition boiling are available in the literature. Figure 8.23 shows the transition boiling data obtained by Dhir [57] during quenching of a 19-mm diameter copper sphere in subcooled water. With increase in subcooling, the transition boiling curves move up and to the right. The maximum heat flux increases with subcooling. Though the maximum heat flux data varied somewhat nonlinearly with subcooling, it was found to correlate with eq. (8.38) when the constant C_1, was chosen to have a value of 6.33. In the transient cooling experiments, the nucleate boiling curves did not collapse on one curve indicating hysteresis in the process as nucleate boiling regime is obtained while traversing the boiling curve from film boiling side. Dhir and Purohit [27] have obtained minimum heat flux data on spheres when liquid was subcooled. In their experiments, water at 1 atm pressure was used as the test liquid. They found the minimum heat flux to increase nearly linearly with subcooling.

8.9 SURFACTANTS AND MIXTURE BOILING

Often, surfactants are added to a liquid to reduce interfacial tension. As given in eq. (3.9) with decrease in σ_{lv}, $\cos\varphi$ increases or φ decreases. With reduced contact angle, fewer cavities nucleate and bubble diameter at departure also decreases. Presence of surfactants affects the interfacial heat transfer as well. However, overall effect of presence of surfactants on nucleate boiling heat transfer is small.

Boiling of mixtures of liquids with different thermophysical properties is a complex process involving understanding of mass transfer across interfaces and has not been addressed in this work.

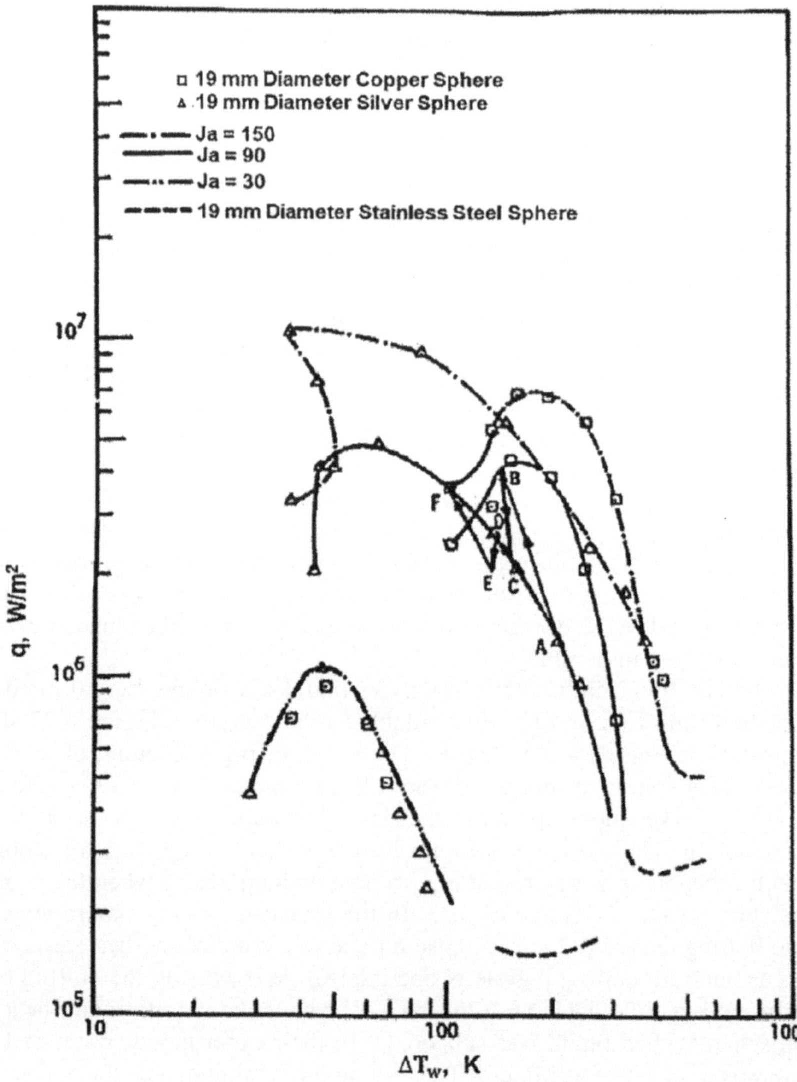

FIGURE 8.23 Dependence of forced convention boiling heat fluxes on sphere surface superheat. Here *Ja* is based on liquid subcooling.

8.10 SUMMARY

In this chapter, we have discussed the effect of various experimental variables on the boiling curve. Experimental variables considered are: Surface Finish; Surface Wettability; Surface Contamination; Heater Size; Geometry and Material Properties and Thickness; System Pressure; Gravity; Mode of the Experiments; and Liquid Subcooling. The description with respect to all modes of boiling is not complete. Effects of surfactants and boiling of mixtures have not been addressed.

EXERCISES

1. Using the data of She plotted in Figure 8.4 as guide, obtain the ratio of maximum heat flux on horizontal plate for contact angles 20° and 90°. Explain your answer. Assume saturated water at 1 atm pressure to be the test liquid.
2. A researcher measures maximum heat flux on a very large horizontal surface and then on a circular disk that is 1.5 cm in diameter. Both surfaces were oxidized and saturated water at 1 atm pressure was the test liquid. The researcher was surprised to find the maximum heat flux on small surface to be about twice that on large surface. Explain why it should not be a surprise?
3. Calculate minimum film boiling heat flux and minimum film boiling temperature for 5 mm diameter cylinder submerged horizontally in water at saturation temperature corresponding to a pressure of 1 atm.
4. Calculate maximum heat flux on a sphere 2 cm in diameter and submerged in saturated water at 1 atm pressure.
5. A heated vertical plate is placed in saturated water at 1 atm pressure. What would be the maximum heat flux if the plate is 5 cm tall?
6. Obtain an expression for film boiling heat transfer coefficient on a long horizontal ribbon of width W placed in a saturated liquid. The bottom surface of the ribbon is insulated. State clearly any assumptions you wish to make.
7. Using data plotted in Figure 8.2 obtain ratio of nucleate boiling heat flux at a contact angles of 90° and 14° for a wall superheat of 20°C. Discuss your result.
8. Obtain maximum heat flux on a large flat plate for water subcooling of 20°, 40°, 60°, and 80°C at 1 atm pressure. Use the correlation proposed by Zuber et al.
9. Repeat problem 8 for pressures of 20, 50, and 100 atm.
10. Calculate maximum heat flux on a 6 mm diameter horizontal cylinder for saturated water at 1 atm pressure.
11. Calculate maximum heat flux on a 1 mm diameter horizontal cylinder submerged in a pool saturated Freon-113 at 1 atm pressure.

NOMENCLATURE

A_h	Heater area
A_v	Area occupied by vapor columns
c	Empirical constant
F	Function
g	Gravitational acceleration
H	Heater width
h	Heat transfer coefficient
h_{fg}	Latent heat of vaporization
Pe	Peclet number
p	Pressure
q	Heat flux
R	Heater radius
U_{vc}	Critical velocity in vapor columns
V	Vapor volume

n Empirical constant
Ja Jakob number $\dfrac{\rho_l c_{pl} \Delta T_{sub}}{\rho_v h_{fg}}$
D Diameter

SUPERSCRIPTS

/	Dimensionless
.	Time rate of change
*	Dimensionless pressure

SUBSCRIPTS

$d2$	Two-dimensional "most dangerous wavelength"
i	Imaginary component
e	Earth normal
l	Liquid
max	Maximum
max, Z	Maximum heat flux given by Zuber's expression
min	Minimum
v	Vapor
sat	Saturation
o	Reference
c	Characteristic
sub	Subcooling
d	Departure
x	Dimensionless group as defined in eq. (8.44)

GREEK LETTERS

α	Thermal diffusivity
β	Coefficient of isobaric expansion
ΔT	Temperature difference
δ	Vapor blanket thickness
λ	Wavelength
μ	Molecular density
ρ	Density
σ	Surface tension
φ	Contact angle
ω	Frequency

REFERENCES

1. Berenson, P.J. (1962), "Experiments on Pool-Boiling Heat Transfer." *Int. J. Heat Mass Transfer*, Vol. 5, pp. 985–999.
2. Bui, T.D. and Dhir, V.K. (1985), "Transition Boiling Heat Transfer on a Vertical Surface." *J. Heat Transfer*, Vol. 107, pp. 756–763.

3. Singh, A., Mikic, B.B., and Rohenow, W.M. (1978). "Relative Behavior of Water and Organics in Boiling." *Proceedings of the 6th International Heat Transfer Conference*, Toronto, Canada, Vol. 1.
4. Shoukri, M. and Judd. R.L. (1978), "On the Influence of Surface Conditions in Nucleate Boiling. The concept of Bubble Flux Density." *ASME, J. Heat Transfer*, Vol. 100, pp. 618–623.
5. Bier, K., Gorenflow, D., Salem, M., and Tanes, Y. (1978), "Pool Boiling Heat Transfer and Size of Active Nucleation Centers for Horizontal Plates with Different Surface Roughnesses, Volume 1." *Proceedings of the 6th International Heat Transfer Conference*, Toronto, ON, Canada, pp. 151–156.
6. Liaw, S.P. and Dhir, V.K. (1986), "Effect of Surface Wettability on Transition Boiling Heat Transfer from a Vertical Surface." *Proceedings of International Heat Transfer Conference*, San Francisco, pp. 2031–2036.
7. Nishikawa, K., Fujita, Y., and Ohta, H. (1984), "Effect of Surface Configuration on Nucleate Boiling Heat Transfer." *Int. J. Heat Mass Transfer*, Vol. 27, pp. 1559–1571.
8. Maracy, M. and Winterton, R.H.S. (1988), "Hyssteresis and Contact Angle Effects in Transition Boiling of Water." *Int. J. Heat Mass Transfer*, Vol. 31, pp. 1443–1449.
9. Hahne, L. and Dieselhorst, T. (1978), "Hydrodynamic and Surface Effects on the Peak Heat Flux in Pool Boiling." *Proceedings of the 6th International Heat Transfer Conference*, Toronto, ON, Canada, Vol. 1, pp. 209–219.
10. Hohl, R., Auracher, H., Blum, J., and Marquardt, W. (1966), "Pool boiling Heat Transfer Experiments with Controlled Wall Temperature Transients." *2nd European Thermal Science and 14th UIT National Heat Transfer Conference*, Rome, pp. 1647–1652.
11. Joudi, K.A. and James, D.D. (1982), "Surface Contamination, Rejuvenation, and Reproducibility of Results in Nucleate Pool Boiling." *J. Heat Transfer*, Vol. 103, pp. 453–458.
12. Ross, T.K. (1967), "Corrosion and Heat Transfer Review." *Ed. Corros. J.*, Vol. 2, pp. 131–142.
13. She, Z. (2021), "Effects of Heater Size, Wettability and Size of Surrounding Vessel on Nucleat Boiling Heat Transfer." PhD Thesis University of California Los Angeles.
14. Lienhard, J.H., Dhir, V.K., and Riherd, D.M. (1974), "Peak Pool Boiling Heat Flux Measurements on Finite Horizontal Flat Plates." *ASME J. Heat Transfer*, Vol. 95, pp. 477–482.
15. Nishikawa, K., Fujita, Y., and Ohta, H. (1974), "Effect of Surface Configuration on Nucleate Boiling Heat Transfer." *Int. J. Heat Mass Transfer*, Vol. 27, pp. 1559–1571.
16. Lienhard, J.H. and Dhir, V.K. (1973), "Hydrodynamic Prediction of Peak Pool Boiling Heat Transfer from Finite Bodies." *ASME J Heat Transfer*, Vol. 95, pp. 152–158.
17. Lance, R.P. and Myers, J.E. (1950), "Local Boiling Coefficients on a Horizontal Tube." *AIChE J.*, Vol. 4, pp. 75–80.
18. Cornwell, K. and Einarsson, J.G. (1989), "The Influence of Fluid Flow on Nucleate Boiling from a Tube." *Proceedings of European Seminar, No. 8: On Advances in Pool Boiling Heat Transfer*, University of Paderborn, May 11–12, 1989, Paderborn, Germany, pp. 28–41.
19. Sun, K.H. and Lienhard, J.H. (1970), "The Peak Pool Boiling Heat Flux on Horizontal Cylinders." *Int. J. Heat Mass Transfer*, Vol. 13, pp. 1425–1439.
20. Lienhard, J.H. and Dhir V.K. (1973), "The Extended Hydrodynamic Theory of the Peak and Minimum Pool Boiling Heat Fluxes." NASA CR2270.
21. Lienhard, J.H. and Dhir, V.K. (1980), "On the Prediction of the Minimum Pool Boiling Heat Flux." *ASME J. Heat Transfer*, Vol. 102, p. 457.
22. Lienhard, J.H., and Wong, P.T.Y. (1964), "The Dominant Unstable Wave Length and Minimum Heat Flux During Film Boiling on a Horizontal Cylinder." *ASME J. Heat Transfer*, Vol. 86, p. 220.

23. Kovalev, S.A. (1966), "An Investigation of Minimum Heat Fluxes in Pool Boiling of Water." *Int. J. Heat Mass Transfer*, Vol. 9, p. 1219.
24. Grigull, U., Abadzic, E. (1967–1968), "Heat Transfer from a Wire in the Critical Region." *Proc. Inst. Mech. Eng.*, Vol. 182, p. 52.
25. Dod, J.S. and Lienhard, J.H. (1972), "The Peak Pool Boiling Heat Flux for a Sphere." *AIChE J.* Vol. 18, pp. 337–342.
26. Dhir, V.K. (1978), "Study of Transient Boiling Heat Fluxes from Spheres Subjected to Forced Vertical Flow." *Proceedings of the 6th International Heat Transfer Conference*, Toronto, Canada, Vol. 4, pp. 451–456.
27. Dhir, V.K., and Purohit, G.P. (1978), "Subcooled Film Boiling Heat Transfer from Spheres." *Nucl. Eng. Design*, Vol. 47, p. 49.
28. Houchin, W.R. and Lienhard, J.H. (1966), "Boiling Burnout in Low Thermal Capacity Heater." *Proceedings of ASME Winter Annual Meeting 1966, New York, (Heat Transfer)*, pp. 1–8.
29. Tachibana, F., Akiyama, M., and Kawamura, H. (1967), "Non-Hydrodynamic Aspects of Pool Boiling Burnout." *J. Nucl. Sci. Technol.*, Vol. 4, pp. 121–130.
30. Fath, H.S. and Judd, R.L. (1978), "Influence of System Pressure on Microlayer Evaporation Heat Transfer." *ASME J. Heat Transfer*, Vol. 100, pp. 49–55.
31. Mikic, B.B. and Rohsenow, W.M. (1969), "A New Correlation of Pool Boiling Data, Including the Effect of Heating Surface Characteristics." *J. Heat Transfer*, Vol. 9, pp. 245–250.
32. Gorenflo, D., Knabe, V., and Bieling, V. (1986), "Bubble Density on Surfaces with Nucleate Boiling—Its Influence on Heat Transfer and Burnout Heat Flux at Elevated Saturation Pressures." *Proceedings of the 8th International Heat Transfer Conference*, Vol. 4, August 1986, San Francisco, CA, pp. 1995–2000. Washington, DC: Hemisphere Publishing Corporation.
33. Merte, H. (1988), "Nucleate Pool Boiling: High Gravity to Reduced Gravity, Liquid Metals to Cryogenics." *Transactions of the 5th Symposium on Space Nuclear Power Systems*, University of New Mexico, January 11–14, 1988, Albuquerque, NM, pp. 437–442.
34. Zell, M., Straub, J., and Vogel, B. (1989), "Pool Boiling Under Microgravity." *Proceedings of European Seminar, No. 8 on Advances in Pool Boiling Heat Transfer*, Paderborn, Germany, pp. 70–74.
35. Siegel, R. and Usiskin, C. (1959), "A Photographic Study of Boiling in the Absence of Gravity." *ASME J. Heat Transfer*, Vol. 81, pp. 230–238.
36. Dhir, V.K., Warner, G., Aktinol, E., Chao, D., Eggers, J., Sheredy, W., and Booth, W. (2012), "Nucleate Pool Boiling Experiment (NPBX) on the International Space Station." *Microgravity Science and Technol.*, Vol. 24, pp. 307–325.
37. Warrier, G., Dhir, V.K., and Chao, D. (2015), "Nucleate Pool Boiling Experiment (NPBX) in Microgravity: International Space Station." *Int. J. Heat. Mass Transfer*, Vol. 83, pp. 781–798.
38. Zuber, N, Tribus, M., Westwater, J.W. (1961), "The Hydrodynamic Crisis in Pool Boiling of Saturated and Subcooled Liquid", *Proceeding of International Heat Transfer, Meeting*, Boulder, CO, p. 130.
39. Merte, H., Lee, H.S., Keller, R.B. (1995), "Report on Pool Boiling Experiment Flown on STS-57 and STS-60." Report No. UM-MEAM-95-091.
40. Kim, J., Benton, J.F., Wisniewski, D. (2002), "Pool Boiling Heat Transfer on Small Heaters: Effect of Gravity and Subcooling." *Int. J. Heat Mass Transfer*, Vol. 45, p. 3919.
41. Straub, J. (2001), "Boiling Heat Transfer and Bubble Dynamics in Microgravity", Adv. Heat Transfer, Vol. 35, p. 58.
42. Parker, J.L. and El-Genk, M.S. (2005), "Enhanced Saturation and Subcooled Boiling of FC-72 and Di-electric Liquid." *Int. J. Heat Mass Transfer*, Vol. 48, p. 3736.

43. Johnson, H.A., (1971), "Transitient Boiling Heat Transfer to Water." *Int. J. Heat Mass Transfer*, Vol. 14, pp. 67–82.
44. Sakurai, A. and Shiotsu, M., (1977), "Transient Pool Boiling Heat Transfer, Part 1: Incipience Boiling Superheat." *ASME J. Heat Transfer*, Vol. 99, pp. 547–553.
45. Sakurai, A. and Shiotsu, M., (1977), "Transient Pool Boiling Heat Transfer, Part 2: Boiling Heat Transfer and Burnout." *ASME J. Heat Transfer*, Vol. 99, pp. 554–560.
46. Kozawa, Y., Innuoe, T., and Okayama, K., (1986), "Enhancement of Bubble Formation and Heat Removal in Transition Boiling." *Proceedings of the 8th International Heat Transfer Conference,* San Francisco, CA, Vol. 4, pp. 2007–2012.
47. Serizawa, A., (1983), "Theoretical Prediction of Maximum Heat Flux in Power Transients." *Int. J. Heat Mass Transfer*, Vol. 26, pp. 921–932.
48. Pasamehmetaglu, K.O., Nelson, R.A., and Gunnerson, F., (1987), "A Theoretical Prediction of Critical Heat Flux in Saturated Pool Boiling during Power Transients." *ASME HTD*, Vol. 77, pp. 57–64.
49. Haramura, Y. and Katto, Y., (1983), "New Hydrodynamic Model of Critical Heat Flux Applicable Widely to Both Pool and Forced Convection Boiling on Submerged Bodies in Saturated Liquids." *Int. J. Heat Mass Transfer*, Vol. 26, pp. 379–399.
50. Bergles, A.E., and Thompson, Jr. W.G., (1970), "The Relationship of Quench Data to Steady State Pool Boiling Data." *Int. J. Heat Mass Transfer*, Vol. 13, pp. 65–68.
51. Veres, D.R., and Florschuetz, L.E., (1971), "A Comparison of Transient and Steady State Pool Boiling Data Obtained Using the Same Heating Surface." *ASME J. Heat Transfer*, Vol. 93, pp. 224–232.
52. Peyaropanakul, W. and Westwater, J.W., (1978), "Evaluation of the Unsteady State Quenching Method for Determining Boiling Curves." *Int. J. Heat Mass Transfer*, Vol. 21, pp. 1437–1445.
53. Lin, D.Y.T. and Westwater, J.W., (1982), "Effect of Metal Thermal Properties on Boiling Curves Obtained by the Quenching Method." *Proceedings of the 7th International Heat Transfer Conference*, Munich, Germany, Vol. 4, pp. 155–160.
54. Irving, M.E. and Westwater, J.W., (1986), "Limitations for Obtaining Boiling Curves by the Quenching Method with Spheres." *Proceedings of the 8th International Heat Transfer Conference*, San Francisco, CA, Vol. 4, pp. 2061–2066.
55. Moissis, R. and Berenson, P. J., (1963), "On the Hydrodynamic Transition in Nucleate Boiling." *ASME J. Heat Transfer*, Vol. 27, p. 221.
56. Elkassabgi, Y. and Lienhard, J.H., (1988), "The Peak Pool Boiling Heat Flux from Horizontal Cylinders in Subcooled Liquids." *ASME J. Heat Transfer*, Vol. 110, pp. 479–488.
57. Dhir, V.K., (1978), "Study of Transient Transition Boiling Heat Fluxes from Spheres Subjected to Forced Vertical Flow." *Proceedings of the 6th International Heat Transfer Conference*, Toronto, ON, Canada, Vol. 4, pp. 451–456.

9 Film and Dropwise Condensation

INTRODUCTION

As we discussed in Chapter 2, vapor can stay in a metastable state at temperatures below the saturation temperature corresponding to pressure in the vapor. However, in most practical situations, phase change occurs when vapor temperature falls below the saturation temperature. Condensation is a phase change process in which vapor in contact with another fluid or a solid at a temperature lower than the vapor condenses to form the liquid phase. Condensation occurring on a solid surface is called surface condensation. However, when vapor mixes with colder air or another species of vapor/gas, condensation can occur leading to formation of tiny droplets (fog). Direct contact condensation on the other hand is the process of condensation of vapor on liquid surface.

In this chapter, we will mainly focus on surface condensation. Since latent heat of vaporization is released during the condensation process, heat must be removed from the surface on which condensation occurs to sustain the process. As such, the process is opposite to that of boiling, where heat must be added to the surface to sustain the process. Surface condensation can be filmwise where as liquid film forms on the surface or dropwise condensation where discreet droplets form on the surface. However, condensation occurring in flow passages or in tubes is called internal condensation. An example of the former is condensation on tubes of a condenser used in power industry, whereas condensation in tubular passages in refrigeration and air conditioning systems is an example of the latter. Film condensation can be laminar or turbulent and may involve pure vapor or vapor containing some non-condensible gas.

9.1 LAMINAR FILM CONDENSATION: INTEGRAL SOLUTIONS

Nusselt [1] was the first one to provide a model for laminar film condensation on a vertical wall and on a horizontal tube. In carrying out his analysis, Nusselt made two key assumptions:

i. Inertia in the condensate film could be neglected even though the film thickness and the condensate velocity increased in the direction of flow.
ii. Neglect of convection terms in the energy equation even though the velocity in the film is nonzero.

Rohsenow [2] offered a correction to Nusselt's solution to account for the second assumption. Using an iterative scheme in an integral approach, Rohsenow accounted for the nonlinearity of the temperature profile in the film, which was assumed to be

 DOI: 10.1201/9781032668437-9

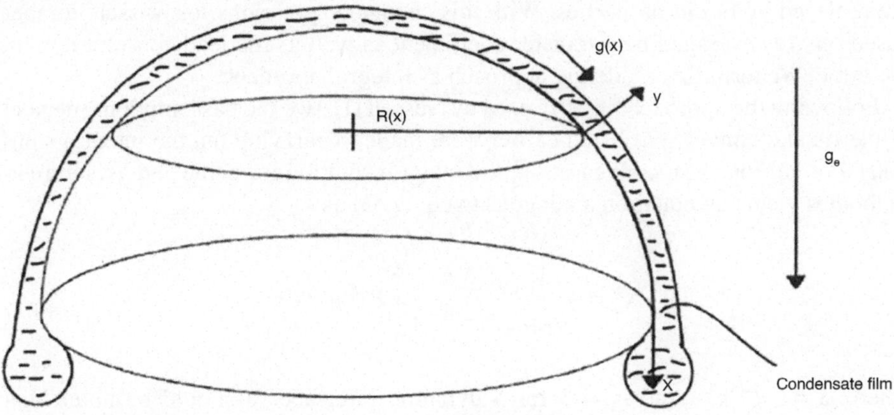

FIGURE 9.1 Film condensation on an axisymmetric body.

linear in Nusselt's analysis. The result was a new definition of the corrected latent of heat of vaporization

$$h'_{fg} = h_{fg} + 0.68c_{pl}\Delta T \tag{9.1}$$

where ΔT is the temperature difference between saturation temperature of vapor and the surface temperature.

As noted in Chapter 7, Dhir and Lienhard [3] developed a generalized analysis for ondensation on plane (e.g., a vertical wall) and for axisymmetric bodies (e.g., cylinders and spheres). Figure 9.1 shows a generalized configuration of axisymmetric bodies from their work. Various assumptions made in carrying out the steady state analysis are as follows:

i. Vapor is at its saturation temperature corresponding to the pressure in the vapor.
ii. Surface is maintained at a constant temperature below saturation temperature.
iii. Condensate film thickness is much smaller than the radius of curvature of the surface.
iv. Fluids are incompressible.
v. Flow in the film is laminar.
vi. Inertia and convection terms can be neglected in the momentum and energy equations, respectively.
vii. The condensate thermophysical properties are evaluated at the mean film temperature.
viii. Interface between vapor and liquid is free of any ripples and vapor imposes no shear on the film.
ix. No non-condensables are present in the vapor.

The formulation and solution procedure is similar to that described in Chapter 7 tailored to film boiling and initially carried out by Dhir and Lienhard [3], for laminar film condensation on plane and axisymmetric bodies except now vapor properties are replaced by liquid properties. With this change, expressions for Nusselt number based on area averaged heat transfer coefficient as well as the lead constant remain the same. We term this modeling approach as integral approach.

Following the approach initially used by Nusselt [1], two tacit assumptions (neglect of inertia and convection) listed earlier were made in carrying out the integral solutions. For laminar film condensation, the integral solution for plane and axisymmetric bodies yields an equation analogous to eq. (7.33) as

$$\Delta^4 = \frac{4C_1\mu k_l \Delta T}{3h'_{fg}\rho_l(\rho_l - \rho_v)} \int_0^x R^{4/3} g^{1/3}\, dx \tag{9.2}$$

where $\Delta = \delta R^{1/3} g^{1/3}$ and $c_1 = 3$ for a dynamic interface. It should be noted that $\Delta = 0$ at $x = 0$ when at least one of the three is zero. For a vertical wall, $\delta = 0$ at $x = 0$ and $g(x) = g_e$. For a horizontal cylinder, $g(x) = g_e \sin\left(\frac{x}{R}\right)$ as such, it is zero at $x = 0$. Also, R normal to the surface is constant. For a sphere, $g(x) = g_e \sin\left(\frac{x}{R}\right)$ and $R(x) = R\sin\left(\frac{x}{R}\right)$, as such $\Delta = 0$ at $x = 0$. An expression for average heat transfer coefficient during laminar film condensation on a vertical wall, similar to eq. 7.36a, is found to be

$$\bar{h} = 0.943\left[\frac{k_l^3 h'_{fg}\rho_l(\rho_l - \rho_v)g_e}{\mu_l \Delta T x}\right]^{1/4} \tag{9.3}$$

Equivalent expression for average heat transfer coefficient for laminar film condensation on horizontal tubes and spheres is:

$$\bar{h} = C\left[\frac{k_l^3 \rho_l(\rho_l - \rho_v)g_e h'_{fg}}{\mu_l \Delta T D}\right]^{1/4} \tag{9.4}$$

where D is the diameter and $C = 0.73$ for a cylinder and 0.81 for a sphere.

Example:

Laminar film condensation occurs on a cone standing at its base. The base of the cone is 2.5 cm in diameter and the cone wall temperature is maintained at 80°C. What is the value of the average heat transfer coefficient if pure saturated steam at a pressure of 1 atm pressure condenses on the cone surface?

Solution:

We first calculate $g(x)$

$$g(x) = g_e \cos 15° = g_e(0.966)$$

Similarly,

$$R(x) = x\sin 15^\circ = x(0.26)$$

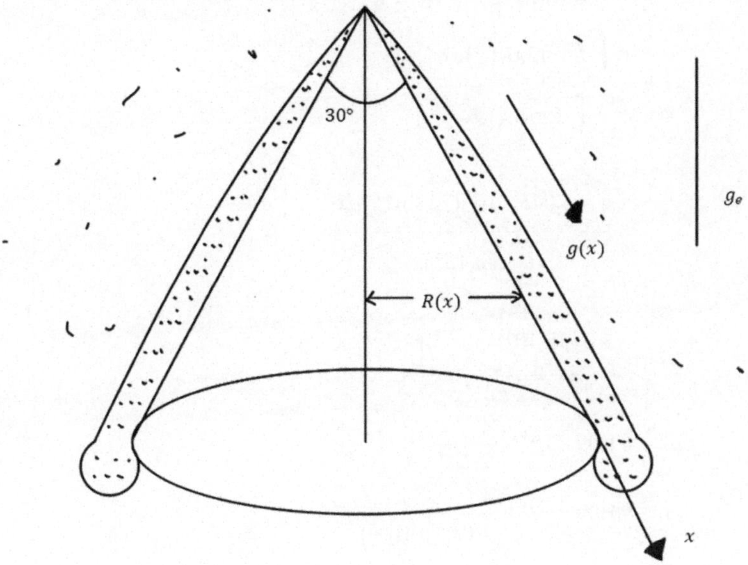

Length of the inclined surface of the cone is

$$L = \frac{1.25}{\sin 15^\circ} = \frac{1.25}{0.26} = 4.8 \text{ cm}$$

With $\Delta = \delta^{1/3} R^{1/3} g(x)^{1/3}$, we find

$$\delta^4 = \frac{4\mu_l k_l \Delta T}{3\rho_l(\rho_l - \rho_v)h'_{fg}} \frac{\int_0^x R(x)^{4/3} g(x)^{1/3}\,dx}{R(x)^{4/3} g(x)^{4/3}}$$

or

$$\delta^4 = \frac{4\mu_l k_l \Delta T}{\rho_l(\rho_l - \rho_v)h'_{fg}} \frac{\int_0^x (x)^{4/3}(\sin 15^\circ)^{4/3} g_e{}^{1/3}(\cos 15^\circ)^{1/3}\,dx}{(x)^{4/3}(\sin 15^\circ)^{4/3} g_e{}^{4/3}(\cos 15^\circ)^{4/3}}$$

$$= \frac{4\mu_l k_l \Delta T}{\rho_l(\rho_l - \rho_v)g_e h'_{fg}} \frac{\int_0^x (x)^{4/3}\,dx}{(x)^{\frac{4}{3}}(\cos 15^\circ)}$$

$$= \left[\frac{12\mu_l k_l \Delta T x}{7\rho_l(\rho_l - \rho_v)g_e h'_{fg}\cos(15^\circ)}\right]$$

$$h(x) = \frac{k_l}{\delta} = \left[\frac{7\rho_l(\rho_l - \rho_v)g_e h'_{fg}\cos(15^\circ)k_l^3}{12\mu_l k_l \Delta T x}\right]^{1/4}$$

$$= \left[\frac{7 \times 958 \times 957.4 \times 9.81 \times 2.3 \times 10^6 \times 0.966 \times (0.68)^3}{12 \times 283 \times 10^{-6} \times 20 \times x} \right]^{1/4}$$

$$= 5.045 \times 10^3 x^{-1/4} \quad \text{W/m}^2\text{K}$$

$$\bar{h} = \frac{\int_0^L h(x) 2\pi R(x) dx}{\int_0^L 2\pi R(x) dx}$$

$$= \frac{\int_0^L \left(5.045 \times 10^3\right) x^{\frac{3}{4}} \sin 15° dx}{\int_0^L x \sin 15° dx}$$

$$= \frac{5.045 \times 10^3 \left[\frac{4}{7} x^{7/4} \right]_0^L}{L^2 / 2}$$

$$= 1.09 \times 10^4 \times \frac{4}{7} \frac{1}{L^{1/4}}$$

$$= 1.09 \times 10^4 \times \frac{4}{7} \frac{1}{\left(4.5 \times 10^{-2}\right)^{1/4}}$$

$$= 12{,}318 \ \text{W/m}^2\text{K}$$

The two assumptions of neglect of inertia in the momentum equation and convection in the energy equation for the condensate film are eliminated in the full boundary layer type of analysis for laminar film condensation on an isothermal vertical wall carried out by Sparrow and Gregg [4]. Subsequently, Dhir and Lienhard [5] carried out analysis similar to that of Sparrow and Gregg but for a variety of surface shapes that admit similar solutions. Vertical surface was one of the special cases of the general formulation. Here we describe their approach. Figure 9.2 shows the coordinate system that was used in the analysis, whereas Figure 9.3 shows the surface shapes that were considered. As we will see later, the gravitational acceleration along these surfaces varies as

$$\frac{g(x)}{g_e} = \left(\frac{x}{L} \right)^n$$

where x is the distance along the surface, L is the length of the plate in the condensate flow direction and exponent $n \geq 0$.

In carrying out the analysis, all of the assumptions listed earlier in this section were made except the assumptions of neglect of inertia in the momentum equation and convection term in the energy equation were not made.

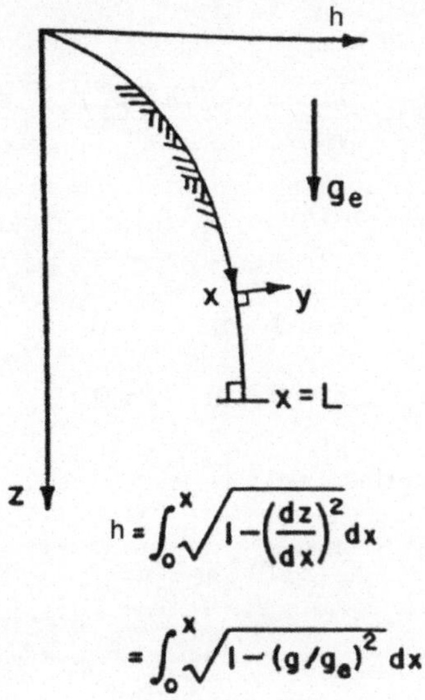

FIGURE 9.2 Coordinates for surfaces experiencing condensation [5].

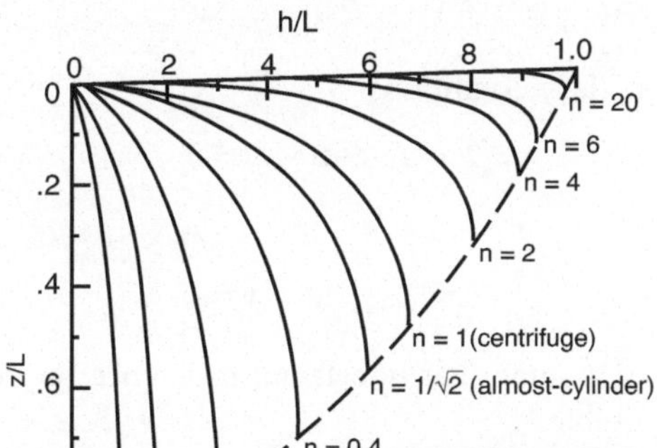

n	at x/L = 1: x/L	h/L
0.05	0.9524	0.0954
0.1	0.9091	0.1690
0.2	0.8333	0.2864
0.4	0.7143	0.4443
0.707	0.5858	0.5858
1	0.5000	0.6667
2	0.3333	0.8000
4	0.2000	0.8889
6	0.1429	0.9231
20	0.0476	0.9755

FIGURE 9.3 The family of body shapes for two-dimensional stagnation point surfaces [5].

9.2 COMPLETE SOLUTIONS FOR LAMINAR FILM CONDENSATION INCLUDING INERTIA AND CONVECTION

The continuity equation and the boundary layer type equation of motion for the film are:

$$\frac{\partial u}{\partial x} + \frac{\partial v}{\partial y} = 0 \tag{9.5}$$

and

$$u\frac{\partial u}{\partial x}+\upsilon\frac{\partial u}{\partial y}=\frac{g(x)(\rho_l-\rho_\upsilon)}{\rho_l}+\nu_l\frac{\partial^2 u}{\partial y^2} \tag{9.6}$$

The boundary conditions for eq. (9.3) are no slip at the wall and no flow into the wall, i.e.,

$$u=0 \quad \text{at} \quad y=0$$

$$\upsilon=0 \quad \text{at} \quad y=0 \tag{9.7}$$

And slip condition at the interface, i.e.,

$$\frac{\partial u}{\partial y}=0 \quad \text{at} \quad y=\delta$$

The energy equation for the film is

$$u\frac{\partial T}{\partial x}+\upsilon\frac{\partial T}{\partial y}=\alpha_l\frac{\partial^2 T}{\partial y^2} \tag{9.8}$$

with the following boundary conditions:

$$T=T_w \quad \text{at} \quad y=0$$

and

$$T=T_{\text{sat}} \quad \text{at} \quad y=\delta \tag{9.9}$$

In writing eq. (9.9), we assume that interfacial resistance for condensation at film–vapor interface is negligible.

To satisfy the continuity equation, we define a stream function ψ such that

$$u=\frac{\partial\psi}{\partial y} \quad \text{and} \quad \upsilon=-\frac{\partial\psi}{\partial x} \tag{9.10}$$

We next seek a similarity variable, η, and a similarity function, $F(\eta)$ [6] in terms of an unknown dimensionless stretching function, *f(x)* and a variable component, *g(x)*, of gravitational acceleration, g_e:

$$\eta = yl/f(x) = y/\delta \quad \text{and} \quad F(\eta)=\psi g_e/\left(\alpha_l g(x) f^3(x)\right) \tag{9.11}$$

where l has dimension of m^{-1} and is defined as

$$l=\left[\frac{g_e(\rho_l-\rho_\upsilon)}{\mu_l\alpha_l}\right]^{1/3} \tag{9.12}$$

The above formulation is similar to that employed in Falkner–Skan solutions [6]. Under these variables, the velocities and their derivatives become

$$u = \frac{\partial \psi}{\partial y} = F'\alpha_l g f^2 l / g_e$$

$$\upsilon = -\frac{\partial \psi}{\partial x} = F'\alpha_l g f^2 f'\eta / g_e - F\alpha_l \frac{d}{dx}\left(gf^3\right) / g_e$$

$$\frac{\partial u}{\partial y} = \frac{\partial^2 \psi}{\partial y^2} = F''\alpha_l g f l^2 / g_e$$

$$\frac{\partial u}{\partial x} = \frac{\partial^2 \psi}{\partial x \partial y} = -F''\alpha_l g f f' l\eta / g_e - \frac{F'\alpha_l g f f' l}{g_e} + F'\alpha_l l \frac{d}{dx}\left(gf^3\right) / f g_e$$

$$\frac{\partial^2 u}{\partial y^2} = \frac{\partial^3 \psi}{\partial y^3} = F'''\alpha_l g l^3 / g_e$$

In the above equations, the primes denote the derivatives of the functions with respect to the single independent variable. Writing eq. (9.6) in terms of new variables and dividing throughout by $\alpha_l \nu_l g l^3 / g_e$, we get

$$\frac{(F')^2}{\Pr_l}\left[f^2 \frac{d}{dx}\left(f^2 g\right) / \lg_e\right] - \frac{FF''}{\Pr_l}\left[f \frac{d}{dx}\left(f^3 g\right) / \lg_e\right] = 1 + F''' \tag{9.13}$$

A successful similar solution requires that if the functions contained in the square parentheses be independent of the position coordinate x, i.e., constant such that

$$f \frac{d}{dx}\left(f^3 g\right) / \left(l g_e\right) \equiv A \tag{9.14}$$

and

$$f^2 \frac{d}{dx}\left(f^2 g\right) / \left(l g_e\right) \equiv B \tag{9.15}$$

We shall soon see the constraint imposed by this requirement on the variability of $g(x)$ with x. Since the constants A and B are interrelated through $f(x)$ and $g(x)$, one of these constants can be set arbitrarily. We set constant $A = 3$ simply on the premise that our formulation should be the same in the limit of constant g as that of Sparrow and Gregg [4]. Thus, eq. (9.10) can be rewritten as

$$F''' + \frac{A}{\Pr_l} FF'' - \frac{B}{\Pr_l}(F')^2 + 1 = 0 \tag{9.16}$$

and the boundary conditions in terms of the new variables are

$$F = F' = 0 \quad \text{at} \quad \eta = 0$$

and

$$F'' = 0 \quad \text{at} \quad \eta = \eta_\delta \tag{9.17}$$

where η_δ is the value of η at $y = \delta$.

Similarly, the energy equation after introduction of a similarity function, θ, for temperature as

$$\theta(\eta) = \frac{T - T_w}{T_{\text{sat}} - T_w} \tag{9.18}$$

becomes

$$\theta'' + AF\theta' = 0 \tag{9.19}$$

With the following boundary conditions:

$$\theta = 0 \quad \text{at} \quad \eta = 0$$

and

$$\theta = 1 \quad \text{at} \quad \eta = \eta_\delta \tag{9.20}$$

Equations (9.16) and (9.19) are fully defined except that we do not know what value of η_δ to use in the boundary conditions. To relate η_δ to physical parameters of the problem, we make an energy balance at the vapor–liquid interface. The heat flux at the interface should be equal to the rate at which enthalpy is added during condensation at the interface. Such an energy balance yields

$$-k_l \left.\frac{\partial T}{\partial y}\right|_{y=\delta} = \rho_l h_{fg}\left[v|_{y=\delta} - u|_{y=\delta} \tan\phi\right] \tag{9.21}$$

where $\tan\phi = \dfrac{d\delta}{dx}$. In terms of the new variables, eq. (9.21) can be rewritten as

$$\begin{aligned}-k_l\theta'(\eta_\delta)\Delta Tl/f(x) = \rho_l h_{fg}\Big[& F'(\eta_\delta)\alpha_l g f^2 f'\eta_\delta/g_e - F(\eta_\delta)\alpha_l \frac{d}{dx}\left(fg^3\right)\Big/g_e \\ & -F'\alpha_l g f^2 l \frac{d\delta}{dx}\Big/g_e\Big]\end{aligned} \tag{9.22}$$

Now we know from eq. (9.11) that

$$\delta \sim f(x)/l \tag{9.23}$$

or

$$\frac{d\delta}{dx} \sim f'(x)\Big/l \tag{9.23a}$$

and

$$\eta_\delta \sim 1 \tag{9.24}$$

Substituting for $\frac{d\delta}{dx}$ and η_δ in eq. (9.22), we find

$$\frac{c_{pl}\Delta T}{h_{fg}}\theta'(\eta_\delta) = f\frac{d}{dx}(gf^3)\Big/(\mathrm{l}g_e)\cdot F(\eta_\delta)$$

or

$$\frac{F(\eta_\delta)}{\theta'(\eta_\delta)} = \frac{1}{A}\frac{c_{pl}\Delta T}{h_{fg}} \tag{9.25}$$

Equation (9.25) provides a relationship between η_δ and the dimensionless group $c_{pl}\Delta T/h_{fg}$. The procedure for numerical integration of eqs. (9.16) and (9.19) will be to assume a value of η_δ and to complete the integration while satisfying the boundary conditions. Once $F(\eta_\delta)$ and $\theta'(\eta_\delta)$ are known, the chosen value can be related to $c_{pl}\Delta T/h_{fg}$.

The magnitude of heat flux at the wall can be simply written as

$$q_w = \left|-k_l\frac{dT}{dy}\right|_{y=0} = k_l\Delta T\theta'(0)\cdot\frac{l}{f(x)} \tag{9.26}$$

or the Nusselt number based on local heat transfer coefficient is

$$\mathrm{Nu} \equiv \frac{hx}{k_l} = \frac{q_w x}{\Delta T k_l} = \frac{\theta'(0)\cdot xl}{f(x)} \tag{9.27}$$

Before going on to the discussion of results, let us determine the surface shapes, which will admit similar solutions. We know from eq. (9.14) that

$$3 = \frac{1}{\mathrm{l}g_e}\left[3f^3f'g + f^4g'\right] \tag{9.14a}$$

and

$$B = \frac{1}{\mathrm{l}g_e}\left[2f^3f'g + f^4g'\right] \tag{9.15a}$$

Solving eqs. (9.14a) and (9.15a) simultaneously, we find that

$$\frac{df^4}{dx} = -\frac{4\mathrm{l}g_e(B-3)}{g} \tag{9.28}$$

and

$$gg'' = \left[\frac{4(B-3)}{3(B-2)}\right](g')^2 \tag{9.29}$$

Integration of eq. (9.29) gives

$$g = (C_1 x + C_2)^n \tag{9.30}$$

where

$$n = \frac{3B-6}{6-B} \tag{9.31}$$

or

$$B = \frac{6(n+1)}{n+3} \tag{9.32}$$

In eq. (9.30), C_1 and C_2 are constants of integration. Since there is no flow at the leading edge, $x = 0$, the stream function ψ or gf^3 should be zero there. This is only possible if $n \geq 0$, or C_2 is zero. If n is equal to zero, g is constant and say equal to g_e. Thus, the appropriate form for g for which similar solutions are possible is

$$g(x) = C_3 x^n \tag{9.33}$$

It must be pointed out that negative values of n are not physically possible. Similarly, integrating eq. (9.28) and after substitution for B from eq. (9.32), we get

$$f(x) = \left[\frac{12 g_e l}{(n+3)} \frac{x}{g(x)}\right]^{1/4} \tag{9.34}$$

Again noting that $gf^3 = 0$ at $x = 0$, the constant of integration must vanish.

Let us next consider the curved surfaces which at earth normal gravity admit similar solutions. Referring to Figure 9.2, the component of gravitational acceleration acting along the surface can be written as

$$\frac{g(x)}{g_e} = \frac{dz}{dx} \tag{9.35}$$

For such a surface with a forward stagnation point, the similar solutions are possible if

$$\frac{g(x)}{g_e} = \left(\frac{x}{L}\right)^n \tag{9.36}$$

where L is value of x at which the surface becomes vertical. Integrating, eq. (9.35) with $z = 0$ at $x = 0$, we obtain

$$\frac{z}{L} = \frac{\left(\frac{x}{L}\right)^{n+1}}{n+1} \tag{9.37}$$

From geometrical consideration, we can also write

$$\frac{dh}{dx} = \sqrt{1 - \left(\frac{dz}{dx}\right)^2} \tag{9.38}$$

Integration of eq. (9.38) after substitution for dz/dx yields

$$\frac{h}{L} = \int_0^{x/L} \sqrt{1 - \left(\frac{x}{L}\right)^{2n}}\, d\left(\frac{x}{L}\right) \tag{9.39}$$

Figure 9.3 displays the family of earth normal gravity surfaces corresponding to the shapes given by eqs. (9.37) and (9.39). Next we consider the heat transfer results for a few such surfaces.

9.2.1 The Vertical Plate

This is the surface that was treated in detail by Sparrow and Gregg [4]. We see from Figure 9.3 that for this surface, $n = 0$ or $B = 2$ and $c_3 = g_e$ in eq. (9.33). Thus, for a vertical plate, our governing equations reduce to exactly the same form as Sparrow and Gregg's [4] original equations. The function $f(x)$ for this case is obtained from eq. (9.34) as

$$f(x) = (4lx)^{1/4} \tag{9.34a}$$

Substituting (9.34a) in eq. (9.27), we get

$$\mathrm{Nu} = \left[\frac{\rho_l g_e (\rho_l - \rho_v) h_{fg} x^3}{4\mu_l k_l \Delta T}\right]^{1/4} \theta'(0) \left[\frac{c_{pl}\Delta T}{h_{fg}}\right]^{1/4}$$

or

$$\frac{\mathrm{Nu}}{\left[\frac{\rho_l g_e (\rho_l - \rho_v) h_{fg} x^3}{4\mu_l k_l \Delta T}\right]^{1/4}} = \theta'(0) \left[\frac{c_{pl}\Delta T}{h_{fg}}\right]^{1/4} \tag{9.27a}$$

The left-hand side of eq. (9.27a) is plotted in Figure 9.4 as a function of $c_{pl}\Delta T / h_{fg}$ or the modified Jakob number for several Prandtl numbers. It is noted that Nusselt's

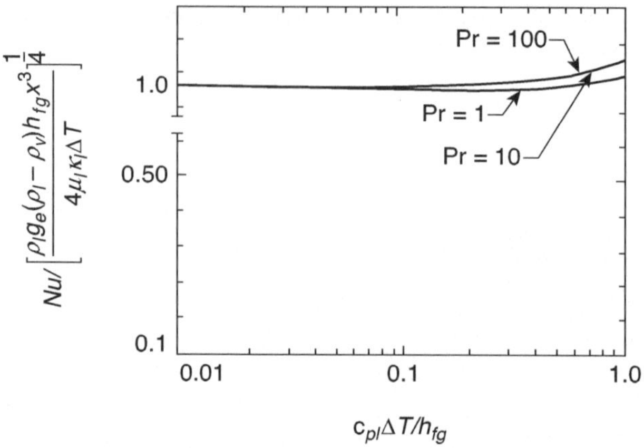

FIGURE 9.4 Heat transfer from a vertical plate [4].

solution is very close to the exact solution as long as $c_{pl}\Delta T/h_{fg} < 10^{-1}$, and the Prandtl number of the condensate is close to unity. For high Prandtl numbers, Rohsenow's correction to Nusselt's solution gives results that match the exact solution even for values of $c_{pl}\Delta T/h_{fg}$ much greater than 10^{-1}. Thus, we find that for high Pr laminar film condensation, Nusselt's solution gives quite accurate results after the latent heat of vaporization has been corrected according to the determination made by Rohsenow [7]. Another point that can be made here is that $\theta'(0)$ varies as $\left[c_{pl}\Delta T/h_{fg}\right]^{-1/4}$ provided $c_{pl}\Delta T/h_{fg}$ values are small. For low Prandtl number fluids, the results start to deviate from Nusselt's solution at relatively low values of $c_{pl}\Delta T/h_{fg}$.

9.2.2 The Synchronously Rotating Plate

When $n=1$, or $B=3$, the gravity varies linearly with the distance from the leading edge, as occurs during condensation on a plate rotating synchronously with the surrounding vapor (i.e., the plate and vapor rotate at the same speed as they would in a closed centrifuge). The component of gravitational acceleration will also vary linearly on a curved surface in an earth normal gravity such as that shown in Figure 9.3, but the rotating plate is of greater practical interest. For a rotating plate with axis located on the line $z=0$, the $g(x)$ and the g_e can be replaced by $\omega^2 x$ and $\omega^2 L$ respectively where ω is the angular speed of the plate. In this case, any static component of external gravity field is considered to be much smaller than $\omega^2 x$, which is not true at small values of x. For this case, the function f from eq. (9.34) becomes

$$f(x) = (3lL)^{1/4} \tag{9.34b}$$

Thus, we see that function is independent of x. Substituting f in eq. (9.23), we find that the thickness of the condensate film on the rotating plate will be constant. Nusselt number from eq. (9.27a) is obtained as

$$\text{Nu} = \theta'(0)\left[\frac{\rho_l\omega^2(\rho_l - \rho_v)h_{fg}x^4}{3\mu_l k_l \Delta T}\right]^{1/4}\left[\frac{c_{pl}\Delta T}{h_{fg}}\right]^{1/4}$$

or

$$\frac{\text{Nu}}{\left[\frac{\rho_l\omega^2(\rho_l - \rho_v)h_{fg}x^4}{3\mu_l k_l \Delta T}\right]^{1/4}} = \theta'(0)\left[\frac{c_{pl}\Delta T}{h_{fg}}\right]^{1/4} \tag{9.27b}$$

We would guess that for low values of $c_{pl}\Delta T/h_{fg}$, the right-hand side of eq. (9.27b) should be equal to unity. This indeed is the case when we carry out the integration. At large values of $c_{pl}\Delta T/h_{fg}$, the exact solutions start to deviate from Nusselt's solution in the same manner as those for a vertical plate and plotted in Figure 9.4. However, Nusselt's solution with Rhosonow's correction does quite well even for high values of $c_{pl}\Delta T/h_{fg}$ provided *Pr* is much greater than unity. The temperature distribution in the condensate film on a rotating plate is plotted in Figure 9.5 for Pr = 1 and for several values of $c_{pl}\Delta T/h_{fg}$. In accordance with the expectation, the profile is found to be nearly linear at low values of $c_{pl}\Delta T/h_{fg}$. The nonlinearity in the profile increases as $c_{pl}\Delta T/h_{fg}$ becomes large. As must be evident by now, larger values of $c_{pl}\Delta T/h_{fg}$ correspond to larger values of film thickness or η_δ.

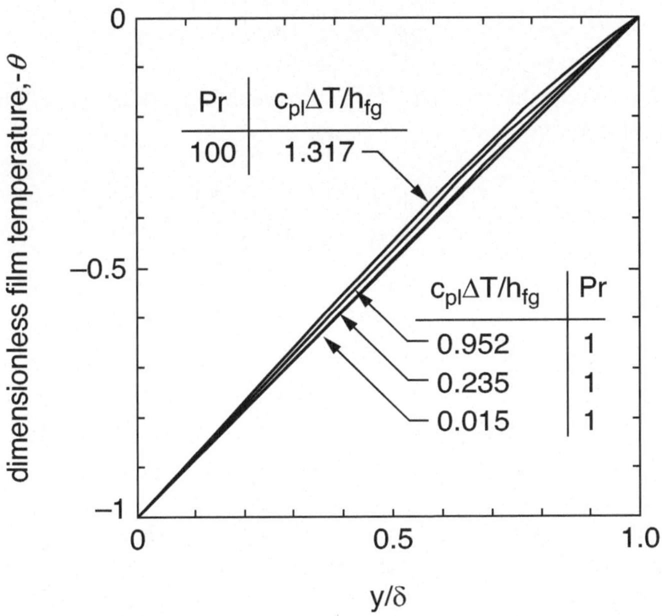

FIGURE 9.5 Temperature distribution in the condensate layer on a rotating plate [5].

9.2.3 Limitations on the Similarity Solution for the Curved Surfaces Shown in Figure 9.2

As n becomes large, the surfaces tend to flatten out and gravitational acceleration acts normal to most of the surface. The limiting case of $n \to \infty$ or condensation on an upward facing horizontal surface was treated by Leppert and Nimmo [8]. If we include the hydrostatic head term in the momentum equation, we can write for the curved surfaces

$$u\frac{\partial u}{\partial x}+\upsilon\frac{\partial u}{\partial y}=\frac{g(x)(\rho_l-\rho_\upsilon)}{\rho_l}-g_e\frac{d\delta}{dx}\frac{dh}{dx}+u_l\frac{\partial^2 u}{\partial y^2} \tag{9.40}$$

In our earlier formulation, we have ignored the second term on the right-hand side of eq. (9.40). This is realistic only if

$$g(x) >> \frac{g_e\rho_l}{(\rho_l-\rho_\upsilon)}\left(-\frac{d\delta}{dx}\right)\sqrt{1-\left(\frac{x}{L}\right)^{2n}} \tag{9.41}$$

The maximum value of the term under the square root sign can be unity. At low pressures, the value of $\dfrac{\rho_l}{\rho_l-\rho_\upsilon}$ will also be close to unity. If we ignore these terms, we can write

$$\frac{g(x)}{g_e} >> \left(-\frac{d\delta}{dx}\right) \tag{9.42}$$

For eq. (9.23) we know that $\delta \sim f/l$. If we substitute for f from eq. (9.34) and for $g(x)/g_e$ from eq. (9.36), we get

$$\left(\frac{x}{L}\right)^n >> -\frac{1}{l}\frac{d}{dx}\left[\frac{12xl}{(n+3)}\left(\frac{L}{x}\right)^n\right]^{1/4}$$

$$\left(\frac{x}{L}\right)^n >> \left[\frac{3(n-1)^4}{64(n+3)(Ll)^3}\right]^{1/4}\left(\frac{x}{L}\right)^{-\frac{(n+3)}{4}} \tag{9.43}$$

$$\left(\frac{x}{L}\right) >> \left[\frac{3(n-1)^4}{64(n+3)(Ll)^3}\right]^{\frac{1}{5n+3}}$$

It is evident from eq. (9.43) that for $n > 1$, the solution without the hydrostatic head term will not be valid over most of the surface. Thus, the usefulness of the similarity solution for surfaces with $n > 1$ is limited. There is an error in the equation given in reference [5]. The correct criterion should be the one given by eq. (9.43).

Next we turn our attention to the reference temperature at which the physical properties should be evaluated. A numerical integration of the governing equations with variable properties has been carried out by several investigators [9,10]. From their work, Minkowycz and Sparrow [9] concluded that Nusselt's solution with Rohsenow's correction will give very accurate results if the properties are evaluated at a reference temperature defined as

$$T_{\text{Reference}} = T_w + 0.31\Delta T \tag{9.44}$$

Since in many practical situations of interest, ΔT is generally much smaller than the wall temperature, little error is made if the properties are evaluated at the wall temperature. Also, as we will discuss next, at higher values of ΔT, which correspond to higher condensation rates, ripples appear at the film surface. The effect of the ripples on heat transfer is much more pronounced than of the variation in properties.

9.3 HEAT TRANSFER DURING WAVEY AND TURBULENT FILM CONDENSATION

In carrying out the laminar film analysis in the preceding section, we had made a number of assumptions including that the liquid–vapor interface of the film is smooth. However, ripples start to appear on the interface as they film thickens and condensate flow rate through the film increases. It has been suggested [11] that ripples start to form when Reynold's number for the condensate flow in the film reaches about 30. With flow in the film being laminar, the ripples dominate up to a Reynold's number of 1800 after which the film becomes turbulent. The rate of heat transfer with ripples and for turbulent film is higher than that based on laminar film analysis. Figure 9.6

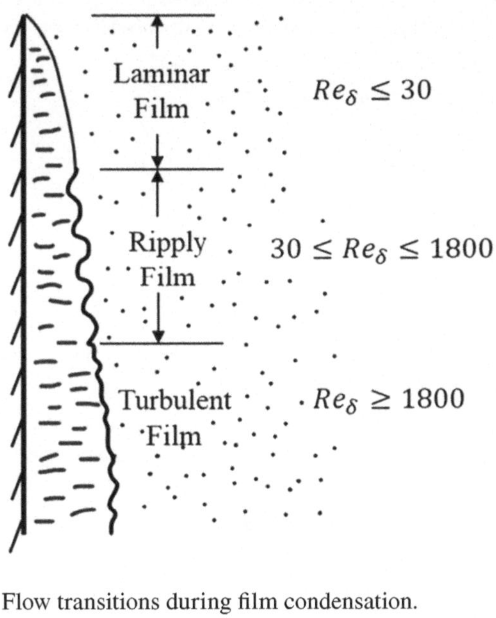

FIGURE 9.6 Flow transitions during film condensation.

shows these transitions during condensation on a vertical plate. Condensate flow Reynold's number based on film thickness is defined as

$$\mathrm{Re}_\delta = \frac{\overline{u_l} D_h}{\nu_l} \tag{9.45}$$

where $\overline{u_l}$ is the average velocity of the condensate in the film, and D_h Is the hydraulic diameter of the film. For a vertical plate with a width, w, normal to the flow direction, D_h is defined as

$$D_h = \frac{4 \cdot \text{flow cross} - \text{sectional area}}{\text{wetted perimeter}}$$

$$= \frac{4 \cdot \delta \cdot w}{w} = 4\delta \tag{9.46}$$

The average condensate velocity through the film is related to the mass flow rate of condensate as

$$\dot{m} = \rho_l \overline{u_l} \cdot \delta \cdot w \tag{9.47}$$

As such, Reynold's number in terms of the average condensate velocity and mass flow rate through the film is written as

$$\mathrm{Re}_\delta = \frac{\overline{u_l} 4\delta}{\nu_l} = \frac{4\dot{m}}{\mu_l w} \tag{9.48}$$

If Q is the total heat transfer rate associated with condensation on a vertical plate of height L and width w, and $\overline{h_l}$ is the average heat transfer coefficient, we can write

$$Q = \overline{h_l} \cdot wL(T_{\text{sat}} - T_w) \tag{9.49}$$

$$\dot{m} = \frac{Q}{h'_{fg}} = \frac{\overline{h_l} \cdot wL(T_{\text{sat}} - T_w)}{h'_{fg}} \tag{9.50}$$

The expression for $\overline{u_l}$ during laminar film condensation on a vertical wall is obtained as

$$\overline{u_l} = \frac{1}{\delta}\int_0^\delta u\, dy = \frac{g_e(\rho_l - \rho_v)\delta^2}{3\mu_l} \tag{9.51}$$

By using $\overline{u_l}$ from eq. (9.51) in (9.48), an expression for Re_δ is obtained as

$$\mathrm{Re}_\delta = \frac{4 g_e \rho_l (\rho_l - \rho_v)\delta^3}{3\mu_l^2} \tag{9.52}$$

Employing $\overline{h_L}$ from eq. (7.36a) and $\delta(L)$ from eq. (7.34), when viscosity and thermal conductivity of vapor are replaced by those of liquid and $\rho_v \ll \rho_l$ (integral solution), we can relate for laminar film, the Nusselt number to Reynold's number for the film as

$$\frac{\overline{h_L}\left(\frac{v_l^2}{g}\right)^{1/3}}{k_l} = 1.47\,\mathrm{Re}_\delta^{-1/3} \tag{9.53}$$

Here characteristic length in Nusselt number is given by $\left(\frac{v_l^2}{g}\right)^{1/3}$. Thus, we see from eq. (9.53) that Nusselt number for laminar film with smooth interface, i.e., $\mathrm{Re}_\delta \leq 30$, decreases with increase in Re_δ.

For laminar wavy film, $30 \leq \mathrm{Re}_\delta \leq 1{,}800$, Kutateladze [11] recommends.

$$\frac{\overline{h_L}\left(\frac{v_l^2}{g}\right)^{1/3}}{k_l} = \frac{\mathrm{Re}_\delta}{1.08\,\mathrm{Re}_\delta^{1.22} - 5.2} \qquad 30 \leq \mathrm{Re}_\delta \leq 1{,}800 \tag{9.54}$$

whereas Labuutsov [12] recommends a correlation for turbulent film as

$$\frac{\overline{h_L}\left(\frac{v_l^2}{g}\right)^{1/3}}{k_l} = \frac{\mathrm{Re}_\delta}{8{,}750 + 58\,\mathrm{Pr}^{-0.5}\left(\mathrm{Re}_\delta^{\frac{3}{4}} - 253\right)} \qquad \mathrm{Re}_\delta \geq 1{,}800 \tag{9.55}$$

It should be noted that in using eq. (9.54) or (9.55), one should first replace $\overline{h_L}$ by Reynold's number using eqs. (9.50) and (9.48) and solve for Reynold's number iteratively. Once Reynold's number is established, use applicable equation to determine $\overline{h_L}$or the average heat transfer coefficient over the length of the plate. Also one should note that according to Lobuntsov's correlation for turbulent film [12,13], Prandtl number is an addition parameter as for single-phase convection.

9.4 DROPWISE CONDENSATION

As described in the previous sections, condensation of vapor on a surface maintained at a temperature below the vapor saturation temperature leads to formation of thin film of condensate on the surface. Continuous film forms on the surface when it is fully wettable by the liquid. If the surface is non-wetting as will be the case for condensation of steam if a thin film of oil existed on the surface, discretely located microdroplets of the condensate will form on the surface. Figure 9.7 shows a picture of dropwise condensation on a vertical surface [14]. The heat transfer coefficients associated with dropwise condensation are much higher than those obtained during film condensation. As such, often surface modifications are made to promote dropwise condensation such as placement of a very thin coating of Teflon or another

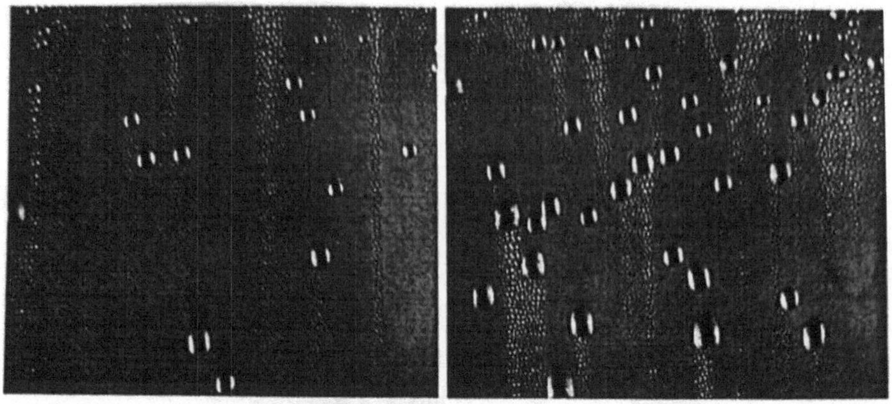

FIGURE 9.7 Dropwise condensation of steam at near-atmospheric pressure on a vertical plane copper surface (5 × 5 cm) promoted with dioctadecyl disulfide (Rose [14]).

material that is not wetted by the condensate. Unfortunately, these surfaces become contaminated with time and dropwise condensation reverts to film condensation. These days a number of researchers are making effort to develop surfaces that do not lose their non-wetting nature with time.

Most of condensation occurs on submicron-size drops. These drops grow with time as condensation proceeds and as a result of merger with neighboring drops. After a drop attains a size such that force due to gravity exceeds that due to surface tension, it rolls down the surface. As it rolls over the surface, any drops in its path merge with the rolling drop. In the wake of the rolling drop, the surface is cleared of any drops and is conducive to formation of new set of drops. Thus, the process is self-sustaining as long as the surface remains non-wetting.

A number of experimental studies of dropwise condensation have been reported in the literature. The data of various investigators show a variability of almost an order of magnitude. There are several reasons for it: (i) accuracy in measurement of small temperature differences between the vapor and the cold surface, (ii) specification of surface condition during condensation, (iii) presence of noncondensables in the vapor.

Figure 9.8 from Rose et al. [15] shows dropwise condensation heat flux data as a function of temperature difference between steam and copper surfaces. The data were obtained at 1 atm pressure. The data show a slight nonlinear dependence of heat flux on temperature difference. The nonlinearity is probably a result of dependence of site density of drops on the temperature difference. The density of micro drops can be as high as $10^7 - 10^9/\text{mm}^2$ [14]. It is noted that for a temperature difference of 4 K, a heat flux of 1 MW/m^2 is obtained. Superheat required in pool nucleate boiling of water at 1 atm pressure to achieve this magnitude of heat flux will be more than five times. The solid line in Figure 9.8 is prediction from the theoretical model of LeFevre and Rose [16], whereas the dotted line is based on the correlation proposed by Rose [14]:

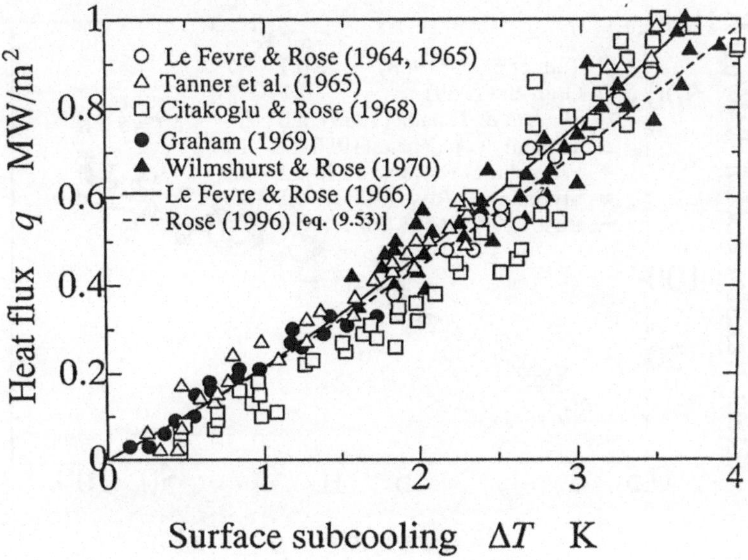

FIGURE 9.8 Dropwise condensation of steam on copper surfaces at atmospheric pressure [15].

$$\bar{h} = T_c^{0.8}(5 + 0.3\Delta T)\ \text{kW/m}^2\text{K} \tag{9.56}$$

In eq. (9.56), T_c is the temperature of saturated vapor in °C, and ΔT is the temperature difference between the vapor and the wall in °C.

Several of the experiments reported in the literature on dropwise condensation have been conducted at steam pressures below 1 atm. Figure 9.9 from Rose et al. [15] shows dependence of heat transfer coefficient on steam pressure varying from 0.01 to 1 atm. The data are well correlated with eq. (9.56). The reduction in heat transfer coefficient with decrease in pressure is reflective of the contribution of interfacial and curvature resistances, which we will discuss soon. The reported data show that thermal properties of the substrate and drop size at departure can also influence the observed heat transfer rates during dropwise condensation. The heat transfer rate under the micro drops is very high in comparison to that under the large drops or on the space between drops. This creates additional thermal resistance known as the constriction resistance. For detailed discussion of constriction resistance, one is referred to the work of Mikic [19]. Tanasawa [20] has shown that larger the size of the drops departing the surface, lower is the rate of heat transfer. This is due to the fact that larger drops sit longer on the cold surface and occupy larger fraction of the surface area. Because of the lower heat transfer rate associated with larger drops, heat transfer rate is observed to decrease with the size of the departing drops.

A number of attempts to model dropwise condensation have been made in the past. The prediction of dropwise condensation heat transfer is complex as was the case for nucleate boiling. It relies on a number of empirical or difficult to control parameters such as number density and size distribution of drops on the surface at

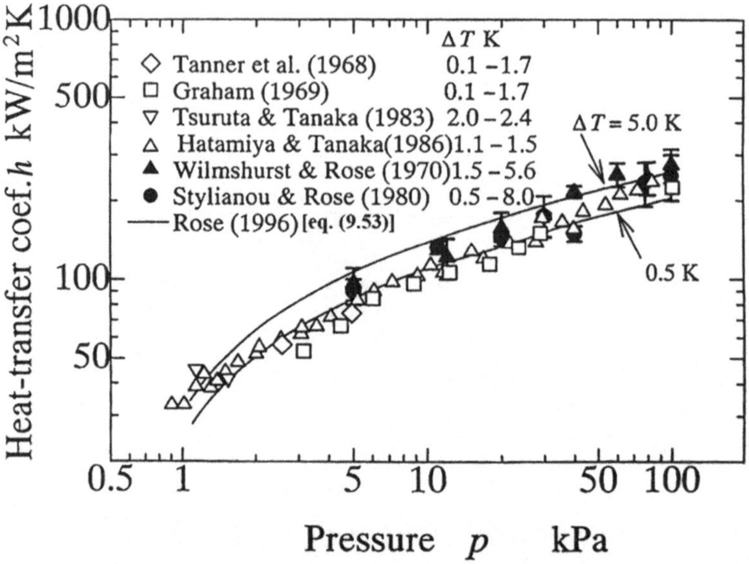

FIGURE 9.9 Dependence of heat transfer coefficient on pressure for steam. Note that the "error bars" indicated for the data of Wilmshurst & Rose [17] and Stylianou & Rose [18] are due to dependence of h on ΔT [15].

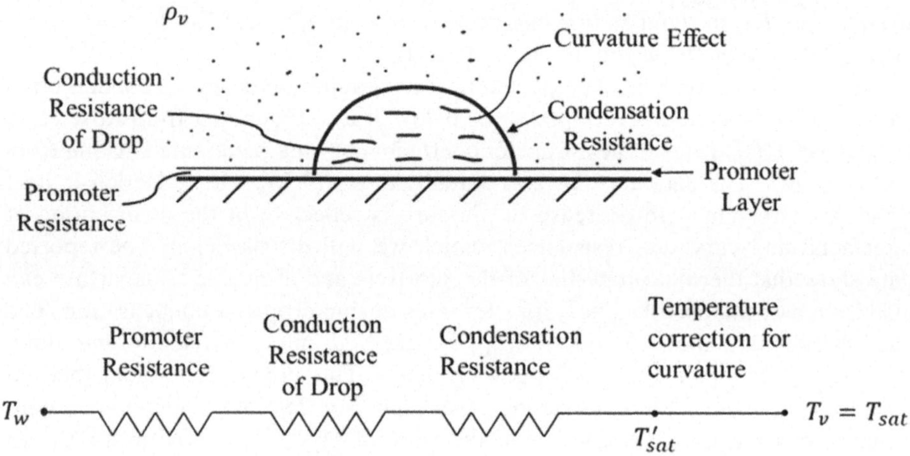

FIGURE 9.10 Various thermal resistances during dropwise condensation.

a given instant. The mechanisms that contribute to the heat transfer through a drop are: (i) promoter thermal resistance; (ii) conduction resistance through drop; and (iii) interfacial condensation resistance. In addition, a correction to the available temperature difference needs to be made to account pressure difference between the drop and the ambient. Figure 9.10 shows a schematic of these mechanisms. In the following, we quantitatively describe these mechanisms.

9.4.1 Promoter Thermal Resistance

Although the promoter layer on the condensing surface is very thin, the resistance imposed by the promoter can be significant as the temperature between the vapor and the cold surface is small. If δ_p Is the thickness of the promoter layer and q_b is the heat flux at the base of the drop and k_p is the conductivity of the promoter material, the temperature drop across the promoter layer is

$$\Delta T_p = q_b \frac{\delta_p}{k_p} \tag{9.57}$$

9.4.2 Conduction Resistance through Drop

Since the drop sits on the cold surface and the temperature of the drop surface is different, a singularity exists in temperatures at the contact line. Also, the surface area of the drop where condensation occurs is different than the base area, a correction factor must be employed to relate the heat flux, q_b, based on the base area of the drop to the surface area. As such, the temperature difference across the drop (i.e., between the curved surface and base of the drop) as a result of conduction resistance can be written

$$\Delta T_c = K_1 \frac{r_b q_b}{k_l} \tag{9.58}$$

where K_1 is a constant that accounts for area change, r_b is the base radius, q_b is the heat flux at the base of the drop, and κ_l is thermal conductivity of the condensate. Constant K_1 is of the order of unity. If a hemispherical drop is replaced by a disc of the same radius as the base of the drop and having the same volume as the drop, K_1, will be equal to 2/3.

9.4.3 Temperature Drop Due to Interfacial Resistance to Mass Transfer by Condensation

As we discussed in Chapter 5, there is evaporation resistance, although very small, at a liquid–vapor interface of a film on a heated surface. Similarly, for condensation at the interface of a drop, there is a condensation resistance. Generally, this resistance is very small. However, since the total resistance across a micro drop is small, contribution of condensation resistance can be a significant fraction of the overall resistance. Similar to the expression for the evaporative resistance, temperature drop due to condensation resistance can be written in terms of the heat flux at the base of the drop as

$$\Delta T_{\text{con}} = \frac{K_2 q_b v_{fg} T_{\text{sat}} \left(2\pi \text{R} \text{T}_{\text{sat}}\right)^{1/2}}{h_{fg}{}^2} \tag{9.59}$$

where constant K_2 accounts for the interface area being larger than the base area. For a hemispherical $(\varphi = 90°)$ drop K_2 will be 0.5, whereas for other contact angle, it will be

$$K_2 = \frac{1}{2}(1+\cos\varphi) \tag{9.60}$$

Accounting for the three resistances and correction for curvature effect, the heat flux, q_b, based on the base area of the drop can be written in terms of the temperature difference between the vapor assumed to be at the saturation temperature at the ambient pressure and the wall, ΔT, as:

$$q_b = \frac{\Delta T - \dfrac{2\sigma T_{\text{sat}} v_{fg} \sin\varphi}{r_b h_{fg}}}{\dfrac{\delta}{k_p} + \dfrac{K_1 \gamma_b}{k_l} + \dfrac{K_2 v_{fg} T_{\text{sat}} (\text{RT}_{\text{sat}})^{1/2}}{h_{fg}^{\,2}}} \tag{9.61}$$

A detailed discussion of the above eq. (9.61) has been given by LeFevre and Rose [15]. The second term in the numerator represents the effect of surface curvature which is described next.

9.4.4 Surface Curvature Effect

Because the drop surface is curved and concave inside, the pressure in the drop will be higher than the pressure in the vapor space. It is similar to pressure in the vapor bubble being higher than that in the ambient. The effect of pressure in the drop being higher than vapor is to effectively increase the saturation temperature of the condensate in the drop above the saturation temperature corresponding to the pressure of the vapor outside. The pressure difference between the drop and the vapor can be simply written as

$$\Delta P = \frac{2\sigma}{r} \tag{9.62}$$

where σ is the interfacial tension, and r is the radius of curvature of the drop surface. If φ is the contact angle r and r_b are related as

$$r = \frac{r_b}{\sin\varphi} \tag{9.63}$$

as $\varphi \to 0, r \to \infty$. This will be the case when liquid spreads on the surface in the form of a film. Using the Clausius–Clapeyron equation, pressure difference can be related to the difference of saturation temperatures ΔT_c, as

$$\Delta T_c = T_{\text{sat}} - T'_{\text{sat}} = \frac{2\sigma T_{\text{sat}} v_{fg}}{r h_{fg}} \tag{9.64}$$

where T_{sat} is the saturation temperature of vapor, v_{fg} is the specific volume change from liquid to vapor, and h_{fg} is the latent heat of vaporization.

Presence of non-condensables in the vapor can significantly reduce the rate of heat transfer during condensation. Analyses of the effect of non-condensables would require some knowledge of mass transfer. Because we have not developed a chapter on mass transfer in this text, that topic has not been included.

Example:

During dropwise condensation on a vertical plate exposed to saturated steam at 1 atm pressure. The surface temperature of the plate is maintained at 90°C. If a hemispherical drop of 5×10^{-6} m radius sits on the surface, calculate the heat flux at the base of the drop. Neglect promoter resistance.

Solution:

To begin with, we first calculate the temperature correction due to curvature of the drop

$$\Delta T_{\text{cur}} = \frac{2\sigma}{r}\cdot\frac{T_{\text{sat}} v_{fg}}{h_{fg}}$$

$$= \frac{2\times 58.9\times10^{-3}\times 373\times 1.67}{5\times10^{-6}\times 2.257\times10^{6}}$$

$$= 6.5°\text{C}$$

As such, available temperature difference is 3.5°C. Next we calculate the conduction resistance of the drop. Converting the drop to a rectangular drop with same base radius as the drop

$$\text{Conduction resistance} = \frac{2}{3}\frac{r}{k_l} = \frac{2}{3}\cdot\frac{5\times10^{-6}}{0.68}$$

$$= 4.9\times10^{-6}\,\text{m}^2\text{K/W}$$

$$\text{Condensation resistance at the vapor liquid interface} = \frac{0.5\times T_{\text{sat}} v_{fg}\,(2\pi R T_{\text{sat}})^{1/2}}{h_{fg}^{\,2}}$$

$$= \frac{0.5\times 373\times 1.67\times\left(2\pi\times\frac{8314}{18}\times 373\right)^{1/2}}{\left(2.257\times10^{6}\right)^{2}}$$

$$= 0.63\times10^{-7}\,\text{m}^2\text{K/W}$$

$$\text{Total resistance} = 4.9\times10^{-6} + 0.63\times10^{-7}$$

$$= 4.96\times10^{-6}\,\text{m}^2\text{K/W}$$

or

$$q_b = \frac{10-6.5}{4.96\times10^{-6}}$$

$$= 7.06\times10^{5}\,\text{W/m}^2$$

9.5 SUMMARY

Laminar and Turbulent film condensation and dropwise condensation are described in this chapter. For laminar film condensation, integral formulation as well as similarity solutions that include both inertia and convection terms in the momentum and energy equations respectively are discussed. For dropwise condensation, various resistances associated with a drop sitting on a cold surface are described. Correlations both for turbulent film condensation and dropwise condensation are included. We have not included the effect of non-condensables, though significant, on rate of heat transfer by condensation.

EXERCISES

1. Calculate heat transfer coefficient for laminar film condensation on a 20 cm tall vertical surface maintained at 90°C. Surface is exposed to pure steam at 100°C and 1 atm pressure.
2. A circular disc of diameter, *D*, spins about its axis at "N" R.P.M. The disc temperature is maintained below the saturation temperature at the imposed pressure such that $T_w < T_{sat}$. The disc rotates synchronously with pure vapor at its saturation temperature and laminar film condensation occurs on the disc. Obtain an expression for the average heat transfer coefficient.
3. Water at temperature of 20°C flows at a velocity of 3 m/s inside a circular aluminum tube 19.1 mm O.D. and 16.1 mm I.D. Saturated steam at a pressure 0.1atm condenses on the outside of the tube. Calculate the condensation rate of steam. Assume that there is no non-condensable present in the steam.
4. Film condensation of steam at saturation temperature and 1 atm pressure occurs on a tall vertical plate. The plate temperature is maintained at 90°C. Calculate the locations at which film will become ripply and turbulent. What will be the rate of heat transfer up to location at which film becomes turbulent?
5. Dropwise condensation of steam at 1 atm pressure occurs on a vertical wall maintained at 95°C. What is the magnitude of total resistance across a hemispherical drop, 10^{-6} m in diameter? What is the wall heat flux? For the latter, use the correlation of Rose.

NOMENCLATURE

A, B	Constants
c_p	Specific heat at constant pressure
F	Similarity function
$f(x)$	Function of x
k	Thermal conductivity
g	Gravitational acceleration
h	Heat transfer coefficient or Coordinate in the horizontal direction
h_{fg}	Latent heat of vaporization

L	Length of the surface
l	Characteristic length defined in eq. (9.9)
R	Radius
r	Radius of a drop
T	Temperature
u	Velocity in the x-direction
v	Velocity in the y-direction
Nu	Nusselt number
n	Constant
q	Heat flux
x	Coordinate along the surface
y	Coordinate normal to the surface
z	Coordinate in the vertical direction

SUPERSCRIPTS

$'$, $''$, $'''$	First, second, and third derivative respectively

SUBSCRIPTS

b	base
e	earth
s	at edge of the film
l	liquid
sat	saturation
w	wall
v	vapor

GREEK LETTERS

α	Thermal diffusivity
ΔT	Temperature difference
δ	Film thickness
η	Similarity variable
ψ	Stream function
μ	Molecular viscosity
ν	Kinematic viscosity
θ	Dimensionless temperature
φ	Contact angle

REFERENCES

1. Nusselt, W. (1916), "Die Oberflächenkondensation des Wasserdamfes." *Z. Ver. Deutsch. Ing.*, Vol. 60, pp. 541–546, 569–575.
2. Rohsenow, W.M. (1956), "Heat Transfer and Temperature Distribution in Laminar Film Condensation." *Trans. ASME*, Vol. 78, pp. 1645–1648.

3. Dhir, V.K. and Lienhard, J.H. (1971), "Laminar Film Condensation on Plane and Axi-Symmetric Bodies in Nonuniform Gravity." *J. Heat Transfer*, Vol. 93, pp. 97–100.
4. Sparrow, E.M. and Gregg, J.L. (1959), "A Boundary-Layer Treatment of Laminar-Film Condensation." *J. Heat Transfer*, Vol. 81, pp. 13–18.
5. Dhir, V.K. and Lienhard, J.H. (1973), "Similar Solutions for Film Condensation with Variable Gravity or Body Shape." *J. Heat Transfer*, Vol. 95, pp. 483–486.
6. Schlichting, H. (1968), Boundary Layer Theory, McGraw Hill Book Company, New York.
7. Rohsenow, W. M. (1965). "Heat Transfer and Temperature Distribution in Laminar-Film Condensation." *J. Heat Transfer*, Vol. 87, pp. 1645–1648.
8. Leppert, G. and Nimmo, B., 1968, "Laminar Film Condensation on Surfaces Normal to Body or Inertial Forces." *J. Heat Transfer*, Vol. 90, pp. 178–179.
9. Minkowycz, W.J. and Sparrow, E.M., 1966, "Condensation Heat Transfer in the Presence of Non-Condensibles, Interfacial Resistance, Superheating, Variable Properties and Diffusion." *Int. J. Heat Mass Transfer,* Vol. 9, pp. 1125–1144.
10. Denny, V.E. and Mills, A.F., 1969, "Non Similar Solutions for Laminar Film Condensation on a Vertical Surface." Int. *J. Heat Mass Transfer*, Vol. 12, pp. 965–979.
11. Kutateladze, S.S. (1963), *Fundamentals of Heat Transfer*, Academic Press, New York.
12. Lobuntsov, D.A. (1957), "Heat Transfer in Film Condensation of Pure Steam on Vertical Surfaces and Horizontal Tubes." *Transenergika*, Vol. 4, pp. 72–80.
13. Incropera, F.P., Dewitt, D.P. (1985), *Fundamentals of Heat and Mass Transfer*, John Wiley and Sons, New York.
14. Rose, J.W. (1964), "Dropwise Condensation of Steam on Vertical Plates", Ph.D. Thesis, London University.
15. Rose, J.W., Utaka. Y., and Tanasawa (1999), *Dropwise Condensation*, Hand Book of Phase Change, Boiling and Condensation, Taylor and Francis.
16. LeFevre, E.J. and Rose, J.W. (1966), "A Theory of Heat Transfer by Dropwise Condensation." *Proceedings of the 3rd International Heat Transfer Conference*, Vol. 2, pp. 362–375.
17. Wilmshurst R., Rose J. W. (1970). Dropwise condensation—further heat transfer measurements . In *Proceedings of 4th International Heat Transfer Conference*, (Versailles), Vol. 4, paper Cs 1.4.
18. Stylianou, S. A. and Rose, J. W.(1980). Dropwise Condensation on Surfaces Having Different Thermal Conductivities, *Journal of Heat Transfer*, vol. 102, pp. 477–482.
19. Mikic, B.B. (1969), "On Mechanism of Dropwise Condensation." *Int. J. Heat Mass Transfer*, Vol. 12, pp. 1311–1323.
20. Tanasawa, I. (1991), "Advances in Condensation Heat Transfer." *Adv. Heat Transfer*, Vol. 21, pp. 55–139.

10 Nucleate and Film Boiling Heat Fluxes in Liquid Metals

INTRODUCTION

Liquid metals are used as coolant in certain types of nuclear fission and fusion reactors. As such, boiling of liquid metals is of interest for both operational and safety concerns. Liquid metals have certain distinct properties in comparison to ordinary liquids. They have high boiling temperatures at very low vapor pressures. They also possess very desirable heat transfer properties: wide range of temperatures at which they remain in liquid state, high temperatures lead to increased power plant efficiency; they have high thermal conductivities and low specific heats, which lead to lower temperature gradients at the heat transfer surface; and they are resistant to nuclear radiation damage. On the other hand, they also possess some undesirable properties: they are chemically active and corrosive, which require the use of costly structural materials and handling techniques—liquid metals react with oxygen to form oxides that can plate out on structural walls—liquid metals pose many safety concerns and are expensive. In this chapter, we discuss distinct features of pool boiling of liquid metals utilizing the material developed earlier with respect to boiling of ordinary liquids.

10.1 INCIPIENT SUPERHEAT FOR BOILING NUCLEATION

As can be deduced from discussion in Chapter 4, theoretically incipient superheat for liquid metals should be less than that for ordinary liquids, but incipient boiling (IB) wall superheats are often much greater. This means that for liquid metals, there may be a large overshoot of the wall temperature at boiling inception. The overshoot is generally higher in pool boiling than in forced convection boiling. Figure 10.1 shows the overshoot behavior on a boiling curve. The main reasons for the overshoot are as follows:

1. Liquid metals usually wet the solid metal surfaces very well, particularly in the range of temperatures encountered in boiling (except Hg).
2. The alkali metals are very reactive chemically and act to reduce surface oxides of many alkali metals and alloys and in turn reduce the contact angle.
3. The solubility of inert gases in liquid metals increases with increase in temperature.
4. The boiling pressures of liquid metals are generally low with respect to their critical pressures, which means that boiling occurs in that part of the vapor pressure curve where the slope $\frac{dp}{dT}$ is low.

DOI: 10.1201/9781032668437-10

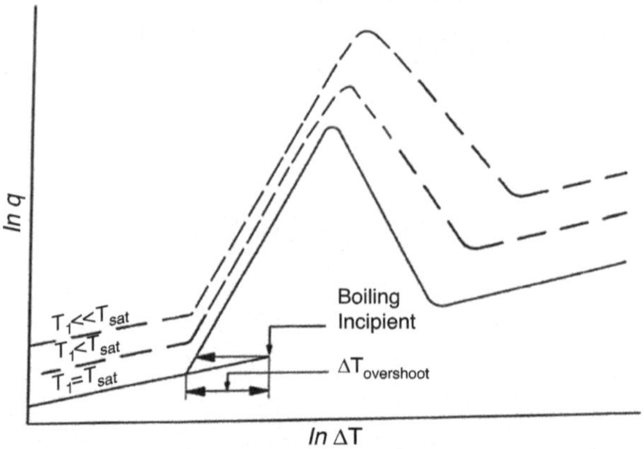

FIGURE 10.1 Overshoot at incipience in liquid metals.

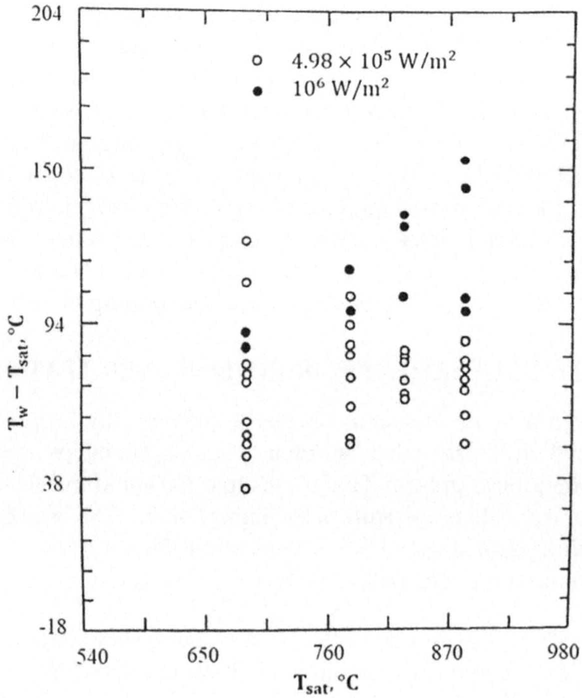

FIGURE 10.2 Incipient-boiling wall superheat data obtained by Pinchera et al. [1] in pool boiling of sodium.

After inception, heat transfer improves, and as a result, wall temperature decreases leading to an overshoot in temperature. The existing experimental data for incipient nucleate boiling show a considerable scatter. Figure 10.2 shows pool boiling data of Pinchera et al. [1] for sodium on an electrically heated rod enclosed by a round pipe. The heater surface had a roughness of 100–400 μm. The data were obtained by setting a heat flux value. As such, the surface temperature went through a rapid transient in temperature. The data show significant scatter. It is also noted that higher heat flux yields higher incipient superheat. Again it can be attributed to rapid increase in surface temperature preceding nucleation. There are a number of variables that could affect the data. The effect of many of these variables is difficult to quantize. However, some of the important variables are as follows:

1. *Natural Convection*: In pool boiling of liquid metals, large temperature fluctuations may exist prior to nucleation; their magnitude depends on such variables as heat flux, smoothness of the heating surface, and system geometry. The main effect of these fluctuations is to produce intermittently lower superheats. In forced flow situations, the temperature fluctuations are generally absent, hence less scatter in data.
2. *Undissolved Gas Bubbles*: In forced flow systems, there is a possibility of entrapping gas bubbles at gas–liquid interfaces in the surge vessel. These bubbles may not completely dissolve in the liquid metal as it heats in the boiler section, and these small bubbles may trigger premature nucleation.
3. *Surface Roughness*: The rougher and more porous the heating surface is, the more reproducible is the incipient superheat. If a sequence of IB superheats is carried out on a smooth surface, the same cavity may not initiate the nucleation each time. Smaller cavities produce larger superheat, which tends to reactivate the neighboring larger cavities by exposing them to vapor. The larger cavities then tend to quench more easily under the equilibrium pressure cycle, and so the variation in size of cavities produces irregular variation in incipient superheats.
4. *Boiling Pressure*: Low boiling pressure tends to give greater scatter to IB superheat partly because the superheats are large and partly because there is greater tendency for different cavities to nucleate from time to time.
5. *Pressure Temperature History*: The pressure and temperature to which system is exposed and the duration of exposure prior to the heat up transient preceding boiling inception can affect the magnitude of IB superheat.
6. *Method of Applying Power to Heater*: When the electrical power is applied to the heater in a series of small steps, the IB data tend to be lower and more precise than when it is applied all at once.
7. *Partial Pressure of Gas in Surface Cavities*: Concentration of inert gas in the cavities affects the incipient boiling behavior. The larger the concentration of gas, the lower will be the IB superheat.
8. *Oxygen Concentration in Liquid*: Concentration of oxygen in the liquid metal would determine the potential for removal of oxides of heater material.

10.2 NUCLEATE BOILING HEAT TRANSFER

In ordinary liquids, the effect of pressure comes through the thermophysical properties of the liquid and vapor. However, in liquid metals, the system pressure may also act as a separate parameter. For example, at low pressures and low heat fluxes, unstable nucleate boiling may occur. In this case, the mode of heat transfer may switch back and forth intermittently between nucleate boiling and natural convection. Another reason for boiling instability with liquid metals is that waiting periods are long and active cavities are frequently quenched by combination of local pressure fluctuations and relatively long exposure to liquid at near saturation pressure. An additional reason is that in the case of liquid metals, considerably more heat is transferred by natural convection and the need for bubble growth to account for a large share of thermal loads does not exist.

Kutateladze [2] developed a correlation for nucleate boiling heat transfer in liquid metals as

$$\mathrm{Nu} = C_1 (\mathrm{Re})^{m1} (\mathrm{Pr})^{m2} (K_P)^{m3} \tag{10.1}$$

where

$$\mathrm{Nu} = \frac{q}{(T_w - T_{\text{sat}})k_l}\left[\frac{\sigma}{g(\rho_l - \rho_v)}\right]^{1/2}$$

$$\mathrm{Re} = \frac{q\rho_l}{\rho_v h_{fg}\mu_l}\left[\frac{\sigma}{g(\rho_l - \rho_v)}\right]^{1/2}$$

$$K_p = p_l/\left[\sigma g(\rho_l - \rho_v)\right]^{1/2}$$

The correlation is similar to Rohsenow's correlation as described in Chapter 5. The last group comes from a consideration of the mass diffusion of material from the liquid interface to the vapor space within a bubble and explicitly represents the effect of pressure. On the basis of experimental data for a number of organic liquids, Kutateladze proposed that

$$C_1 = 7\times10^{-4};\; m_1 = 0.7;\; m_2 = 0.35-0.7;\quad m_3 = 0.7$$

Kutateladze's equation seems to work well for alkali metals boiling on commercially smooth stainless-steel surfaces and reduced pressures $\frac{p_l}{p_c}$ greater than 0.001. The value of constant C_1, which normally varies with liquid metal/heating surface system, must be estimated from experimental data.

Example:

Calculate using Kutateladze's correlation, nucleate boiling heat flux at a wall superheat of 5°C for pool boiling of sodium at saturation temperature corresponding to 1 atm pressure. Relevant properties of sodium are:

$$p_{cr} = 408 \text{ atm} \qquad c_{pl} = 1{,}217 \text{ J/kgK}$$

$$T_{cr} = 2{,}733 \text{ K} \qquad h_{fg} = 3.86\times10^6 \text{ J/kg}$$

$$\sigma = 113 \times 10^{-3}\ \text{N/m} \qquad \mu_l = 143 \times 10^{-6}\ \text{kg/ms}$$

$$\rho_l = 740\ \text{kg/m}^3 \qquad k_l = 58.7\ \text{W/mK}$$

$$\rho_v = 0.27\ \text{kg/m}^3$$

Solution:

Using eq. (10.1)

$$\frac{q\sqrt{\dfrac{\sigma}{g(\rho_l-\rho_v)}}}{\Delta T k_l} = 7\times10^{-4}\left(\frac{q\rho_l\sqrt{\dfrac{\sigma}{g(\rho_l-\rho_v)}}}{\rho_v h_{fg}\mu_l}\right)^{0.7}(\text{Pr}_l)^{0.35}\left(\frac{p_l}{\left[\sigma g(\rho_l-\rho_v)\right]^{1/2}}\right)^{0.7}$$

After substitution for the numerical values of the desired properties, we obtain

$$q = 53744\Delta T^{3.33}\ \text{W/m}^2$$

For a superheat of 5°C

$$q = 1.14\times10^7\ \text{W/m}^2$$

As we will see later, this value is significantly above the critical or maximum heat flux. Since superheat of 5°C is relatively small, the above calculation highlights the sensitivity of the calculated value to the empirical constants employed in correlating the data, especially the effect of liquid to vapor density ratio.

Subbotin et al. [3] have investigated nucleate boiling of sodium or cesium over pressures varying from 0.015 to 0.5 atm, and potassium over the range of 0.65–1.05 atm. Their data are plotted in Figure 10.3 and were correlated as

$$h = c\left(\frac{k_l h_{fg}\rho_l}{\sigma T_{\text{sat}}{}^2}\right)^{1/3} q^{2/3}\left(\frac{p_l}{p_{cr}}\right)^s \tag{10.2}$$

where $c = 8.0$ and $s = 0.45$ for $\frac{p_l}{p_{cr}} < 0.001$, and $c = 1.0$, $s = 0.15$ for $\frac{p_l}{p_{cr}} > 0.001$. In plotting the data in Figure 10.3, the critical pressures of 337, 165, and 116 atm for sodium, potassium, and cesium, respectively, have been used as proposed by Dwyer [4]. A transition in the data is seen to occur at $\frac{p_l}{p_{cr}} \simeq 10^{-3}$. The reason for this is not clear. It should be noted that according to eq. (10.2), $q \propto \Delta T^3$ as was the case for ordinary liquids.

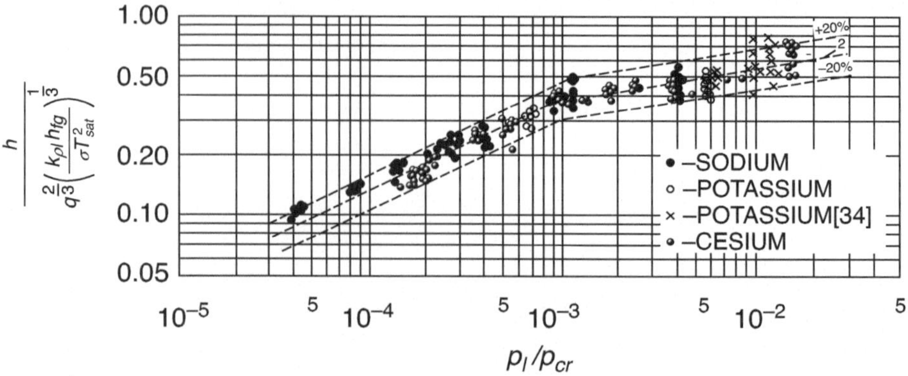

FIGURE 10.3 Correlation of experimental heat transfer results for stable nucleate pool boiling of alkali metals on a smooth, flat stainless-steel surface [4].

10.3 PEAK OR CRITICAL POOL BOILING HEAT FLUX

The observed peak pool boiling heat fluxes with liquid metals are generally much higher than predicted by Zuber's hydrodynamic model or by Kutateladze's semiempirical correlations. Various investigators have reasoned that the higher heat fluxes are due to relatively large contribution of conduction-convection to the total CHF in liquid metal boiling. Assuming that the considerable amount of energy from the heater was removed by conduction or convection, Noyes and Lurie [5] correlated their CHF results for sodium along with data of other investigators by an equation of the type

$$q_{cr} = q_{c-c} + q_{evap} \tag{10.3}$$

$$q_{cr} = 4\times10^5 + 0.16\sqrt{\rho_v}\,h_{fg}\left[\sigma g(\rho_l - \rho_v)\right]^{1/4} \tag{10.4}$$

Equation (10.4) correlated the sodium data within ±20%. Equation (10.4) is also equivalent to

$$q_{cr} = 4\times10^5 + 1.76\times10^5\,p_l^{\,0.42}$$

or

$$q_{cr} = 560{,}000\,p_l^{\,0.20} \tag{10.5}$$

where q_{cr} is in W/m^2 and p_l is in N/m^2.

Equation (10.5) correlates sodium data from pressures of $3.43\times10^3\,\frac{\text{N}}{\text{m}^2}$ to $3.43\times10^5\,\frac{\text{N}}{\text{m}^2}$.

Another empirical equation has been proposed by Caswell and Balzhiser [6]:

$$\frac{q_c c_{pl}\sigma}{h_{fg}{}^2 \rho_v k_l} = 1.18\times10^{-8}\left(\frac{\rho_l-\rho_v}{\rho_v}\right)^{0.71} \tag{10.6}$$

Equation (10.6) correlates data for potassium, sodium, and rubidium within ±6%. This is very good, but the authors caution that it is probably risky to apply the equation to systems and conditions appreciably different from those for which the data were obtained. For example, the equation does not show a dependence on *g*. Thus, it may not be applicable for systems in which body force is different than earth's normal gravity. Figure 10.4 shows their correlation of data.

Kirillov [7] proposed that critical heat flux was proportional to the rate of growth of a vapor bubble and concluded that

$$q_{cr} = 7.2\times10^5 k_l{}^{0.6}\left(\frac{p_l}{p_{cr}}\right)^{1/6} \tag{10.7}$$

In the reduced pressure range 10^{-4}–3×10^{-2}, he correlated the data for sodium, potassium, and cesium. In eq. (10.7), k_l is expressed in W/mK, and q_{cr} is W/m^2. Equation (10.7) again does not depend on *g*, and the same caution should be observed as for eq. (10.6) while using it.

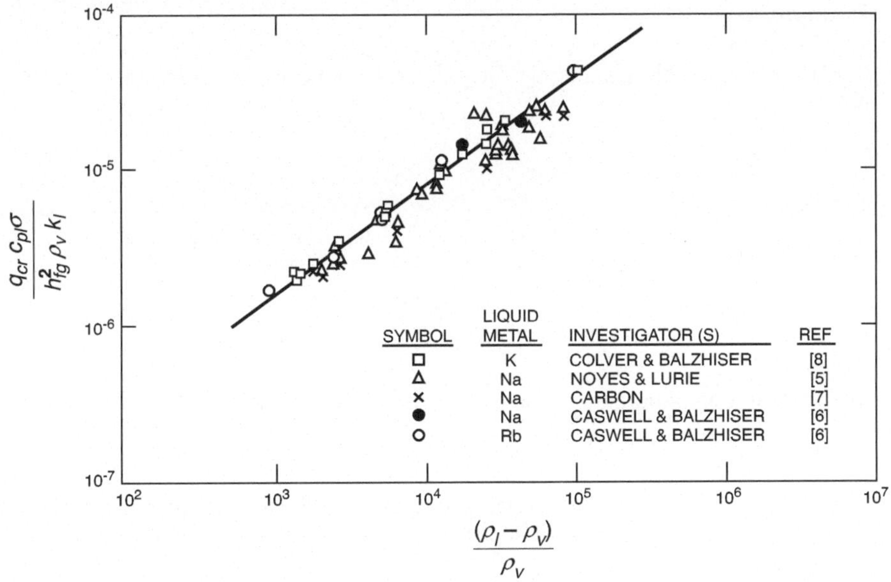

FIGURE 10.4 Correlation of CHF data by eq. (10.6) for stable nucleate pool boiling of liquid metals on horizontal cylindrical heaters [6],

Subbotin et al. [8] also proposed an equation similar to Noyes and Lurie [5], i.e.,

$$q_{cr} = q_{\text{evap}} + q_{cc}$$

or

$$q_{cr} = \left(1 + \frac{q_{cc}}{q_{\text{evap}}}\right) q_{\text{evap}}$$

Assuming that

$$\frac{q_{cc}}{q_{\text{evap}}} = \frac{45}{p_{cr}}\left(\frac{p_{cr}}{p_l}\right)^{0.4}$$

Subbotin et al. [8] suggested a final correlation as

$$q_{cr} = 0.14 h_{fg} \rho_v \left[1 + \frac{45}{p_{cr}}\left(\frac{p_{cr}}{p_l}\right)^{0.4}\right]\left[\frac{\sigma g(\rho_l - \rho_v)}{\rho_v^{\;2}}\right]^{1/4} \tag{10.8}$$

where p_{cr} is in atmospheres, and expression for q_{evap} is the same as obtained by Zuber for q_{cr} of ordinary liquids with a slightly different lead constant.

Example:

Calculate the critical or maximum heat flux for pool boiling of sodium at 1 atm pressure.

Solution:

From eq. (10.8)

$$q_{\max} = 0.14 \rho_v h_{fg} \left[1 + \frac{45}{p_{cr}}\left(\frac{p_{cr}}{p_l}\right)^{0.4}\right]\left[\frac{\sigma g(\rho_l - \rho_v)}{\rho_v^{\;2}}\right]^{1/4}$$

After substitution of numerical values for various parameters, we obtain

$$q_{\max} = 3.32 \times 10^6 \text{ W/m}^2$$

If we use the nucleate boiling correlation evaluated earlier to determine the dependence of wall heat flux on wall superheat, we obtain the wall superheat at the maximum heat flux as

$$\Delta T_{\max} = 3.5°\text{C}$$

This is a rather low value as noted earlier because of uncertainties in employed empirical constants.

The law of corresponding states was very successfully applied by Bonshernsky [9] to predict peak heat flux in organic liquids.

$$\frac{q_{cr}}{q_{cr}^*} = f\left(\frac{p_l}{p_{cr}}\right) \tag{10.9}$$

where q_{cr}^* is the value of the critical heat flux at a particular value of the reduced pressure $\frac{p_l^*}{p_{cr}}$. Subbotin et al. used the above equation to liquid metals with excellent results. Their value of q_{cr}^* was arbitrarily taken at $\frac{p_l^*}{p_{cr}} = 0.003$. They found that data for sodium, potassium, and cesium as well as rubidium could be correlated by

$$\frac{q_{cr}}{q_{cr}^*} = 2.20\left(\frac{p_l}{p_{cr}}\right)^{0.125} \tag{10.10}$$

The data covered the range of reduced pressure 5×10^{-4} to 3×10^{-2}. The limitation of eq. (10.10) is that a dependable value of q_{cr}^* at the reference value of reduced pressure must be available for predicting q_{cr} at some other value of reduced pressure. Equation (10.10) shows only 0.125 power dependence on the boiling pressure when the data for the four alkali metals are considered. This dependence is similar to that reported by Kirillov [7].

As we see a number of correlations have been developed for critical heat flux in pool boiling of liquid metals. Two reasons for this are that: (i) conduction and convection heat transfer plays a much stronger role in liquid metals than in ordinary liquids; (ii) contribution of mass diffusion at the vapor liquid interface of bubbles is much higher in liquid metals as opposed to ordinary liquids. The different approaches taken to incorporate their contribution are reflected in the formulation of the correlations.

10.4 FILM BOILING

Liquid metal vapor, particularly those of alkali metals, consists of not only gaseous atoms but also molecules

$$\begin{aligned} &\mathrm{nM} \leftrightarrow \mathrm{Mn} \\ &\rightarrow \text{Pressure} \\ &\leftarrow \text{Temperature} \end{aligned} \tag{10.11}$$

At saturation condition, since temperature and pressure are coupled, and they increase or decrease together, the influence of increase of pressure and increase of temperature is thus opposite to each other in the equilibrium. However, the pressure effect outweighs the temperature effect, so that an increase in saturation temperature actually produces an increase in dimer and tetramer molecules concentration. Figure 10.5 shows typical dependence of dimer and tetramer fractions as a function of saturation temperature for potassium.

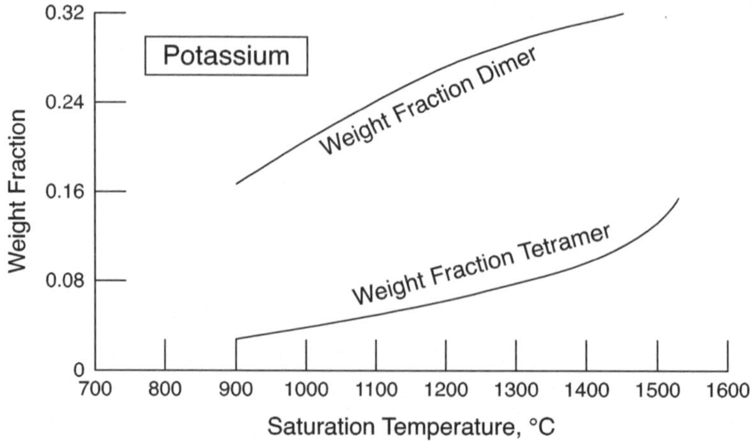

FIGURE 10.5 Weight fraction of dimers and tetramers for potassium.

Ewing et al. [10] have suggested a correlation to calculate dimer and tetramer concentrations. This correlation is of the type

$$\log_{10} k_P = a + \frac{b}{T(°\text{K})} \tag{10.12}$$

where k_P is the chemical equilibrium constant in terms of partial pressures, and a and b are empirical constants. Use of this equation shows that sodium vapor at 1 atm. will have 20% (by weight) dimers while potassium and cesium will have about 16% dimers by weight under equilibrium conditions. However, under these conditions, concentrations of tetramers are negligible. At constant pressure, the above percentages drop sharply with temperature increases comparable to the temperatures encountered in film boiling. For example, at 1 atm, the dimer concentration in potassium vapor falls to about 4% (by weight) at wall temperature of 1,089°C, which corresponds to $(T_w - T_{sat})$ of 316°C.

Now, let us consider the above phenomena in connection with film boiling of an alkali metal, where only dimerization reaction is significant. Because of relatively large temperature difference across the film, the reaction represented by eq. (10.11) tends to go to the left as the heating surface is approached and to the right as the vapor liquid interface is approached, Therefore, the concentration of monomers decreases with distance from the heating surface, while that of the dimers increases. These concentration gradients promote diffusion of the monomers toward the relatively cold vapor/liquid interface and the dimer molecules toward the relatively hot heating surface/vapor interface. As the monomer atoms diffuse toward the liquid/vapor interface, they react exothermically to produce dimer molecules, and as the dimer molecules formed (plus those generated) at the liquid/vapor interface diffuse toward the heating surface, they decompose endothermically to produce monomer atoms. The net result of this process is an additional mode of heat transport from the heating surface to the boiling liquid. The easiest way to deal with it is to combine it with the conduction mode by employing an effective thermal conductivity, one that includes polymerization effect.

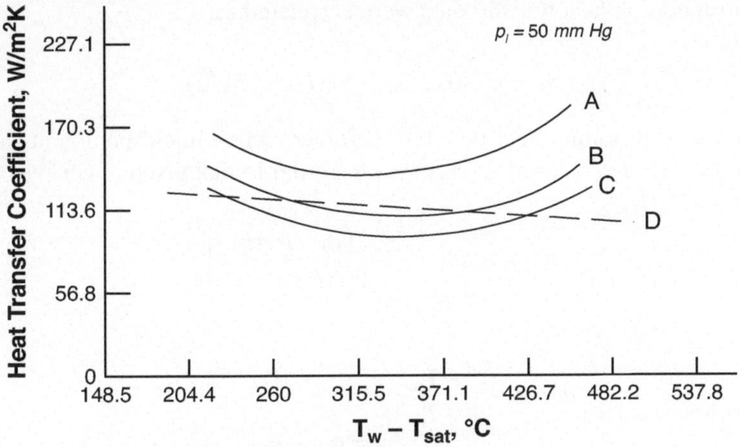

FIGURE 10.6 Film boiling heat transfer coefficient for potassium [4,11].

Few observations of film boiling heat transfer in liquid metals have been made mainly because of very high heater temperatures needed to sustain stable film boiling. Observations of film boiling of potassium on a flat horizontal stainless steel plate have been made by Padilla [11]. His results for an ambient pressure of 50 mm Hg are shown in Figure 10.6 by curve A; while curve B is curve A minus the contribution of radiation heat transfer. Curve C is curve B minus the contribution of dimerization heat transport effect. Curve D represents Berenson's prediction with constant of proportionality increased to 0.84.

In this figure, the experimental curve shows considerable curvature while theoretical curve is essentially a straight line. Padilla's results for potassium show that the heat transfer increased more rapidly at high temperature differences than the theoretical correlations would indicate even when due allowance is made for the effect of radiation. The upswing in the heat transfer may be caused by increased turbulence in the vapor film because of larger volume flux of vapor and by presence of ripples on the vapor–liquid interface. The turbulence and ripples could conceivably produce local random and intermittent thinning of the vapor film. Geometrical features of Padilla's test setup may have also played some role. Evidently, many more experimental results are needed before a satisfactory understanding of film boiling heat transfer to alkali metals can be achieved.

Subcooled film boiling of sodium on tantalum, spheres has been studied by Farahat [12]. He found the observed minimum boiling temperature to be much higher than that predicted by Zuber–Berenson equation for ordinary liquids. They correlated their data for minimum heat flux and minimum film boiling temperature as:

$$q_{\min} = 1.99 \times 10^5 + 1.08 \times 10^4 \Delta T_{\text{sub}}\,(^\circ\text{C})\ \text{W/m}^2 \tag{10.13}$$

$$\Delta T_{\min} = 438.9 + 22\Delta T_{\text{sub}}\ \ ^\circ\text{C} \tag{10.14}$$

where $\Delta T_{\text{sub}} = T_{\text{sat}} - T_l$, °C.

The film boiling heat transfer data were correlated as:

$$h = h_{\text{sat}} + 0.88h_r + Kh_c\left(\Delta T_{\text{sub}}/\Delta T_w\right) \tag{10.15}$$

where h_{sat} is film boiling heat transfer coefficient when liquid pool is at saturation temperature. For h_{sat}, he used an expression similar to that given in Chapter 7 as

$$h_{\text{sat}} = 0.68\left[\frac{k_v^{\,3} g\rho_v\left(\rho_l - \rho_v\right)h'_{fg}}{\mu_v \Delta T_w D}\right]^{1/4} \tag{10.16}$$

and for

$$h_r = \sigma\frac{T_w^{\,4} - T_{\text{sat}}^{\,4}}{T_w - T_{\text{sat}}}\frac{1}{\dfrac{1}{\varepsilon_1\left(T_w\right)} + \dfrac{1}{\varepsilon_2\left(T_{\text{sat}}\right)} - 1} \tag{10.17}$$

where ε_1 and ε_2 are emissivities of sphere surface and liquid sodium, respectively. The natural convection heat transfer coefficient, h_c, at the vapor–liquid interface was obtained as:

$$h_c = \frac{k_l}{D}\left[0.75(\text{Gr}\,\text{Pr})^2\right]^{1/4} \tag{10.18}$$

and

$$K = \frac{11.86}{\left(\Delta T_{\text{sub}}\right)^{0.7}} \tag{10.19}$$

$$\Delta T_w = T_w - T_{\text{sat}}\ ^\circ\text{C}, \quad \Delta T_{\text{sub}} = T_{\text{sat}} - T_l\ ^\circ\text{C} \tag{10.20}$$

In the above equation, K is an empirical constant. It should be noted that in writing eq. (10.15), different modes of heat transfer were simply added together. This is strictly not true since convection at the interface influences the thickness of the film itself. This fact is reflected through parameter K.

10.5 SUMMARY

In this chapter, we have discussed pool boiling of liquid metals. Various correlations developed for nucleate boiling and critical heat flux have been included. Film boiling phenomenon is also described and a correlation for film boiling on spheres with sodium as test liquid is given.

EXERCISES

1. Compare nucleate boiling heat transfer coefficients obtained from the correlation of Kutateladze and Subbotin et al. for sodium at a pressure of 1 atm. Discuss your results.

2. Calculate maximum or critical heat fluxes for sodium, potassium, and cesium boiling at 1 atm pressure.
3. Film boiling of sodium occurs on a 1 cm diameter sphere. Sodium is at 1 atm pressure and has a subcooling of 50°C. Calculate film boiling transfer coefficient when wall superheat is 300°C. Include the radiation contribution.

NOMENCLATURE

a, b	Constants
c_p	Specific heat
D	Diameter
Gr	Grashof number
g	Gravitational acceleration
h	Heat transfer coefficient
K_p	Dimensionless pressure
K	Empirical constant
k	Thermal conductivity
M	Metal vapor atoms
Nu	Nusselt number
Pr	Prandtl number
p	Pressure
q	Heat flux
Re	Reynolds number
T	Temperature, °K

SUBSCRIPTS

cr	Critical or maximum
fg	Liquid to vapor
l	Liquid
v	Vapor
sat	Saturation
sub	Subcooled

GREEK LETTERS

ΔT	Temperature difference
μ	Molecular viscosity
ν	Kinematic viscosity
ρ	Density
σ	Surface tension

REFERENCES

1. Pinchera, G.C., Tomassetti, G., Gambardella, G., and Farello, G.E. (1967), "Experimental Boiling Studies Related to the Fast Reactor Safety." *Proceedings of the International Conference on Safety of Fast Reactors*, Aix-en-Provence, France.
2. Kutateladze, S.S. (1952), *Heat Transfer during Condensation and Boiling*, Second Edition, State Scientific and Technical Publishing House of Literature on Machinery, Moscow.
3. Subbotin, V.I., Sorokin, D.N., and Kudryavtsev, A.P. (1970), "Generalized Relationship for Calculating Heat Transfer in the Developed Boiling of Alkali Metals", (Transl.) *Sov. At. Energy*, Vol. 29, p. 45.
4. Dwyer, O.E. (1976), *Boiling Liquid-Metal Heat Transfer*, American Nuclear Society. Hinsdale, Illinois.
5. Noyes, R.C., and Lurie, H. (1966), "Boiling Sodium Heat Transfer." *Proceedings of the 3rd International Conference on Heat Transfer*, August 7–12, Chicago, IL, Vol. 5, p. 92.
6. Caswell, B.F., and Balzhiser, R.E. (1966), "The Critical Heat Flux for Boiling Liquid Metal Systems." *Chem. Eng. Prog. Symp. Ser.*, Vol. 62, No. 64, p. 41.
7. Kirillov, P.L. (1968), "A Generalized Functional Relationship between the Critical Heat Flux and Pressure in the Boiling of Metals in Large Quantities" (Transl.) *Sov. At. Energy*, Vol. 25, p. 143.
8. Subbotin, V.I., Orechkin, D.M., Sorokin, D.N., and Kundryavtsev, A.P. (1966), "Critical Heat Flux in Alkali Metal Boiling Under Conditions of Free Convection." (Transl.), *Symposium on Alkali Metal Coolants*, Viena, November 29–December 2, International Atomic Energy Agency, Viena.
9. Borishansky, V.M. (1961), "Allowing for the Influence of Pressure on the Heat Transfer and Critical Thermal Loads During Boiling, in Accordance with the Theory of Thermodynamic Similarity" (Transl.), *Voprosy teplootdachi I gidravliki dvukhfaznykh sred*, Gosenergoizdat, Moscow.
10. Ewing, C.T., Stone, J.P., Spann, J.R., and Miller, R.R. (1967), "Molecular Association in Sodium, Potassium, and Cesium Vapors at High Temperatures." *J. Phys. Chem.*, Vol. 47, p. 473.
11. Padilla, A. (1966), "Film Boiling of Potassium on a Horizontal Plate." Ph.D. Thesis, University of Michigan, Ann Arbor, MI.
12. Farahat, M. (1971), "Transient Boiling Heat Transfer From Spheres to Sodium." Ph.D. Thesis, Northwestern University, Cook County, IL.

11 Two-Phase Flow

INTRODUCTION

In pool boiling, vapor bubbles generated at the heated surface leave the surface as a result of buoyancy. As vapor bubbles rise up, liquid in the pool flows down, and a circulatory flow pattern is created. If there is an imposed forced flow either parallel or normal to the heated surface, bubble detachment from the surface and subsequent trajectory will be affected by the drag imposed by the liquid on the bubbles. Outflow of vapor bubbles can be in the same direction as the imposed liquid flow or in the opposite direction. The former is called parallel flow, whereas the latter is countercurrent flow. If the heater surfaces are placed in a large chamber, the composition of the flow may not change significantly in the flow direction. Such a situation is called external flow boiling. However, if the heated surface is confined, such as flow in a heated tube, or an annulus, the process is called internal flow boiling. In this chapter, we describe basic parameters, flow regimes, and pressure drop under two-phase flow condition.

11.1 VOID FRACTION, QUALITY, SUPERFICIAL VELOCITIES, AND FLOW REGIMES

11.1.1 Void Fraction

In a static two-phase system, the volume fraction of the lighter phase (i.e., vapor or gas) is called the void fraction. For two-phase flow in a boiling channel, the void fraction will vary over the cross section of the tube and along the tube. In a transient situation, it can be a function of axial, radial, and angular position as well as time. As an example, at onset of boiling because of concentration of vapor bubbles on the heated surface, void fraction will be high there as compared with the bulk. In this work, however, we will assign for steady-state conditions a single value over the cross section of the tube or, void fraction, , will be equal to

$$\bar{\alpha} = \frac{A_v}{A} \tag{11.1}$$

where A_v is the cross-sectional area occupied by vapor, and A is the cross-sectional area of the tube. In the same vein, the fractional area occupied by the liquid will be

$$\begin{aligned} \frac{A_l}{A} &= \frac{A - A_v}{A} = 1 - \frac{A_v}{A} \\ &= 1 - \bar{\alpha} \end{aligned} \tag{11.2}$$

DOI: 10.1201/9781032668437-11

11.1.2 Quality

For a flowing system, cross-sectionally averaged quality is defined as the mass flow rate of vapor divided by the total mass flow rate, or

$$\bar{x} = \frac{\dot{m}_v}{\dot{m}_t} \tag{11.3}$$

where

$$\dot{m}_t = \dot{m}_v + \dot{m}_l \tag{11.4}$$

or

$$\dot{m}_l = \dot{m}_t - \dot{m}_v$$

or

$$\frac{\dot{m}_l}{\dot{m}_t} = 1 - \frac{\dot{m}_v}{\dot{m}_t} = 1 - \bar{x} \tag{11.5}$$

If V_v is the local average velocity of vapor over the vapor occupied area and V_l is the local average velocity of liquid in the liquid occupied area at a given location, we can write

$$\dot{m}_v = \rho_v A_v V_v \tag{11.6}$$

$$\dot{m}_l = \rho_l A_l V_l \tag{11.7}$$

$$\bar{x} = \frac{\dot{m}_v}{\dot{m}_v + \dot{m}_l} = \frac{\rho_v A_v V_v}{\rho_v A_v V_v + \rho_l A_l V_l}$$

Dividing throughout by $\rho_v A V_v$, we obtain

$$\bar{x} = \frac{\bar{\alpha}}{\bar{\alpha} - \frac{\rho_l}{\rho_v}(1-\bar{\alpha})\frac{V_l}{V_v}} \tag{11.8}$$

For a homogeneous flow, the liquid and vapor velocities are taken to be equal. This is true only for high pressures and high velocities. However, in general, because of buoyancy, vapor velocity is higher than the liquid velocity. The ratio between the local vapor and liquid velocity is known as a slip factor S and is simply

$$S = \frac{V_v}{V_l} \tag{11.9}$$

In the early studies of two-phase flow with application to thermal hydraulics of nuclear reactors, empirical relations were developed for slip factor S. Figure 11.1 from [1] shows dependence of slip factor on system pressure.

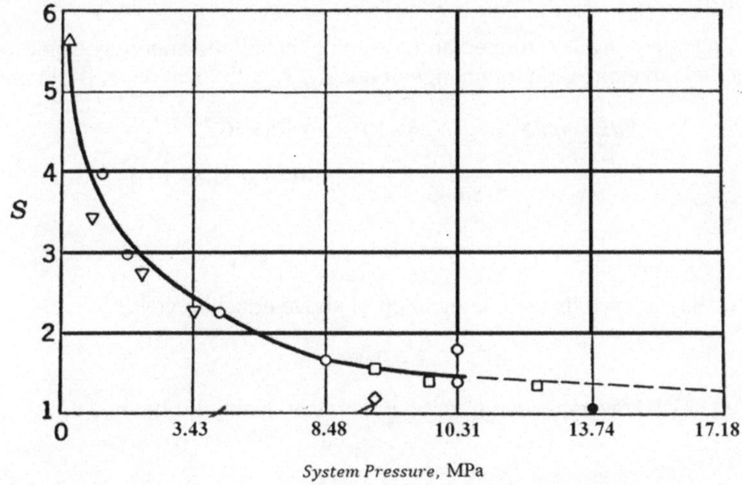

FIGURE 11.1 Slip ratio as a function of pressure for $V_l \approx 0.61$ m/s and $x \approx 0.05$ [1].

In terms of slip factor S, one can rewrite eq. (11.8) for average void fraction, $\bar{\alpha}$, as

$$\bar{\alpha} = \frac{\bar{x}}{(1-\bar{x})\dfrac{\rho_v}{\rho_l}S + \bar{x}} \tag{11.10}$$

$$= \frac{1}{1 + \dfrac{(1-\bar{x})}{\bar{x}}\dfrac{\rho_v}{\rho_l}S} \tag{11.11}$$

As an example for steam water flow at 1 atmosphere pressure, the void fraction for a quality, $\bar{x} = 0.01$ and slip factor $S = 2$, will be

$$\bar{\alpha} = \frac{1}{1 + \dfrac{0.99}{0.01} \times \dfrac{0.596}{958} \times 2}$$

$$= 0.91$$

We find that for a fixed slip factor, the density ratio or system pressure plays a major role in determining the void fraction corresponding to a given quality.

Example:

Saturated water at 7 MPa pressure enters 1.5 cm diameter electrically heated tube. The tube is 3.5 m long. The wall heat flux is 5×10^5 W/m^2, and the mass velocity of water entering the tube is 5×10^6 kg/hr m^2. Obtain the axial variation of quality and void fraction for slip factors of 1&2.

Solution:

With liquid entering as saturated and assuming that all the energy goes into vapor production, an expression for change of quality, $\bar{x}$, with location, z, is obtained as

$$\frac{d\bar{x}}{dz} = \frac{q\pi D}{\dot{m}h_{fg}} = \frac{5\times10^5\times\pi\times1.5\times10^{-2}}{\frac{5\times10^6}{3,600}\times\frac{\pi}{4}\left(1.5\times10^{-2}\right)^2\times\left(1.5\times10^6\right)}$$

$$= 0.064$$

Since quality at $z = 0$, is zero, integration of above equation yields

$$\bar{x} = 0.064z$$

From eq. (11.11), average void fraction in terms of quality is obtained as

$$\bar{\alpha} = \frac{1}{1+\frac{(1-\bar{x})}{\bar{x}}\frac{\rho_v}{\rho_l}S}$$

For $S = 1$

$$\bar{\alpha} = \frac{1}{1+\frac{(1-0.064z)}{0.064z}\frac{37.17}{738}\cdot1}$$

For $S = 2$

$$\bar{\alpha} = \frac{1}{1+\frac{(1-0.064z)}{0.064z}\frac{37.17}{738}\cdot2}$$

The following table gives quality and void fraction at various locations along the tube.

z, m	0	1	2	3	3.5
$\bar{x}$	0	0.064	0.128	0.192	0.224
$\bar{\alpha}$ $(S=1)$	0	0.576	0.745	0.825	0.851
$\bar{\alpha}$ $(S=2)$	0	0.404	0.593	0.702	0.741

As expected, local void fraction decreases as slip between phases increases.

11.1.3 Superficial Velocities

Before proceeding further, it is important to define superficial mass velocities. The mass velocity of vapor G_v is defined as

$$G_v = \frac{\dot{m}_v}{A} = \frac{\dot{m}_t\bar{x}}{A}\ \left(\text{kg/m}^2\text{s}\right) \tag{11.12}$$

Similarly, mass velocity of the liquid, G_l, is:

$$G_l = \frac{\dot{m}_l}{A} = \frac{\dot{m}_t(1-\bar{x})}{A}\ \left(\text{kg/m}^2\text{s}\right) \tag{11.13}$$

and mass velocity based on total flow rate is

$$G_m = \frac{\dot{m}_t}{A}\left(\text{kg/m}^2\text{s}\right) \tag{11.14}$$

Superficial velocities of vapor or gas and liquid respectively based on tube cross-sectional area, A, become

$$\bar{j}_v = \frac{\dot{m}_v}{A\rho_v} = \frac{\dot{m}_t\bar{x}}{A\rho_v}\ (\text{m/s}) \tag{11.15}$$

and

$$\bar{j}_l = \frac{\dot{m}_l}{A\rho_l} = \frac{\dot{m}_t(1-\bar{x})}{A\rho_l}\ (\text{m/s}) \tag{11.16}$$

Local area averaged vapor or gas and liquid velocities in terms of superficial velocities can be written as

$$V_v = \frac{\bar{j}_v}{\bar{\alpha}} \tag{11.17}$$

and

$$V_l = \frac{\bar{j}_l}{(1-\bar{\alpha})} \tag{11.18}$$

and superficial velocity of the mixture, $\bar{j}$, based on mixture density, ρ_m, becomes

$$\bar{j} = \frac{\dot{m}_t}{A\rho_m} = \frac{G_m}{\rho_m} \tag{11.19}$$

Since average quality, $\bar{x}$, and void fraction, $\bar{\alpha}$, increase in the direction of flow along a heated channel, mixture density decreases in the flow direction.

For a differential volume of length, dz, in a channel, local mixture mass can be written in terms of the volumes of the two fluids as

$$m = \rho_v v_v + \rho_l v_l$$

where v_v and v_l are the volumes of vapor and liquid respectively.

$$m = \rho_v A\bar{\alpha}\ dz + \rho_l A(1-\bar{\alpha})\ dz$$

or

$$\rho_m A dz = \rho_v A\bar{\alpha}\ dz + \rho_l A(1-\bar{\alpha})\ dz$$

or

$$\rho_m = \rho_v \bar{\alpha} + \rho_l (1 - \bar{\alpha}) \tag{11.20}$$

Substituting for local void fraction, $\bar{\alpha}$, in terms of quality, we obtain

$$\begin{aligned}\rho_m &= \rho_v \frac{1}{1 + \frac{(1-\bar{x})}{\bar{x}} \frac{\rho_v}{\rho_l} S} + \rho_l \left[1 - \frac{1}{1 + \frac{(1-\bar{x})}{\bar{x}} \frac{\rho_v}{\rho_l} S} \right] \\ &= \frac{\rho_v}{1 + \frac{(1-\bar{x})}{\bar{x}} \frac{\rho_v}{\rho_l} S} + \frac{\frac{1-\bar{x}}{\bar{x}} \rho_v S}{1 + \frac{(1-\bar{x})}{\bar{x}} \frac{\rho_v}{\rho_l} S}\end{aligned} \tag{11.21}$$

For $S = 1$, eq. (11.21) reduces to

$$\rho_m = \frac{\rho_l}{\bar{x} \frac{\rho_l}{\rho_v} + 1 - \bar{x}} \tag{11.21a}$$

Zuber and Findley [2] proposed that slip between the phases was reflected in the drift that exists between the rate at which vapor bubbles are moving away relative to the vapor–liquid mixture.

Writing the identity for local vapor velocity, V_v

$$V_v = j + (V_v - j) \tag{11.22}$$

and using the relation given by eq. (11.17) based on local rather than average over the cross section, the above equation becomes

$$j_v = \alpha j + \alpha (V_v - j) \tag{11.23}$$

By averaging across the cross section of the tube, the above relation can be written as

$$\bar{j}_v = \frac{\bar{x} G}{\rho_v} = C_o \bar{j}\ \bar{\alpha} + V_{vj} \bar{\alpha} \tag{11.24}$$

where

$$C_o = \frac{\overline{j\alpha}}{\bar{j}\ \bar{\alpha}} \tag{11.25}$$

and

$$V_{vj} = \frac{\overline{(V_v - j)\alpha}}{\bar{\alpha}} \tag{11.26}$$

In eqs. (11.24)–(11.26), the upper bars represent averaging over the cross section. From eq. (11.24), $\bar{\alpha}$ can be written as

$$\bar{\alpha} = \frac{\bar{j_v}}{C_o \bar{j} + V_{vj}} \tag{11.27}$$

Parameter C_o accounts for the radial void fraction and velocity distribution over the cross section and can be termed as global slip. The drift velocity V_{vj} is a measure of the local slip as represented by the velocity of vapor relative to the mixture. For example, in Chapter 4, we determined the terminal velocity of a vapor bubble rising in a pool of liquid. Drift velocity V_{vj} has been taken to be approximately equal to the terminal rise velocity of the bubble:

$$V_{vj} = V_t = K\left[\frac{(\rho_l - \rho_v) g\sigma}{\rho_l^2}\right]^{1/4} \tag{11.28}$$

where K is an empirical constant.

Lahey [3] has shown that eq. (11.28) works well for churn-turbulent bubbly flow. However, eq. (11.28) has been frequently used for other flow regimes as well.

Substituting for $\bar{j_v}$ and $\bar{j}$ from eqs. (11.15) and (11.19) respectively in eq. (11.27), we obtain

$$\bar{\alpha} = \frac{\dfrac{\bar{x}}{\rho_v} G_m}{C_o\left[\dfrac{\bar{x}}{\rho_v} + \dfrac{(1-\bar{x})}{\rho_l}\right] G_m + V_{vj}}$$

$$= \frac{\bar{x}}{C_o\left[\bar{x} + \dfrac{\rho_v}{\rho_l}(1-\bar{x})\right] + \dfrac{\rho_v V_{vj}}{G_m}} \tag{11.29}$$

Comparing eq. (11.29) with eq. (11.11), we find slip factor S to be

$$S = C_o + \frac{(C_o - 1)\rho_l}{\rho_v(1-\bar{x})} + \frac{\rho_l V_{vj}}{G_m(1-\bar{x})} \tag{11.30}$$

The first two terms result from averaging of the void fraction over the cross section and represent integral slip, whereas the last term represents the local slip between the phases. Parameter C_o and drift velocity V_{vj} depend on the flow regime, which we will discuss next. However, this dependance is not strong. For example, Dix [4] has shown from experiments that value of C_o depends on system pressure but is close to unity for water at pressures between 1 and 140 atm and average quality greater than 0.1. Although constant K in eq. (10.28) depends on flow regime, it is usually taken to be independent of the flow regime. The constant K in the expression for drift velocity is assigned to be +2.9 for upflow and −2.9 for downflow.

11.1.4 Flow Regimes

As vapor/gas flow rate in a tube increases for a fixed liquid flow rate, the configuration of the phases in the tube changes. Figure 11.2 shows these flow configurations or flow regimes in a vertical adiabatic tube. For low vapor/gas flow rate, the lighter phase breaks down into discreetly located bubbles. The bubbles can be of different sizes and shapes. This flow regime is called bubbly flow. As the vapor/gas flow rate is increased, bubbles start to merge with each other leading to formation of long vapor bubbles or slugs. These slugs are separated by liquid layer or plugs that may contain isolated bubbles. This flow regime is known as the slug flow. With further increase in vapor/gas flow, vapor interface between slugs and liquid becomes unstable, and some vapor slugs may break up and merge with other bubbles leading to formation of large bubbles. The corresponding flow regime is called churn turbulent flow. Finally, with further increase in vapor/gas flow rate, liquid is pushed to the walls to form a liquid film. Vapor/gas is confined to the middle of the tube as a continuous phase. This is called annular flow regime. Some liquid droplets in the vapor/gas may be entrained from the liquid film and even be deentrained from the flow. At very high vapor/gas flow rates, liquid film may break up and liquid ligaments/drops are entrained in the vapor/gas phase. This is called entrained flow regime. Again, some deentrainment may occur up stream. For adiabatic walls, thin liquid film may or may not continue to persist on the wall. Similar flow regimes for a horizontal tube are depicted in Figure 11.3. For relatively high liquid flow rates and low flow rates of gas/vapor, individual bubbles of lighter fluid flow through the tube. This flow regime is called bubbly flow. As liquid flow rate is decreased with moderate gas flow

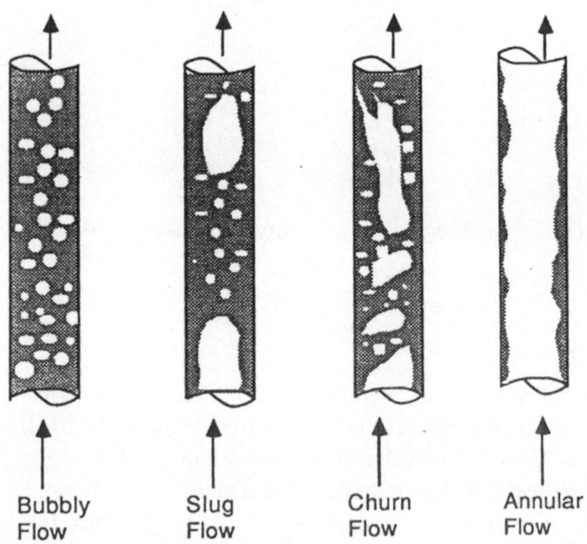

FIGURE 11.2 Flow patterns in vertical flow [5].

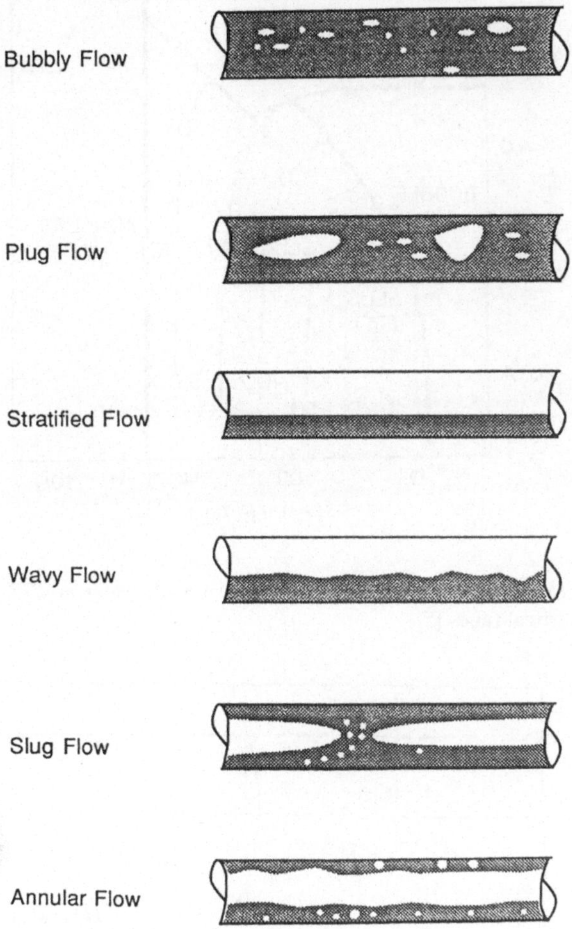

FIGURE 11.3 Flow patterns in horizontal flow [5].

rates, large vapor slugs with some smaller bubbles dominate the flow field. This flow regime is called plug flow. With further decrease in liquid flowrate, stratified flow condition develops where continuous layer of gas/vapor flows over a liquid layer underneath. With increase in gas flow, the interface between gas/vapor and liquid may become wavy. However, with increase in liquid flow rate, the stratified flow regime gives way to slug flow in which large vapor slugs move through the tube with liquid being the continuous phase filling the cross section of the tube. For higher gas/vapor velocities, annular flow condition develops in which liquid is confined to the walls of the tube with lighter phase filling the core.

Figure 11.4 shows a map of flow regimes in a 55 mm diameter vertical tube with air and water at 1 atm pressure and 20°C [5]. The figure is from the work of Taitel et al. [6].

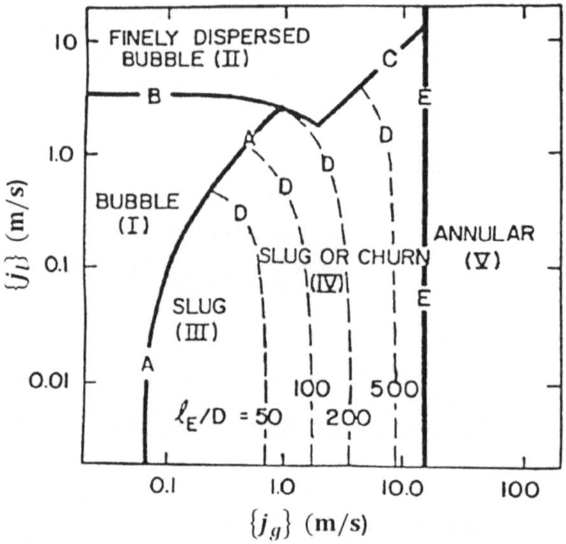

FIGURE 11.4 Flow regime aap of Taitel et al. [6] for air-water at 25°C and 0.1 MPa in 55 mm diameter vertical tubes [5].

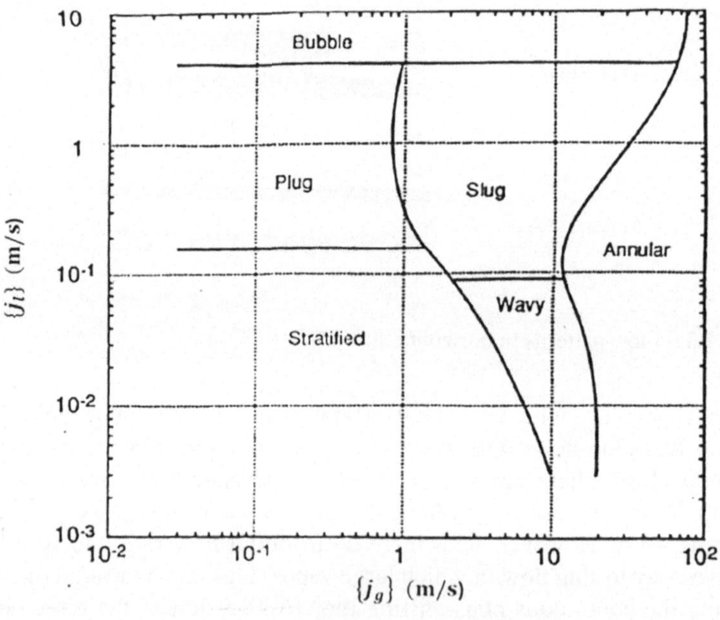

FIGURE 11.5 Flow map of Mandhane et al. for horizontal flow [5].

Existence and transitions of various regimes are influenced by the tube diameter. Detailed discussion of the physical basis for these transitions is given in references [7] and [8]. For horizontal tubes, map proposed by Mandhane et al. [9] is given in Figure 11.5. For both orientations, the maps are presented in terms of the superficial velocities of vapor/gas and liquid through the tube. It should be emphasized that the maps are based on the assumption of one-dimensional flow. Detailed visual observations are required to develop maps that account for changes in the flow patterns in the circumferential and radial directions.

11.2 TWO-PHASE PRESSURE DROP

In a heated tube with boiling on the tube inner wall, two-phase pressure drop can be significantly higher than that for single-phase flow. Key reasons are: increase in friction pressure drop, and inclusion of acceleration pressure drop as vapor is added from the wall to the two phase mixture in the tube. Figure 11.6 shows a differential volume in the channel. Again only one dimensional flow is considered.

Rate of momentum inflow at the lower face of the differential volume

$$= \dot{m} V_m \tag{11.31}$$

$$= \frac{\dot{m}_t \dot{m}_t}{\rho_m A} = \frac{\dot{m}_t^{\,2}}{\rho_m A} \tag{11.32}$$

where $\dot{m}_t$ is the total mass flow rate, ρ_m is the mixture density, V_m is the mixture velocity, and A is the tube area of cross section. Rate of momentum outflow at the upper boundary is

$$= \frac{\dot{m}_t^{\,2}}{\rho_m A} + \frac{d}{dz}\left(\frac{\dot{m}_t^{\,2}}{\rho_m A}\right) dz \tag{11.33}$$

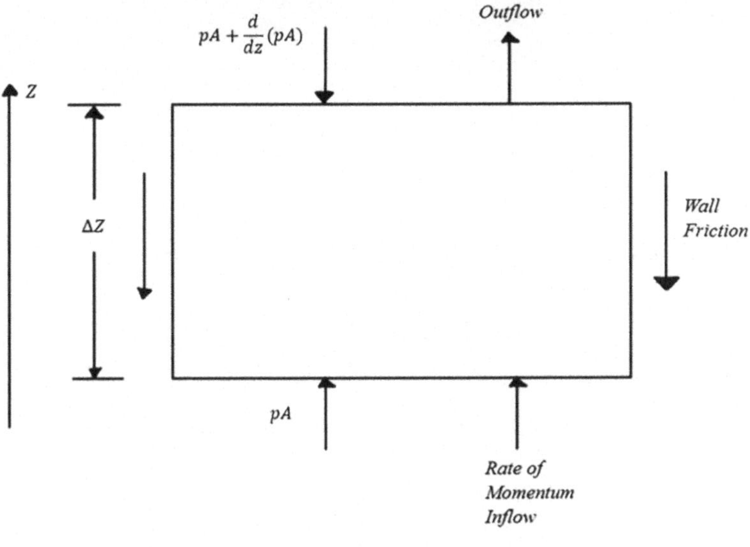

FIGURE 11.6 Two-phase pressure drop configuration over a control volume.

Forces due to pressure, p, acting at inlet and outlet faces of the differential volume are

$$F_{p_i} = pA \tag{11.34}$$

$$F_{p_o} = -\left[p + \frac{dp}{dz}dz\right]A \tag{11.35}$$

Wall to fluid friction force in terms of two phase friction factor, f_{tp}, and mixture velocity and density is written as

$$F_f = -f_{tp}\frac{1}{2}\rho_m\left(\frac{\dot{m}_t}{\rho_m A}\right)^2 P_w dz \tag{11.36}$$

\

where P_w is the tube perimeter.

Hydrostatic force in terms of the mixture density is written as

$$F_{\text{hydro}} = -\rho_m A dz g \tag{11.37}$$

Equating the net rate of momentum influx with the sum of the forces acting on the differential volume, we find

$$\frac{\dot{m}_t^2}{\rho_m A} + \frac{d}{dz}\left(\frac{\dot{m}_t^2}{\rho_m A}\right)dz - \frac{\dot{m}_t^2}{\rho_m A} = F_{p_i} + F_{p_o} + F_f + F_{\text{hydro}} \tag{11.38}$$

$$= pA - \left[p + \frac{dp}{dz}dz\right]A - \frac{1}{2}f_{tp}\rho_m\left(\frac{\dot{m}_t}{\rho_m A}\right)^2 P_w dz - \rho_m A dz g \tag{11.39}$$

or

$$\frac{dp}{dz} = \underbrace{-\frac{1}{A}\frac{d}{dz}\left(\frac{\dot{m}_t^2}{\rho_m A}\right)}_{\text{Acceleration pressure drop}} \underbrace{-\frac{1}{2}f_{tp}\frac{\dot{m}_t^2}{\rho_m A^2}\frac{P_w}{A}}_{\text{Friction pressure drop}} - \underbrace{\rho_m g}_{\text{Hydrostatic head}} \tag{11.40}$$

In terms of the mass velocity, G_{m}, for a constant area tube eq. (11.40) can be written as

$$\frac{dp}{dz} = -G_m^2\frac{d}{dz}\left(\frac{1}{\rho_m}\right) - \frac{4f_{tP}}{D_h}\left(\frac{G_m^2}{2\rho_m}\right) - \rho_m g \tag{11.41}$$

In writing eq. (11.41), the hydrostatic diameter, D_h, is defined as

$$D_h = \frac{4A}{P_w}$$

and

$$G_m = \frac{\dot{m}_t}{A}$$

Friction factor for single-phase flow is generally related to the Reynold's number of the flow, i.e.,

$$f = C\,\mathrm{Re}^{-n} \tag{11.42}$$

where C is 16 and $n = 1$ for laminar flow and $C = 0.366$ and $n = 0.25$; and $C = 0.182$ and $n = 0.2$ have been used in the literature for turbulent flow. Reynold's number for single-phase flow is defined as

$$\mathrm{Re} = \frac{\bar{U} D_h \rho}{\mu}$$

where $\bar{U}$ is the mean velocity, ρ is the density, and μ is the molecular viscosity. If such a functional dependence of friction factor is assumed to apply for two-phase flow as well, Reynold's number in terms of two-phase mixture properties can be written as

$$\mathrm{Re}_{tp} = \frac{\dot{m}_t}{\rho_m A}\frac{D_h}{\mu_m}\rho_m = \frac{\dot{m}_t D_h}{A\mu_m} = \frac{G_m D_h}{\mu_m} \tag{11.43}$$

A number of relations for molecular viscosity of two-phase mixtures have been proposed in the literature. McAdam et al. [10] have proposed

$$\mu_m = \mu_l \left[1 + \bar{x}\left(\frac{\mu_l}{\mu_v}\right) - 1\right]^{-1} \tag{11.44}$$

Cichitti et al. [11] have suggested

$$\mu_m = \mu_l \left[1 + \bar{x}\left(\frac{\mu_v}{\mu_l}\right) - 1\right] \tag{11.45}$$

and according to Dukler et al. [12]

$$\mu_m = \mu_l \left[1 + \beta\left(\frac{\mu_v}{\mu_l}\right) - 1\right] \tag{11.46}$$

where β is the ratio of superficial velocity of vapor $\bar{j}_v$ averaged over the cross section to the superficial velocity of the mixture defined as

$$\bar{j} = \bar{j}_v + \bar{j}_l \tag{11.47}$$

As such for two-phase mixture, two-phase friction factor, f_{tp}, from eq. (11.42) becomes

$$f_{tp} = C\left(\frac{G_m D_h}{\mu_m}\right)^{-n} \tag{11.48}$$

From eq. (11.42) to (11.48), we find that

$$\frac{f_{tp}}{f_{lo}} = \left(\frac{\mu_m}{\mu_l}\right)^{n} \tag{11.49}$$

where f_{lo} represents liquid only.

Using McAdams et al.'s eq. (11.44) for μ_m in the above, we find

$$\frac{f_{tp}}{f_{lo}} = \left[1+\bar{x}\left(\frac{\mu_l}{\mu_v}-1\right)\right]^{-n} \tag{11.50}$$

From eq. (11.40), we know that

$$\left(\frac{dp}{dz}\right)_{\text{acc}} = -G_m^2 \frac{d}{dz}\left(\frac{1}{\rho_m}\right) \tag{11.51}$$

For homogeneous flow $S = 1$, and using ρ_m from eq. (11.21a), eq. (11.51) can be written as

$$\left(\frac{dp}{dz}\right)_{\text{acc}} = -G_m^2 \frac{d}{dz}\left(\frac{1+\bar{x}\left(\frac{\rho_l}{\rho_v}-1\right)}{\rho_l}\right)$$

or

$$\left(\frac{dp}{dz}\right)_{\text{acc}} = -G_m^2 \frac{d}{dz}\left(\frac{1}{\rho_l}+\frac{\bar{x}}{\rho_v}-\frac{\bar{x}}{\rho_l}\right) \tag{11.52}$$

In terms of specific volumes, eq. (11.52) can be written as

$$\left(\frac{dp}{dz}\right)_{\text{acc}} = -G_m^2 \frac{d}{dz}\left(v_l+\bar{x}(v_{vl})\right) \tag{11.53}$$

In eq. (11.53), v_l is the specific volume of liquid and v_{vl} is the change in specific volume as liquid converts into vapor. Rewriting eq. (11.53)

$$\left(\frac{dp}{dz}\right)_{\text{acc}} = -G_m^2 \left(\frac{dv_l}{dz}+\frac{d\bar{x}}{dz}v_{vl}+\frac{\bar{x}dv_{vl}}{dz}\right) \tag{11.54}$$

For liquids $\frac{dv_l}{dz}$ is very small and can be neglected. As such,

$$\left(\frac{dp}{dz}\right)_{\text{acc}} \simeq -G_m^2\left(v_{vl}\frac{d\bar{x}}{dz}+\bar{x}\frac{dv_{vl}}{dp}\frac{dp}{dz}\right) \tag{11.55}$$

As noted above, change in specific volume of liquid with pressure is small. Noting this eq. (11.55) can be further approximated as

$$\left(\frac{dp}{dz}\right)_{\text{acc}} = -G_m^2\left(v_{vl}\frac{d\bar{x}}{dz}+\bar{x}\frac{dv_v}{dp}\frac{dp}{dz}\right) \tag{11.56}$$

We know from eq. (11.40) and eq. (11.41),

$$\left(\frac{dp}{dz}\right)_{\text{fri}} = \frac{-4f_{tp}}{D_h}\left(\frac{G_m^2}{2\rho_m}\right) \tag{11.57}$$

and

$$\left(\frac{dp}{dz}\right)_{\text{hyd}} = -\rho_m g \tag{11.58}$$

Combining the contributions of acceleration, friction, and hydrostatic pressure, we find

$$\frac{dp}{dz} = -G_m^2\left(v_{vl}\frac{d\bar{x}}{dz}+\bar{x}\frac{dv_v}{dp}\frac{dp}{dz}\right)-\frac{4f_{tp}}{D_h}\left(\frac{G_m^2}{2\rho_m}\right)-\rho_m g \tag{11.59}$$

Or

$$-\frac{dp}{dz} = \frac{G_m^2 v_{vl}\frac{d\bar{x}}{dz}+\frac{4f_{tp}}{D_h}\left(\frac{G_m^2}{2\rho_m}\right)+\rho_m g}{\left(1+G_m^2\bar{x}\frac{dv_v}{dp}\right)} \tag{11.60}$$

It should be noted that $\frac{dv_v}{dp}$ is a negative quantity.

Thus, knowing variation of $\bar{x}$, ρ_m, f_{tp}, and $\frac{dv_v}{dp}$, two-phase pressure gradient along the boiling channel can be calculated. It should be noted that in arriving at eq. (11.60), we have invoked the assumption of homogeneous flow ($S = 1$). This is generally not true, and one must rederive eq. (11.60) for $S > 1$.

11.3 PRESSURE DROP DURING FLOW BOILING IN A CHANNEL

As we will discuss flow boiling in Chapter 12, during flow boiling in a channel, pressure gradient in the channel varies with the axial position. If subcooled liquid enters the channel, the location where liquid will become saturated can be simply obtained by energy balance along the flow direction, z, as

$$\dot{m}_t c_{pl} \frac{dT}{dz} = q_w P_w \tag{11.61}$$

Integrating eq. (11.61), we obtain

$$\dot{m}_t c_{pl} \left(T_{\text{sat}} - T_{li}\right) = \int_0^{l_b} q_w P_w \, dz \tag{11.62}$$

where T_{sat} is the liquid saturation temperature at the system pressure, T_{li} is the liquid temperature at inlet, q_w is the wall heat flux, P_w is the tube perimeter, and l_b is the length of the tube where bulk liquid reaches saturation temperature. Since pressure drop in single-phase region is small, T_{sat} can be taken to be the saturation temperature corresponding to the pressure at inlet. Although depending on the magnitude of heat flux and development of the thermal boundary layer, boiling may begin on the wall prior to bulk liquid reaches saturation temperature, its effect on overall pressure drop through the channel is small. For a constant heat flux case, the location, l_b, where boiling begins can be obtained from eq. (11.62) as

$$l_b = \frac{\dot{m}_t c_{pl} \left(T_{\text{sat}} - T_{li}\right)}{q_w P_w} \tag{11.63}$$

In the region from entrance to where bulk boiling begins, correlations for single phase pressure drop can be used under the assumption that flow developing layer thickness is small. Similarly, hydrostatic heat can be simply calculated as due to liquid alone. After bulk boiling begins, again energy balance can be used to calculate the increase in equilibrium quality, $\overline{x}$, in the flow direction:

$$\dot{m}_t h_{fg} \frac{d\overline{x}}{dz} = q_w P_w \tag{11.64}$$

For constant wall heat flux, after integrating eq. (11.64) we obtain

$$\overline{x} = \frac{q_w P_w \left(z - l_b\right)}{\dot{m}_t h_{fg}} \tag{11.65}$$

The dependence of $\overline{x}$ on z obtained from eq. (11.65) can be used to obtain two-phase pressure gradient as described in Section 11.2.

11.4 SUMMARY

In this chapter, we have described the concepts of void fraction and quality in two-phase flow through tubes. Relation between them has been developed. Flow regimes in both vertical and horizontal tubes have been described. An expression for two-phase pressure drop through vertical tubes has been derived.

EXERCISES

1. For the flow boiling example given in the text, obtain the flow regime that is expected to exist at the midplane of the tube.
2. For the flow boiling example given in the text, calculate the two phase pressure drop at the mid-plane of the tube when slip factor $S = 2$.

NOMENCLATURE

A	Channel or tube cross-sectional area
Co	Constant
c_{pl}	Specific heat of liquid
f	Friction factor
G	Mass velocity based on cross-sectional area
h_{fg}	Latent heat of vaporization
J	Superficial velocity based on cross-sectional area
l	Length
$\dot{m}$	Mass flow rate
p	Pressure
P_w	Wetted perimeter
q	Heat flux
S	Slip factor
T	Temperature
V	Velocity
v	Specific volume
x	Quality
$\overline{x}$	Area and time averaged quality
z	Coordinate in the direction of flow

SUBSCRIPTS

b	Boiling
l	Liquid
v	Vapor
m	Mass
$\dot{m}$	Mass flow rate
t	Total

REFERENCES

1. El-Wakil, M.M. (1962), *Nuclear Power Engineering*, McGraw-Hill Book Co. New York.
2. Zuber, N. and Findley, J.A. (1965), "Average Volumetric Concentration in Two Phase Flow Systems." *J. Heat Transfer*, Vol. 87, p. 453.
3. Lahey, R.T. (1974), *Two Phase Flow in Boiling Water Nuclear Reactors*, MEDP-13388 General Electric Company
4. Dix, G.E. (1971), *Vapor Void Fraction for Forced Convection with Subcooled Boiling at Low Flow Rates*, NEDO-10491, General Electric Company. San Francisco, CA.
5. Todreas, N. and Kazimi, M.S. (1989), *Nuclear Systems I Thermal Hydraulic Fundamentals*, Hemisphere Publishing Corporation, New York.
6. Taitel, Y., Bornea, D., and Dukler, A.E. (1980), "Modeling Flow Pattern Transition for Steady Upward Gas-Liquid Flow in Vertical Tubes", *AIChE J.*, Vol. 26, p. 345.
7. Taitel, Y. and Dukler, A.E. (1977), "Flow Regime Transitions for Vertical Upward Gas-Liquid Flow: A Preliminary Approach", Program at AIChE 70th Annual Meeting
8. Taitel, Y. and Dukler, A.E. (1976), "A Model for Predicting Flow Regime Transition in Horizontal and Near Horizontal Gas-Liquid Flow." *AIChE J.*, Vol 22, p. 47.
9. Mandhane, J.M., Gregory, G.A., and Aziz, K. (1974), "A Flow Map for Gas-Liquid Flow in Horizontal Pipes", Intl. J. Multiphase Flow, Vol. 1, p. 537.
10. McAdams, W.H., Woods, W. K., and Heroman, Jr., L. C., (1942), "Vaporization Inside Horizontal Tubes II. Benzene-Oil Mixtures." *Trans. ASME*, Vol. 64, p. 193.
11. Cicchitti, A, Lombardi, C, Silvestri, M, Soldaini, G, & Zavattarelli, R (1960), "Two Phase Cooling Experiments – Pressure Drop, Heat Transfer and Burnout Measurements." *Energ. Nucl.* Vol. 7, p. 407.
12. Dukler, A.E., Moye W., and Cleveland, R. G. (1964), "Pressure Drop and Holdup in Two Phase Flow: Part A – A Comparison of Existing Correlations. Part B – An Approach through Similarity Analysis." *AIChE J.*, Vol. 10, p. 38.

12 Flow Boiling Heat Transfer

INTRODUCTION

As noted in Chapter 11, flow boiling can be divided into two categories: external flow boiling and internal flow boiling. In external flow boiling, liquid flows across heated surfaces such as tubes or rod bundles. The flow rate of liquid is such that change in quality as the liquid flows over the heated surfaces is small. Thus, little effect of physical structure of external flow is considered on heat transfer from the heated surfaces. In internal flow boiling, however, fluid moves through a heated tube/enclosure and significant changes in the flow configuration occur as boiling on the heated surface takes place and vapor is added to the flowing fluid. Depending on the flow regime, heat removal processes from the heated surface can be different. In this chapter, we discuss both external and internal flow boiling.

12.1 EXTERNAL FLOW BOILING

In addressing external flow boiling, we traverse the boiling curve from onset of nucleate boiling to partial and fully developed nucleate boiling; critical or maximum heat flux; and film boiling.

12.1.1 Onset of Nucleate Boiling

Wall superheat at which nucleate boiling begins can be determined by using eq. (4.30) developed in Chapter 4. The heat flux in eq. (4.30) assuming that bulk liquid is saturated can be written as

$$q = \frac{k_l}{\delta}(T_w - T_{\text{sat}}) \tag{12.1}$$

where δ is the thermal layer thickness. Knowing the thickness of the thermal layer along the surface (for a developing or a fully developed flow), eq. (12.1) can be used in conjunction with eq. (4.30) to determine the wall superheat or heat flux at onset of nucleate boiling.

12.1.2 Partial and Fully Developed Nucleate Boiling

As was noted in Chapter 5, in partial nucleate boiling, the rate of heat transfer by convection continues to play an important role. This in turn makes the rate of heat transfer dependent on the orientation of the surface with respect to direction of gravity (e.g., Figure 5.9 for pool boiling on a flat surface) and flow velocity and geometry of the surface (e.g. Figure 8.8 for heat transfer on a horizontal cylinder subjected to forced flow).

DOI: 10.1201/9781032668437-12

Because of the interaction of the heater geometry and flow velocity, a reliable prediction of partial nucleate boiling heat fluxes has to rely on experimental data.

In fully developed nucleate boiling, the contribution of heat transfer by convection becomes relatively small (e.g. Figure 5.10), and one can use pool boiling correlations to predict dependence of fully developed nucleate boiling heat flux on wall superheat. Some examples are: correlation of Rohsenow's eq. (5.27) and correlations of Stephan and Abdelsalam eq. (5.28) and Cooper eq. (5.33).

12.1.3 Critical or Maximum Heat Flux

Similar to pool boiling, maximum heat flux condition under external flow boiling for well-wetting surfaces occurs when vapor removal rate just starts to fall short of vapor production rate. This leads to accumulation of vapor at the surface and degradation of rate of heat transfer.

Hardly any mechanistic models have been reported in the literature for forced flow maximum heat flux across surfaces. The maximum heat flux data have generally been correlated with several dimensionless groups. These dimensionless groups are formed by recognizing various independent parameters on which maximum heat flux for a saturated liquid depends and by using the Buckingham Π theorem. For inviscid fluids, the maximum heat flux can be written as

$$q_{\max} = \left(\rho_v, \rho_l, U_l, \sigma, h_{fg}, l, g\left(\rho_l - \rho_v\right)\right) \tag{12.2}$$

In the above equation, ρ_v and ρ_l are the vapor and liquid densities, respectively, U_l is the free stream liquid velocity, σ is the surface tension, h_{fg} is the latent heat of vaporization, l is the characteristic length of the heater, and $g\left(\rho_l - \rho_v\right)$ is the buoyancy term. From eq. (12.2), the dimensionless maximum heat flux is written as

$$\frac{q_{\max}}{\rho_v h_{fg} U_l} = f_1\left(\frac{\rho_v}{\rho_l}, \frac{\rho_l U_l^2 l}{\sigma}, \frac{h_{fg}}{U_l^2}, \frac{\rho_l U_l^2}{g\left(\rho_l - \rho_v\right) l}\right) \tag{12.3}$$

The second dimensionless group on the right-hand side is the Weber number, whereas the last group is Froude number. If liquid inertia dominates the buoyancy, inverse Froude number is not an important parameter. Similarly, $\frac{h_{fg}}{U_l^2}$ is generally of little importance. As such, eq. (12.3) reduces to

$$\frac{q_{\max}}{\rho_v h_{fg} U_l} = f_2\left(\frac{\rho_v}{\rho_l}, \frac{\rho_l U_l^2 l}{\sigma}\right) \tag{12.4}$$

A horizontal cylinder is the most investigated geometry for external flow boiling. The maximum heat flux data on a horizontal cylinder at high velocities and free from gravity effect have been correlated by Hasan et al. [1] as

$$\Phi = 0.000919\left[1 + \frac{16.8}{\mathrm{We}_f^{1/3}}\right] \tag{12.5}$$

where

$$\Phi = \frac{\pi q_{\max}}{\rho_v h_{fg} U_l} \tag{12.6}$$

$$\mathrm{We}_f = \frac{\rho_l U_l^2 D}{\sigma} \tag{12.7}$$

$$\gamma = \frac{\rho_l}{\rho_v} \tag{12.8}$$

The effect of density ratio γ is absorbed in the two empirical constants in eq. (12.5).

Figure 12.1 shows the correlation and the high-speed data that is free of gravity. It is seen that most of the data are correlated within ±20% of predictions from eq. (12.5). It should be noted that as shown by Lance and Meyer [2] that at low velocities, interaction between gravity and inertia can create a dead region for bubble movement and thereby can lead to a premature burnout.

Not much work has been done in correlating the minimum heat flux during external flow boiling. However, several attempts have been made to model forced flow film boiling heat transfer coefficient. The model requires solution of momentum and energy equation for the vapor film and only momentum equation for the liquid if

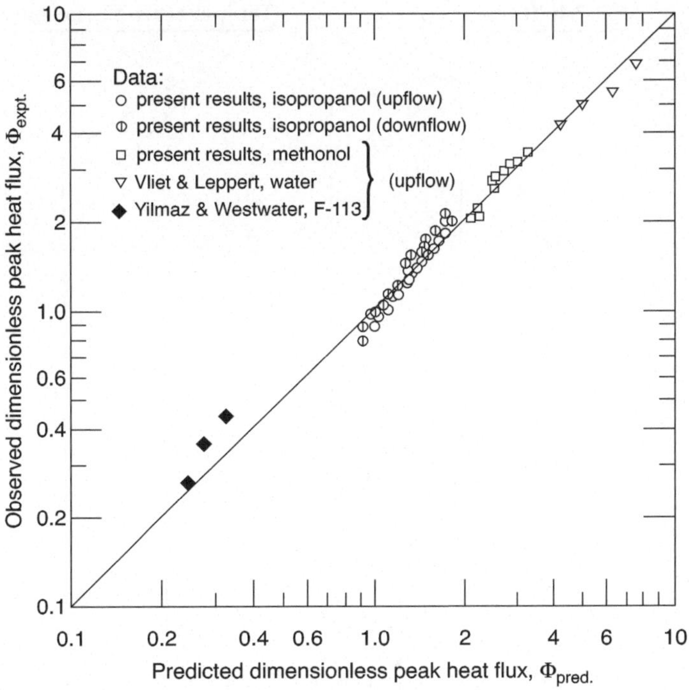

FIGURE 12.1 Comparison of high-speed burnout prediction with experimental data [1].

the liquid is saturated. Solutions either by matching both shear stress and tangential velocity at the interface and/or by assuming vapor velocity at the interface being equal to the liquid velocity have been reported in the literature. At high flow velocities, generally the film boiling heat transfer coefficient varies as square root of the imposed velocity.

12.2 INTERNAL FLOW BOILING

The most common geometry involving internal flow boiling is a tube. Since the enthalpy and state of the fluid flowing through the heated tube are continuously changing, the bulk flow conditions are not constant along the direction of flow. Figures 12.2 and 12.3 depict flow boiling and development of flow patterns at low-to-moderate and high heat fluxes, respectively. For subcooled liquid at inlet, nucleation begins at some distance downstream of the inlet where required nucleation superheat condition is satisfied. Prior to nucleation, the heat transfer from the solid wall is by developing single-phase forced convection. Subcooled nucleate boiling occurs when the bulk liquid is below saturation temperature. In subcooled boiling, bubbles mostly remain attached to the surface. As bulk liquid becomes saturated, bubbles start to slide along the surface before leaving the heated surface. This is onset of saturated nucleate boiling. Bubbles grow to a certain size on the tube wall before detachment

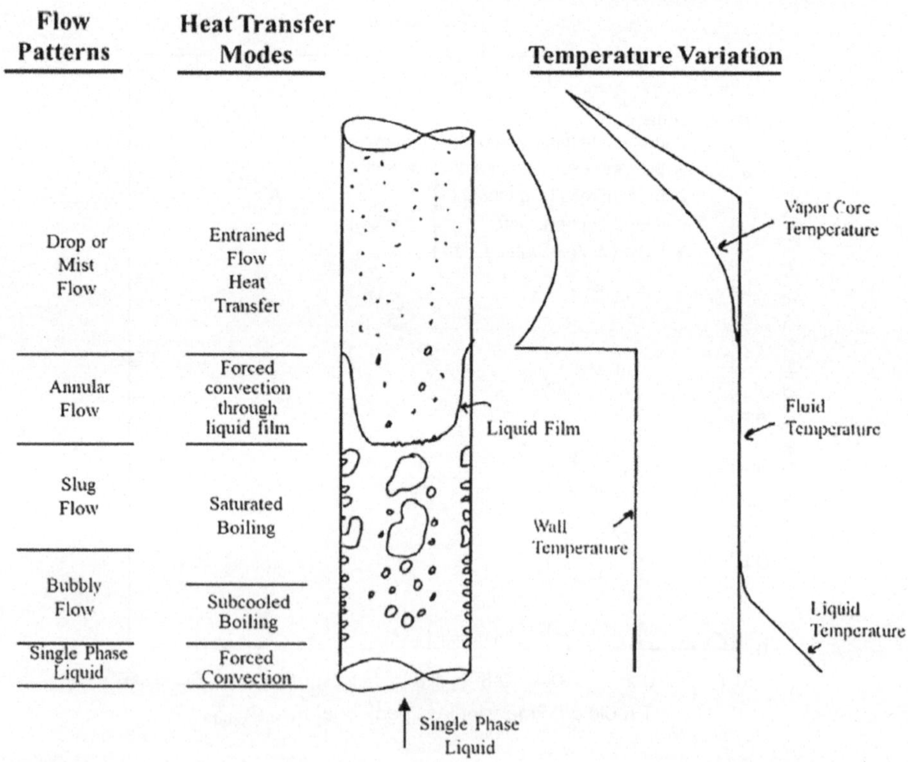

FIGURE 12.2 Flow boiling in a tube at low to moderate heat fluxes and low flow rates.

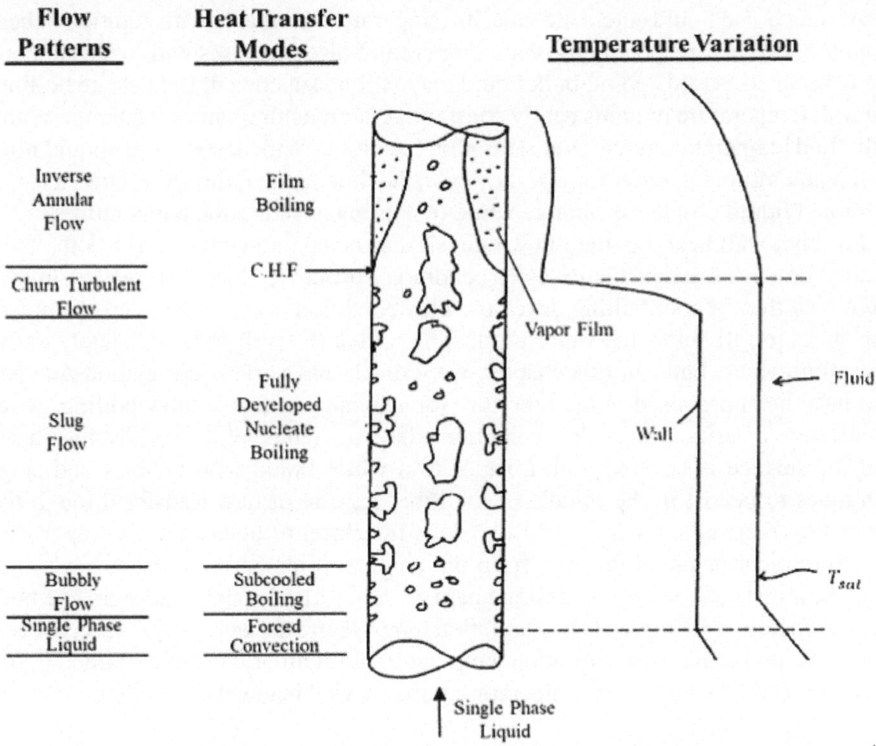

FIGURE 12.3 Flow boiling in a tube at high flow rates and high heat fluxes.

and gradually move to the central portion of the tube. For an upward flowing liquid, surface tension, buoyancy, lift, inertia and liquid drag are the forces that determine the bubble size at departure.

Bubbles may slide along the wall for a short distance before moving away from the wall. Lift is the main force pushing the bubbles away from the wall. As bubbles merge in the bulk, large bubbles form leading to slug flow, while nucleate boiling occurs on the heated wall. As more and more vapor is added and liquid is depleted, for low liquid flow rates and low-to-moderate heat fluxes, a vapor core forms with liquid in the form of a film confined to the tube wall. This is annular flow. Heat transfer from the wall occurs via forced convection in the film. Liquid droplets from the film interface can also be entrained in the vapor core, leading to entrained flow regime. As the film dries out with deficiency of liquid, a rapid rise in wall temperature occurs for a constant wall heat flux condition. This is termed as the dryout heat flux. Subsequently, wall temperature might decrease in the flow direction as entrained liquid droplets hit the solid surface and evaporate while forced convection to vapor occurs from the wall. As concentration of liquid droplets in the flow diminishes, heat transfer from the wall is mainly to vapor and a turn around in temperature occurs. Wall and bulk fluid temperatures along the flow direction are also shown in Figures 12.2 and 12.3.

As subcooled liquid enters the tube, its temperature increases with addition of heat from the wall. Correspondingly, wall temperature also increases until nucleate boiling is initiated. At this point, bulk liquid may still be subcooled. In nucleate boiling, the wall temperature remains nearly constant. After reaching saturation temperature, bulk fluid temperature remains at saturation condition. With dryout of the liquid film, wall temperature increases rapidly; however, the fluid temperature rise shows a lag as entrained liquid droplets evaporate while remaining at saturation temperature.

For high wall heat flux high flow rate, similar behavior occurs except as the bulk quality or void fraction increases, a condition similar to that of critical or maximum heat flux in pool boiling develops. This condition occurs when vapor removal rate starts to fall below the vapor production rate at the wall or liquid supply to the wall diminishes. Later in this chapter, we will discuss various correlations/models that have been proposed in the literature for critical heat flux in flow boiling. After occurrence of critical heat flux condition, the wall temperature rapidly increases, and the surface is covered with a vapor film while liquid with bubbles and slugs continues to persist in the middle of the tube. As rate of heat transfer through the vapor film increases due to instability of the interface, turbulence in the vapor film or entrainment of liquid droplets from the core, wall temperature decreases somewhat with distance before increasing again as vapor film becomes thicker. The bulk fluid temperature increases from saturation temperature as vapor in the film is superheated. Liquid remains at saturation temperature until all or most of entrained liquid is evaporated. The bulk fluid temperature remains well below the vapor temperature.

12.2.1 Onset of Nucleate Boiling

With forced convection cooling at the inlet of the heated tube, inception of boiling occurs when nucleation superheat criterion is satisfied. The inception superheat depends on the morphology (size and shape of cavities available on the surface) and wettability of a surface. The procedure outlined in section 12.1.1 can be used to predict for a given heat flux, the inception superheat or the distance from where onset of boiling on the wall occurs when the bulk liquid is still subcooled. Basu et al. [3] have investigated onset of nucleated boiling during subcooled forced flow boiling.

Assuming spherical cavities to be present on the surface, they, as was done in Chapter 4, related the wall superheat at inception to the size of the nucleating cavity as

$$\Delta T_{ONB} = \frac{4\sigma T_{\text{sat}}}{\rho_v h_{fg} D_c} \tag{12.9}$$

The size, D_c, of the nucleating cavity was related to the size of the cavity, D_{co}, corresponding to the minimum wall superheat needed for nucleation and contact angle, φ, measured in degrees as

$$D_{co} = \left[\frac{8\sigma T_{\text{sat}} k_l}{\rho_v h_{fg} q_w}\right]^{1/2} \tag{12.10}$$

$$D_c = F(\varphi) D_{co} \tag{12.11}$$

where

$$F(\varphi) = 1 - \exp\left[-\left(\frac{\pi\varphi}{180}\right)^3 - 0.5\left(\frac{\pi\varphi}{180}\right)\right] \tag{12.12}$$

There are a number of other correlations reported in the literature. For example, Bergles and Rohsenow [4] proposed a correlation for onset of nucleate boiling with water at pressures varying from 0.1 to 13.6 MPa. Their correlation is

$$q_w = 1{,}277.3 p^{1.156} (T_w - T_{\text{sat}})^{\frac{2.3}{p^{0.0234}}} \tag{12.13}$$

Where p is in MPa, $(T_w - T_{sat})$ is in °C and q_w is in $\frac{\text{W}}{\text{m}^2}$.

12.2.2 Bubble Departure

In later stages of subcooled boiling, bubbles start to detach from the surface, slide along the wall, and move into the bulk. As noted earlier, hydrodynamics forces, i.e., liquid drag, inertia, buoyancy, surface tension, and lift force all play a role in determining the location at which bubbles leave the wall. Because of condensation at the upper portions of the bubble, thermal conditions affect the bubble growth rate and thereby influence the bubble departure location. A number of correlations for bubble departure in flow boiling have been reported in the literature. One of these correlations is by Saha and Zuber [5]. Their correlation distinguishes the effect of thermal conditions near the wall and the flow hydrodynamics. They have proposed that for low mass flow rates, thermal processes near the wall affect the condensation at the top region of the bubble and in turn the departure of bubbles. They used Peclet number as the criterion to distinguish between thermally controlled and thermally and hydrodynamically controlled bubble departure. Peclet number is defined as

$$\text{Pe} = \frac{\text{Nu}}{\text{St}} \tag{12.14}$$

where

$$\text{Nu} = \frac{qD_e}{k_l (T_{\text{sat}} - T_b)} \tag{12.15}$$

and

$$\text{St} = \frac{q}{Gc_{pl} (T_{\text{sat}} - T_b)} \tag{12.16}$$

In the above q is wall heat flux, D_e is equivalent tube diameter, k_l is thermal conductivity of the liquid, T_w is wall temperature, T_b is the bulk temperature, and G is the mass velocity.

Thermally controlled departure occurred for $\mathrm{Pe} < 7\times 10^4$. For this case,

$$\mathrm{Nu} = 455 \tag{12.17}$$

or

$$T_{sat} - T_b = 0.0022\left(\frac{qD_e}{k_l}\right) \tag{12.18}$$

For thermally and hydrodynamically controlled departure, which occurs at high mass velocities, i.e., $\mathrm{Pe} > 7\times 10^4$ and

$$\mathrm{St} = 0.0065 \tag{12.19}$$

or

$$T_{sat} - T_b = 154\left(\frac{q}{Gc_{pl}}\right) \tag{12.20}$$

Their correlation covered data for water for pressures 0.1–13.8 MPa and mass velocity G from 95 to 2760 kg/m^{2}s and heat flux, q in the range $0.29 - 1.89\times 10^6 \ \frac{\mathrm{W}}{\mathrm{m}^2}$. In developing the correlation, they also included water and Freon data of other investigators as well [6].

12.2.3 Nucleate Boiling Heat Transfer

Transition from single-phase forced convection to subcooled/saturated nucleate boiling leads to stronger increase in heat flux with wall superheat. This behavior is similar to that for pool boiling. The distance over which this transition occurs is generally small and can be neglected. However, any correlation that includes transition data and saturated nucleate boiling data will generally show a weaker dependence of heat flux on wall superheat. There are a number of correlations developed in the literature for fully developed nucleate boiling. McAdams et al. [7] proposed a correlation for subcooled water flowing in a tube at pressures between 0.2 and 0.62 MPa as

$$q = 21.2\left(T_w - T_{sat}\right)^{3.8} \tag{12.21}$$

where heat flux is in $\frac{\mathrm{W}}{\mathrm{m}^2}$ and $\left(T_w - T_{sat}\right)$ is in °C. For pressures between 3.43 and 13.74 MPa, Jens and Lottes [8] have suggested the following relationship

$$q = \frac{\exp\left(4p/6.2\right)}{(25)^4}\left(T_w - T_{sat}\right)^4 \tag{12.22}$$

However, at higher pressures, Jens and Lottes correlation has been noted to give lower wall temperatures than data. Thom et al. [9] have suggested a similar correlation based on data from 5.96 to 13.74 MPa as

$$q = \frac{\exp(2p/8.7)}{(22.7)^2}(T_w - T_{sat})^2 \tag{12.23}$$

In the above relations (eqs. (12.22) and (12.23)), q is in $\frac{\text{MW}}{\text{m}^2}$, p is in MPa and $(T_w - T_{sat})$ is in °C. Weaker dependence of heat flux on wall superheat in Thom et al. [9] correlation is due to their inclusion of data for transition from single-phase forced convection to fully developed nucleate boiling. It should be noted that these correlations are independent of quality or flow regimes that exit in the tube.

Example:

For the example given in Chapter 11, calculate the wall superheat in nucleate boiling using the correlation of Thom et al.

Solution:

According to the correlation of Thom et al.,

$$q = \frac{\exp\left(\frac{2p}{8.7}\right)}{(22.7)^2}(T_w - T_{sat})^2$$

$$0.5 = \frac{\exp\left(\frac{2 \times 7}{8.7}\right)}{(22.7)^2}(T_w - T_{sat})^2$$

or

$$T_w - T_{sat} = 7.18°\text{C}$$

Chen [10] using a large variety of data has developed a correlation that includes contributions of both forced convection and nucleate boiling heat transfer. Chen's correlation includes data for quality, x, up to 0.7. As such, it could correlate data up to the early part of annular flow when nucleate boiling also occurs in the liquid film. Chen's correlation combines contributions of single-phase forced convection and nucleate boiling as

$$h = h_{nb} + h_{fc} \tag{12.24}$$

The correlation for h_{nb} is a modification of Forster and Zuber [11] equation and is

$$h_{nb} = S(0.00122)\left[\frac{k_l^{0.79} c_{pl}^{0.45} \rho_l^{0.49}}{\sigma^{0.5} \mu_l^{0.29} h_{fg}^{0.24} \rho_v^{0.24}}\right]\Delta T^{0.24} \Delta p^{0.75} \tag{12.25}$$

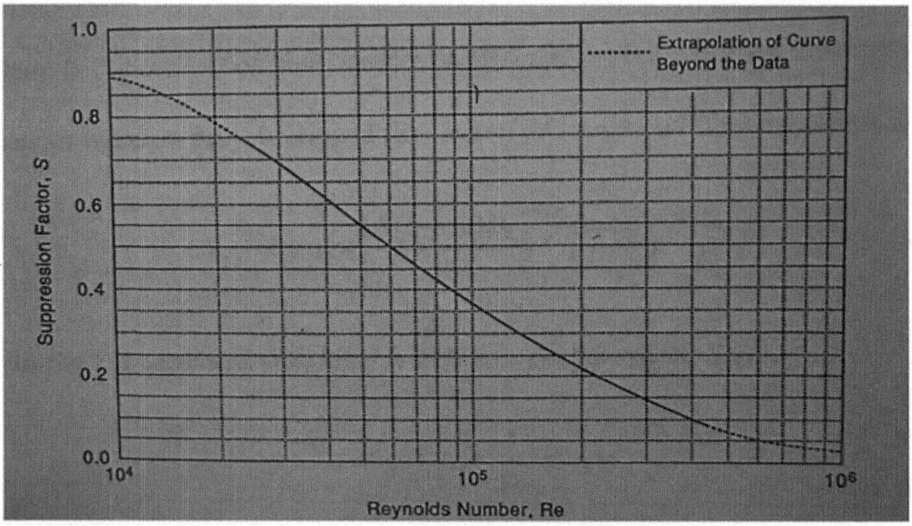

FIGURE 12.4 Suppression factor (S) used in Chen's correlation [10].

where ΔT is $T_w - T_{sat}$ and Δp is difference between pressures for which T_w and T_{sat} are the saturation temperatures respectively, i.e.,

$$\Delta p = p(T_w) - p(T_{sat}) \tag{12.26}$$

In eq. (12.25), if SI units are used, Δp and ΔT are in N/m^2 and $°K$, respectively. S is the suppression factor and is a function of Reynolds number based on total liquid mass flow rate. It is plotted in Figure 12.4 and is correlated as [10]

$$S = \frac{1}{1 + 2.53 \times 10^{-6}\,\mathrm{Re}^{1.17}} \tag{12.27}$$

where

$$\mathrm{Re} = \mathrm{Re}_l\, F^{1.25} \equiv \left(\frac{GD_h}{\mu_l} \right) F^{1.25} \tag{12.28}$$

Single-phase forced convection heat transfer coefficient, h_{fc}, is a modified Dittus Boelter correlation for fully developed turbulent flow based on local mass flow rate of liquid, i.e.,

$$h_{fe} = 0.023 \left[\frac{G(1-x)D_h}{\mu_l} \right]^{0.8} \left(\frac{\mu_l c_{pl}}{k_l} \right)^{0.4} \frac{k_l}{D_h} F \tag{12.29}$$

The parameter F is correlated with inverse of Martinelli parameter, X_{tt}, as

$$F = 1 \quad \text{for} \quad \frac{1}{X_{tt}} < 0.1 \tag{12.30}$$

$$= 2.33\left(0.213 + \frac{1}{X_{tt}}\right) \quad \text{for} \quad \frac{1}{X_{tt}} > 0.1$$

where

$$\frac{1}{X_{tt}} = \left(\frac{\bar{x}}{1-\bar{x}}\right)^{0.9}\left(\frac{\rho_l}{\rho_v}\right)^{0.5}\left(\frac{\mu_v}{\mu_l}\right)^{0.1} \tag{12.31}$$

Chen's correlation is based on water as test liquid and for

$$\text{Pressure} = 0.17 - 3.5 \text{ MPa}$$

$$\text{Liquid Velocity at Inlet} = 0.06 - 0.45\frac{\text{m}}{\text{s}}$$

$$\text{Quality}, \bar{x} = 0 - 0.7$$

$$\text{Heat Flux up to } 2.4\,\frac{\text{MW}}{\text{m}^2}$$

A number of correlations have been developed for the upper region of annular flow when heat transfer from the liquid film mainly occurs by forced convection with evaporation at the liquid–vapor interface. These correlations mainly utilize the inverse of Martinelli parameter, X_{tt}, to correlate the data. Dengler and Addoms [12] correlated their data as

$$h = 3.5\left(\frac{1}{X_{tt}}\right)^{0.5} h_{fc} \tag{12.32}$$

where

$$h_{fc} = 0.023\left(\frac{GD_h}{\mu_l}\right)^{0.8}\left(\frac{\mu_l c_{pl}}{k_l}\right)^{0.4}\frac{k_l}{D_h}$$

It should be noted that in the above heat transfer coefficient is evaluated assuming liquid alone is flowing in the tube. Benett et al. [13] also obtained a similar correlation with the constant 3.5 replaced by 2.4 and the exponent on $\frac{1}{X_{tt}}$ being 0.66.

12.2.4 Dryout Heat Flux

As described earlier in section 12.2, at low-to-moderate flow rates and heat fluxes, dryout heat flux represents the condition for a given liquid mass flow rate and liquid temperature at inlet at which a liquid film on the wall dries out in annular flow. In the absence of any entrained liquid, this condition will correspond to a local quality $x = 1$. Simple energy balance along the tube length for constant heat flux yields

$$q_{\text{dryout}} = \frac{GD}{4L}\left(c_{pl}\left(T_{\text{sat}} - T_{li}\right) + h_{fg}\right) \tag{12.33}$$

where G is the liquid mass velocity at inlet, D is the diameter of the heated tube, L, is the tube length at which dryout occurs (actual length of the tube can be much longer), c_{pl} is the liquid specific heat, T_{li} is the liquid temperature at inlet, and h_{fg} is the latent heat of vaporization. In arriving at eq. (12.33), pressure drop through the tube is considered to be much smaller than the system pressure. According to eq. (12.33), for a fixed tube length at dryout, the dryout heat flux will increase with mass velocity and liquid subcooling. In practice, because of entrainment of liquid from the interface of the liquid film and existence of some liquid in the vapor core, quality at dryout is less than unity. Since it is the total energy deposited into the liquid stream that determines the dryout location in a tube, the profile of the heat flux applied to the tube wall has little effect on the quality at dryout.

At high flow rates and heat fluxes, the occurrence of critical heat flux on tube wall is similar to that for pool boiling. Now the blanketing of the surface with vapor with rapid rise in wall temperature occurs even when the core is full of liquid. This condition develops when the vapor removal rate from the surface falls short of the vapor production rate. The phenomenon responsible for the critical heat flux, which is often referred to as "Departure from Nucleate Boiling" or q_{DNB} is the result of imbalance of vapor removal rate and liquid supply rate to the surface. We discuss it next. However, it should be noted that to assess the safety margin, departure from nucleate boiling ratio or DNBR is evaluated. It is simply the ratio of q_{CHF} to local q.

12.2.5 Critical or Maximum Nucleate Boiling Heat Flux

A number of correlation for critical heat flux in tubes and annuli have been reported in the literature. These correlations have been developed by nuclear reactor vendors and academic researchers. One of these correlations is by Tong [14] developed at Westinghouse. The correlation has been developed for uniform heat flux distribution along the tube but can be extended to situations when heating is nonuniform. It can be applied to tubes, annuli, and rod bundles. The correlation is developed for water. For uniformly heated channels, Tong's correlation for critical heat flux as given in [6] is:

$$\begin{aligned} q_{CHF,\,EU} = \Big[&(2.022 - 0.06238p) + (0.1722 - 0.001427p) \\ &\exp(18.177 - 0.5987p)x\Big] \\ &\cdot\Big[(0.1484 - 1.596x + 0.1729x|x|)\cdot 2.326G + 3271\Big] \\ &\cdot[1.157 - 0.869x]\cdot\big[0.2664 + 0.837\exp(-1.241D_h)\big] \\ &\cdot\big[0.8258 + 0.0003413(h_f - h_{li})\big] \end{aligned} \tag{12.34}$$

where q is in $\frac{\text{kW}}{\text{m}^2}$, p is in MPa, G is in $\frac{\text{kg}}{\text{m}^2\text{s}}$ and D_h is in cm, h_{li} is liquid enthalpy at inlet, h_f is enthalpy of liquid at saturation temperature in kJ/kg, x is the equilibrium quality, and p is the pressure in MPa. The correlation is based on data for

$$p = 5.58 - 13.8 \text{ MPa}$$

$$G = 1.356 \times 10^3 - 6.78 \times 10^3 \frac{\text{kg}}{\text{m}^2\text{s}}$$

$$D_h = 0.51 - 1.78 \text{ cm}$$

$$x = -0.15 \text{ to } 0.15$$

$$L = 0.254 - 3.658 \text{ m}$$

$$\frac{\text{Heated Perimeter}}{\text{Wetted Perimeter}} = 0.88 - 1.0$$

For nonuniform heating of the channels, Tong [14] suggested a correction factor for relating CHF in uniform to nonuniform heated channels as

$$q_{CHF,\,EU} = q_{CHF,\,NU} \cdot F \tag{12.35}$$

where F is defined as

$$F = \frac{C}{q_{\text{local}}\left[1 - \exp\left(-CL_{CHF,\,EU}\right)\right]} \int_0^{L_{CHF}} q(z)\exp(-C(L_{CHF} - z))dz \tag{12.36}$$

$$C = \frac{4.23 \times 10^6 (1 - x_{CHF})^{7.9}}{G^{1.72}} \text{ m}^{-1} \tag{12.37}$$

L_{CHF} is the length at which critical heat flux occurs in a uniformly heated tube.

Tong's correlation was developed for use in pressurized water reactors (PWRs), which have normal reactor vessel pressure of about 13.8 MPa.

For boiling water reactors (BWRs) conditions $p \simeq 7$ MPa, Janssen and Levy [15] have proposed

$$q = 10^6\left[2.22 + 0.55\frac{G}{10^3}\right] \quad \text{for} \quad x < x_1 \tag{12.38}$$

$$= 10^6\left[5.147 - \frac{0.628G}{10^3} - 4.71x\right] \quad \text{for} \quad x_1 < x < x_2 \tag{12.39}$$

$$= 10^6\left[1.906 - \frac{0.381G}{10^3} - 0.653x\right] \quad \text{for} \quad x_2 < x \tag{12.40}$$

where x is the quality and

$$x_1 = 0.197 - 0.0797\frac{G}{10^3} \tag{12.41}$$

$$x_2 = 0.254 - 0.019\frac{G}{10^3} \tag{12.42}$$

For pressures between 4.2 and 9.8 MPa , the correlation has been modified as

$$q_{CHF} = q_{CHF}(13.8\,\text{MPa}) + 1.004\times 10^5(13.8 - p) \tag{12.43}$$

In the above heat flux is in $\frac{\text{W}}{\text{m}^2}$, G is in $\frac{\text{kg}}{\text{m}^2\text{s}}$ and pressure, p, is in MPa. The range of parameters over which the correlations have been developed are:

$$p = 8.28 - 20\,\text{MPa}$$

$$G = 5.44\times 10^2 - 8.16\times 10^3\ \frac{\text{kg}}{\text{m}^2\text{s}}$$

$$x = -\text{ve (subcooled liquid) to } 0.45$$

$$D_h = 0.622\ -\ 3.17\,\text{cm}$$

$$L = 0.737 - 2.74\,\text{m}$$

Bernath [16] has developed a correlation based on heat transfer coefficient in nucleate boiling and the wall temperature difference. According to him,

$$q_{CHF} = h_{CHF}\left(T_{w,CHF} - T_b\right) \tag{12.44}$$

where T_b is bulk temperature in °C and

$$h_{CHF} = 3.43\times 10^4\left(\frac{D_h}{D_h \pm D_i}\right) + \frac{77.79V}{D_h^{0.6}}\ \frac{\text{W}}{\text{m}^2\text{K}} \quad \text{for} \quad D_h \le 0.03\,\text{m} \tag{12.45}$$

$$T_{W,CHF} = \left(203.9\ln p - 54\left(\frac{p}{p + 0.103}\right) - 0.82V\right) + 17.78\ ^\circ\text{C} \tag{12.46}$$

And + sign is used when $D_i > D_h$ and − sign when $D_i < D_h$

$$p = \text{System Pressure in MPa } (0.158\ -\ 20.619\,\text{MPa})$$

$$V = \text{Liquid velocity at inlet, m/s } (1.22 - 16.46\,\text{m/s})$$

T = Temperature in °C
h = heat transfer coefficient in $\frac{\text{W}}{\text{m}^2\text{K}}$

D_h = Hydraulic diameter in m (0.0036 – 0.0168 m)
D_i = inner diameter of an annulus, m

Several other correlations for critical heat flux in tubes and annuli have been reported in the literature. Some of these correlations have been extended to flow over rod bundles. Todreas and Kazimi [6] in their text list a number of such correlations.

12.2.6 Post-Critical Heat Flux Heat Transfer

After occurrence of critical heat flux, on a tube wall, vapor film covers the tube surface, whereas two-phase mixture prevails in the core. A number of correlations for post-critical heat transfer or for flow film boiling have been reported in the literature, and most reprehensive of them is the correlation of Groeneveld [17]. In his work as reported in [6], Groeneveld correlated water-steam data from tubes, annuli, and rod bundles. His correlation is

$$\mathrm{Nu} = a\left[\mathrm{Re}_v\left[\bar{x}+\frac{\rho_v}{\rho_l}(1-\bar{x})\right]\right]^b (\mathrm{Pr}_v)^c\, Y \tag{12.47}$$

where $\mathrm{Nu} = \dfrac{hD}{k_v}$ and $\mathrm{Re}_v = \dfrac{GD}{\mu_v}$

and

$$Y = \left[1-0.1\left(\frac{\rho_l-\rho_v}{\rho_v}\right)^{0.4}(1-\bar{x})^{0.4}\right]^d \tag{12.48}$$

The values of constants a, b, c, and d as well as range of parameters are given below:

Constant	a	b	c	d
Tubes	1.09×10^{-3}	0.989	1.41	–1.15
Annuli	5.2×10^{-2}	0.688	1.26	–1.06

Range of Parameters		
Parameter	**Tubes (Vertical or Horizontal)**	**Annuli (Vertical)**
D (mm)	2.5 – 25	1.5 – 6.3
p (MPa)	6.8 – 21.5	3.4 – 10.0
G (kg/m^{2}s)	700 – 5,300	800 – 4,100
q (kW/m^2)	120 – 2,100	450 – 2,250
$\bar{x}$	0.1 – 0.9	0.1 – 0.9

12.3 SUMMARY

In this chapter, we have described both external and internal flow boiling. For external flow boiling, a correlation for maximum heat flux during cross flow over cylinders is given. Internal flow boiling correlations include those for onset of nucleate boiling, nucleate boiling heat flux, critical heat flux, and post-critical heat flux heat transfer. The information developed is introductory.

EXERCISES

1. Subcooled water at a temperature of 570 K enters an annulus, with inner tube diameter of 1.5 cm and outer tube diameter of 2.0 cm. The pressure at inlet is 7 MPa. Inner tube is electrically heated uniformly with a wall heat flux of 10^5 W/m^2, whereas the outer tube is insulated. Mass velocity of water entering the annulus is 10^6 kg/hr m^2. Calculate the location at which water will reach its saturation temperature. What is the quality at exit and what would be the critical heat flux?
2. Saturated water at 7 MPa enters a 1.5 cm diameter electrically heated tube. The tube is 3.5 m long. The wall heat flux is 5×10^5 W/m^2 and the mass velocity of the water entering the tube is 5×10^6 kg/hr m^2. Obtain axial variation of the tube wall temperature. To obtain the heat transfer coefficient, use correlation of Jens and Lottes.
3. Use the expression of Janssen and Levy to determine the critical heat flux in Problem 2. What is the DNBR?

NOMENCLATURE

c_p	Specific Heat
D	Diameter
D_c	Diameter of a nucleate cavity
Fr	Froude number
G	Mass velocity
g	Gravitational acceleration
h	Heat transfer coefficient
h_{fg}	Latent heat of vaporization
k	Thermal conductivity
l	Characteristic length of the heater
Nu	Nusselt number
p	Pressure
Pe	Peclet Number
q	Heat Flux
R	Radius of the cylinder
St	Stanton number
S	Suppression factor
T	Temperature

U	Imposed velocity
We	Weber number
x	Quality

SUBSCRIPTS

e	Equivalent
f	Saturation
h	Hydraulic diameter
l	Liquid
li	Liquid at inlet
max	Maximum
sat	Saturation
v	Vapor
w	Wall

GREEK LETTERS

Γ	Solute concentration
δ	Thermal layer thickness
ε	Characteristic energy scale
ρ	Density
σ	Surface tension
Φ	Potential function
ϕ	Angular position or contact angle

REFERENCES

1. Hasan, M.Z., Hasan, M.M., Eichhorn, R., and Lienhard, J.H. (1981), "Boiling Burnout during Cross Flow Over Cylinders, Beyond the Interface of Gravity." *ASME, J. Heat Transfer*, Vol. 103, pp. 478–484.
2. Lance, R.P. and Meyer, J.E. (1950), "Local Boiling Coefficients on a Horizontal Tube." *AIChE J.*, Vol. 4, pp. 75–80.
3. Basu, N., Warrier, G.R., and Dhir, V.K. (2002), "Onset of Nucleate Boiling and Active Nucleation Site Density during Flow Boiling." *J. Heat Transfer*, Vol. 124, pp. 717–728.
4. Bergles, A.E. and Rohsenow, W.M. (1964), "The Determination of Forced Convection Surface Boiling Heat Transfer". *Journal of Heat Transfer*, Vol. 86, pp. 365.
5. Saha, P. and Zuber, N. (1974), "Point of Net Vapor Generation and Vapor Void Fraction in Subcooled Boiling", *Proceedings of the 5th International Heat Transfer Conference*, Vol. IV, Tokyo.
6. Todreas, N.E. and Kazimi, M.S. (1989), *Nuclear Systems – Thermal Hydraulics Fundamentals*. Hemisphere Publishing Corporation, New York.
7. McAdams, W.H., Kennel, W.E., Minden, G.S., Carl, R., Picernell, P.M., and Dew, H.E. (1949), "Heat Transfer at High Rates of Water with Surface Boiling." *Ind. Chem. Eng.*, Vol. 41, p. 1945.
8. Jens, W.H. and Lottes, P.A. (1951), "Analysis of Heat Transfer, Burnout, Pressure Drop, and Density Data for High Pressure Water." USAEC Report ANL-4627.

9. Thom, J.R.S., Walker, W.M., Fallon, T.A., and Reising, G.F.S. (1966), "Boiling in Subcooled Water Flowing in Tubes and Annuli", *Proc. Inst. Mech. Eng.*, pp. 180–226.
10. Chen, J.C. (1963), "A Correlation for Boiling Heat Transfer to Saturated Fluids in Convective Flow." Paper No. 63-HT-34, ASME.
11. Forster, H.K. and Zuber, N. (1955), "Dynamics of Vapor Bubbles and Boiling Heat Transfer", *AIChE J.*, Vol. 1, pp. 531–535.
12. Dengler, C.E. and Addoms, J.M. (1956), "Heat Transfer Mechanism for Vaporization of Water in a Vertical Tube." *Chem. Eng. Prog. Ser.*, Vol. 52, p. 95.
13. Bennett, J. A. R.; Collier, J. G.; Pratt, H. R. C.; and Delaney, T. J. (1961), "Heat transfer to two-phase gas-liquid systems. Part I: Steam-water mixtures in the liquid-dispersed region in an annulus", Trans. Inst. Chem. Eng., Vol. 39, PP 113–126.
14. Tong, L.S. (1972), "Boiling Crisis and Critical Heat Flux." IID-25887, AEC Critical Review Series, USAEC.
15. Janssen, E. and Levy, S. (1962), *Burn Out Limit Curves for Boiling Water Reactors.*, APED-3892, General Electric Company. New York.
16. Bernath, L. (1960), "A Theory of Local Boiling and Its Application to Existing Data." *Chem. Eng. Prog. Symp. Ser.*, Vol. 56, No. 30, p. 95.
17. Groeneveld, D.C. (1973), *Post-Dryout Heat Transfer at Operating Conditions*, AECL-4309, Atomic Energy of Canada Ltd, Chalk River Canada.

13 Numerical Simulations of Pool Boiling

INTRODUCTION

In this chapter, we describe in detail the approach that has been used to simulate pool nucleate and film boiling numerically. The technique used in carrying out the simulations is described in detail, and the results are validated with data from experiments for partial and fully developed nucleate boiling and for film boiling on both a horizontal plate and a cylinder held horizontally. Film boiling at very high pressures is also investigated.

13.1 NUMERICAL ANALYSIS

Although the conservation equations for both vapor and liquid phases in nucleate and film boiling are the same, the physics of the two processes is quite different. As such in the following, we describe each process separately.

13.2 NUCLEATE BOILING SIMULATION

In partial nucleate boiling, the nucleation sites are sparsely populated, and as a result, bubble–bubble interactions are not important. As such, one can focus only on the thermal and hydrodynamic processes at and around a single nucleation site. Because of the symmetry that is expected to exist around the cavity, axisymmetric analysis can be carried out for single bubbles in partial nucleate boiling. However, in fully developed nucleate boiling, when bubbles start to merge in the lateral direction, one needs to carry out transient three-dimensional analysis. In analyzing the growth of a single bubble, Son et al. [1] divided the computational domain into micro and macro regions, as shown in Figure 13.1. The micro region contains the thin film that forms underneath the bubble, whereas the macro region consists of the bubble and the liquid surrounding the bubble away from the wall. In carrying out the analysis, the flow is assumed to be laminar, and the fluid properties, including density, viscosity, and thermal conductivity, are evaluated at a mean temperature in each phase.

13.2.1 Micro Region

In the microlayer, liquid flows radially inward, and mass flow rate of liquid in the thin film decreases as evaporation takes place at the vapor–liquid interface. For the microlayer, lubrication theory is employed, and it has been used and validated by a

DOI: 10.1201/9781032668437-13

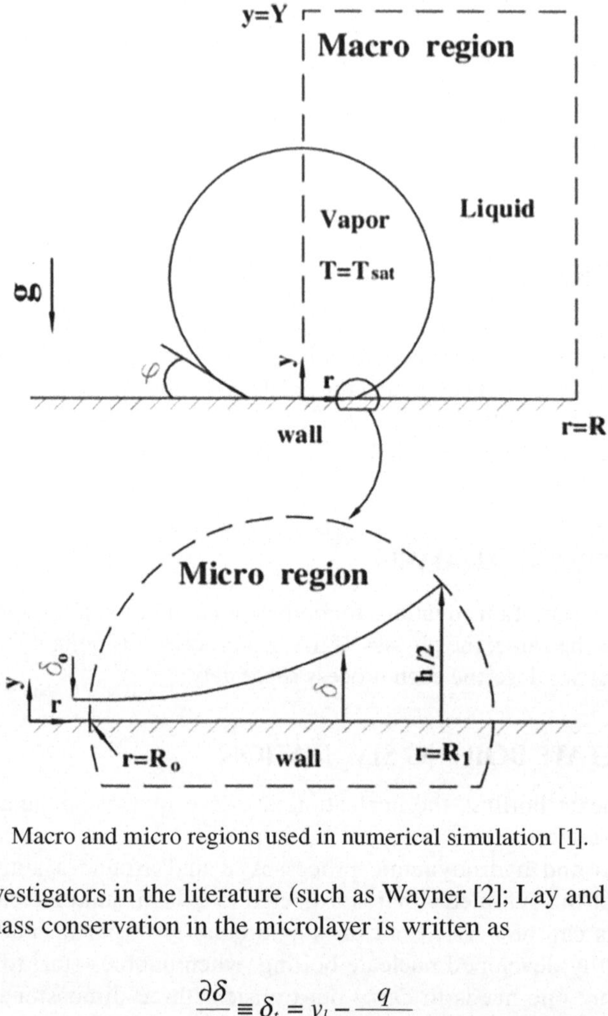

FIGURE 13.1 Macro and micro regions used in numerical simulation [1].

number of investigators in the literature (such as Wayner [2]; Lay and Dhir [3]). The equation of mass conservation in the microlayer is written as

$$\frac{\partial \delta}{\partial t} \equiv \delta_t = v_l - \frac{q}{\rho_l h_{fg}} \tag{13.1}$$

In eq. (13.1), the liquid velocity normal to the liquid–vapor interface, v_l, is obtained from the continuity equation as

$$v_l = -\frac{1}{r}\int_0^{\delta} r u_l \, dy \tag{13.2}$$

where u_l is the liquid velocity in the radial direction. Neglecting inertia, the momentum equation for the microlayer neglecting inertia is written as

$$\frac{\partial p_l}{\partial r} = \mu_l \frac{\partial^2 u_l}{\partial y^2} \tag{13.3}$$

The energy conservation equation for the film yields

$$q = \frac{k_l (T_w - T_{sat})}{\delta} \tag{13.4}$$

The inertial terms in the momentum equation and convection terms in the energy equation have been neglected because the Reynolds number based on the liquid film thickness and the liquid velocity in the micro region is less than unity.

Using the modified Clausius–Clapeyron equation, (e.g., Wayner [2]), the evaporative heat flux is written as

$$q = h_{ev}\left[T_{int} - T_v + \frac{(p_l - p_v)T_v}{\rho_l h_{fg}}\right] \tag{13.5}$$

where

$$h_{ev} = \left(\frac{1}{2\pi R_v T_v}\right)^{0.5} \frac{\rho_v {h_{fg}}^2}{T_v} \tag{13.6}$$

and T_v is taken to be T_{sat} corresponding to p_v. The second term in eq. (13.5) results from the fact that pressure in the liquid can be less than that in the vapor. The pressures in the vapor and liquid phases are related [3] as

$$p_l = p_v - \sigma\kappa - \frac{A}{\delta^3} + \frac{q^2}{\rho_v {h_{fg}}^2} \tag{13.7}$$

where surface tension, σ, was taken to be a function of temperature, and A is the dispersion constant relating disjoining pressure to the film thickness. Dispersion constant represents intramolecular force between liquid molecules and solid atoms of the substrate. Typical value of A used in this work is -8.5×10^{-21} J. In eq. (13.7), the second term on the right-hand side accounts for the capillary pressure, the third term for the disjoining pressure, and the last term originates from the recoil pressure due to momentum of vapor molecules leaving the interface being much higher than the liquid molecules approaching the interface. The curvature of the interface is defined as

$$\kappa = \frac{1}{r}\frac{\partial}{\partial r}\left[\frac{r\dfrac{\partial \delta}{\partial r}}{\sqrt{1+\left(\dfrac{\partial \delta}{\partial r}\right)^2}}\right] \tag{13.8}$$

The integration of eq. (13.3) in the y direction with the boundary conditions of $u = 0$ at $y = 0$ and $\dfrac{\partial u_l}{\partial y} = 0$ at $y = \delta$ yields

$$\frac{dp_l}{dr} = -\frac{3\mu_l \bar{u}_l}{\delta^2} \tag{13.9}$$

where $\bar{u}_l$ the liquid velocity averaged across over the microlayer. Assuming that the vapor state is at $p_v = p_{sat}$ and $T_v = T_{sat}$, the combination of eqs. (13.4) and (13.5) yields

$$p_l - p_{sat} = -\frac{\rho_l h_{fg}}{T_{sat}}\left(T_{int} - T_{sat} - \frac{q}{h_{ev}}\right) \tag{13.10}$$

Assuming the microlayer is quasi-steady $\left(\frac{\partial \delta}{\partial t} = 0\right)$, the combination of the above equations yields four first-order differential equations for δ, δ', T_{int} and $\bar{u}_l$:

$$\frac{d\delta}{dr} = \delta' \tag{13.11}$$

$$\frac{d\delta'}{dr} = -\frac{\delta'\left(1+\delta'^2\right)}{r} + \frac{\left(1+\delta'^2\right)^{1.5}}{\sigma}\left[-\left(p_l - p_{sat}\right) - \frac{A}{\delta^3} + \frac{q^2}{\rho_v {h_{fg}}^2}\right] \tag{13.12}$$

$$\frac{dT_{int}}{dr} = \frac{3\mu_l T_{sat} h_{ev} \bar{u}}{\rho_l h_{fg} \delta\left(k_l + h_{ev}\delta\right)} - \frac{q\delta'}{k_l + h_{ev}\delta} \tag{13.13}$$

$$\frac{d\left(r\delta\bar{u}_l\right)}{dr} = -\frac{rq}{\rho_l h_{fg}} \tag{13.14}$$

where $'$ denotes $\frac{d}{dr}$, q is obtained from eq. (13.4), and $p_l - p_{sat}$ from eq. (13.10).

The boundary conditions for the above equation are as follows:

At $r = R_0$,

$$\delta = \delta_0; \quad \delta' = \bar{u}_l = 0 \tag{13.15}$$

where δ_o is of the order of molecular size [3], for example, $\delta_o = 6\times10^{-10}$ m. Since no evaporation occurs at $r = R_0$, δ' and $\bar{u}_l$ are zero there.

At $r = R_1$,

$$\delta = \frac{h}{2}; \quad \delta' = \tan\varphi; \quad \delta'' = 0 \tag{13.16}$$

where $\frac{h}{2}$ is the distance to the first computational node for the LS function ϕ from the wall and φ is an apparent contact angle that can be measured experimentally. In implementing the above boundary conditions, the radius R_1, radial location of the liquid–vapor interface at $y = h/2$, was determined from the solution of macro region. For a given contact angle φ, eqs. (13.11)–(13.14) are integrated using six boundary conditions, eqs. (13.15) and (13.16), under the condition that the dispersion constant A and R_0 are unknown. The practical procedure to determine A and R_0 for given φ and R_1 is as follows:

i. guess R_0,
ii. guess A,
iii. guess T_{int} at $r = R_0$,
iv. integrate eqs. (13.11)–(13.14) from $r = R_0$ and $\delta = \delta_0$, using a fourth-order Runge–Kutta method with adaptive grid sizes, to reach $\delta = h/2$,
v. repeat steps (iii)–(iv) if $\delta'' \neq 0$ at $\delta = h/2$,
vi. repeat steps (ii)–(v) if $\delta' \neq \tan\varphi$ at $\delta = h/2$,
vii. repeat steps (i)–(vi) if $r \neq R_1$ at $\delta = h/2$.

Considering that the averaged $\bar{q}_{\text{micro}}$, which is defined as $\bar{q}_{\text{micro}} = 2\int_{R_0}^{R_1} qrdr/\left(R_1^2 - R_0^2\right)$, is nearly independent of R_1 as long as $(R_1 - R_0)/R_1 << 1$, a significant amount of computing time can be saved on the above iterative calculations.

13.2.2 Macro Region

To analyze the macro region numerically, the Level Set (LS) formulation [4] modified by Son et al. [1] to accommodate the effect of phase change at the interface is used. The interface separating the two phases is captured by ϕ, which is defined as a signed distance from the interface. The negative sign is chosen for the vapor phase and the positive sign for the liquid phase. In the analysis, it is assumed that the vapor phase is maintained at the saturation temperature corresponding to pressure in the vapor space.

The equations governing conservation of momentum, energy, and mass for the liquid–vapor region are written in terms of Heaviside function, H, as

$$\hat{\rho}\left(\frac{\partial \boldsymbol{u}}{\partial t} + \boldsymbol{u}\cdot\nabla\boldsymbol{u}\right) = -\nabla p + \hat{\rho}g - \hat{\rho}\beta_T\left(T - T_{\text{sat}}\right)g - \sigma\kappa\nabla H + \nabla\cdot\hat{\mu}\nabla u + \nabla\cdot\hat{\mu}\nabla\boldsymbol{u}^T \quad (13.17)$$

$$\rho_l c_{pl} H\left(\frac{\partial T}{\partial t} + \boldsymbol{u}\cdot\nabla T\right) = \nabla\cdot\hat{k}\nabla T \quad \text{for } H > 0$$

$$T = T_{\text{sat}}\left(p_v\right) \quad \text{for } H = 0 \quad (13.18)$$

$$\nabla\cdot u = \frac{\dot{\mathbf{m}}}{\hat{\rho}^2}\cdot\nabla\hat{\rho} + \dot{V}_{\text{micro}} \quad (13.19)$$

where

$$\hat{\rho} = \rho_v + \left(\rho_l - \rho_v\right)H$$

$$\hat{\mu}^{-1} = \mu_v^{-1} + \left(\mu_l^{-1} - \mu_v^{-1}\right)H$$

$$\hat{k}^{-1} = k_l^{-1}H$$

$$H = 1 \text{ in the liquid, if } \phi \geq +1.5h$$

$$= 0 \text{ in the vapor, if } \phi \leq -1.5h$$

$$= 0.5 + \frac{\phi}{3h} + \sin\left[\frac{2\pi\phi}{3h}\right]/(2\pi) \quad \text{if } |\phi| \leq 1.5h$$

where h is a grid spacing. This equation implies that the interface separating two phases is replaced by a transition region of finite thickness ($3h$). The interface thickness can be maintained as constant for all the time while keeping the LS function as a distance function.

The volume source term included in eq. (13.19) due to liquid–vapor phase change is derived from the conditions of the mass continuity and energy balance at the interface:

$$\dot{\mathbf{m}} = \hat{\rho}(\mathbf{U} - \mathbf{u}) = \hat{k}\nabla T / h_{fg} \tag{13.20}$$

where $\boldsymbol{U}$ is the interface velocity.

In eq. (13.19), $\dot{V}_{\text{micro}}$ is obtained from the microlayer solution as

$$\dot{V}_{\text{micro}} = \int_{R_0}^{R_1} \frac{k_l (T_w - T_{\text{int}})}{\rho_v h_{fg} \delta \Delta V_{\text{micro}}} r dr \tag{13.21}$$

where ΔV_{micro} is a vapor-side control volume near the micro region. In the LS formulation, the liquid–vapor interface is described as $\phi = 0$. The zero LS of ϕ is advanced by the interfacial velocity while solving the following equation:

$$\frac{\partial \phi}{\partial t} = -\boldsymbol{U} \cdot \nabla \phi \tag{13.22}$$

where

$$\mathbf{U} = \mathbf{u} + \dot{\mathbf{m}}/\rho \tag{13.23}$$

where ρ is the interpolated density. It was noted by Sussman et al. [4] that the LS function should be maintained as a distance function at all times to prevent excessive numerical errors. Therefore, at each time step, the LS function is reinitialized as a distance function, $|\nabla \phi| = 1$, from the interface by obtaining a steady-state solution of the equation:

$$\frac{\partial \phi}{\partial \tau} = \frac{\phi_o}{\sqrt{\phi_o^2 + h^2}} \left(1 - |\nabla \phi|\right) \tag{13.24}$$

where ϕ_o is a solution of eq. (13.22).

The boundary conditions for the governing equations for the macro region are as follows:

at the wall ($y = 0$),

$$u = v = 0,\ T = T_w,\ \frac{\partial \phi}{\partial y} = -\cos\varphi \tag{13.25}$$

at the planes of symmetry ($r = 0, R$),

$$u = \frac{\partial v}{\partial r} = \frac{\partial T}{\partial r} = \frac{\partial \phi}{\partial r} = 0 \tag{13.26}$$

at the top of computational domain (free surface, $y = Y$),

$$\frac{\partial u}{\partial y} = \frac{\partial v}{\partial y} = \frac{\partial \phi}{\partial y} = 0,\ T = T_{\text{sat}} \tag{13.27}$$

When discretizing the governing equations temporally, the diffusion terms are treated by a fully implicit scheme and the convection and source terms by a first-order explicit method. With that, the discretized governing equations are expressed as

$$\frac{\phi_o - \phi^n}{\Delta t} = -U^n \cdot \nabla \phi^n \tag{13.28}$$

$$\frac{\partial \phi^{n+1}}{\partial \tau} = \frac{\phi_o}{\sqrt{\phi_o^{\,2} + h^2}}\left(1 - \left|\nabla \phi\right|^{n+1}\right) \tag{13.29}$$

$$\rho_l c_{pl} H \frac{T^{n+1} - T^n}{\Delta t} = -\rho_l c_{pl} H \boldsymbol{u}^n \cdot \nabla T^n + \nabla \cdot \hat{k} \nabla T^{n+1} \tag{13.30}$$

$$\hat{\rho}\frac{\boldsymbol{u}^{n+1} - \boldsymbol{u}^n}{\Delta t} = -\hat{\rho}\boldsymbol{u}^n \cdot \nabla \boldsymbol{u}^n - \nabla p^{n+1} - \sigma\kappa\nabla H + \hat{\rho} g + \nabla \cdot \hat{\mu} \nabla \boldsymbol{u}^{n+1} + \nabla \cdot \hat{\mu} \nabla \boldsymbol{u}^{nT} \tag{13.31}$$

$$\nabla \cdot \boldsymbol{u}^{n+1} = \frac{\dot{\boldsymbol{m}}^{n+1}}{\hat{\rho}^2} \cdot \nabla \hat{\rho} + \dot{V}_{\text{micro}} \tag{13.32}$$

where superscripts n and $n+1$ represent n and $n+1$ time steps, respectively. Fluid properties and the step function are evaluated from ϕ^{n+1}.

In order to obtain the governing equation for pressure that achieves mass conservation, the fractional-step method, or projection method, is used. This method has been developed by Chorin [5], Kim and Moin [6], and Bell and Colella [7]. The projection method is formulated in a staggered grid system in which the locations for velocity components are displaced from those for pressure and temperature (refer to Figure 13.2). This is done to avoid the difficulty caused by implementation

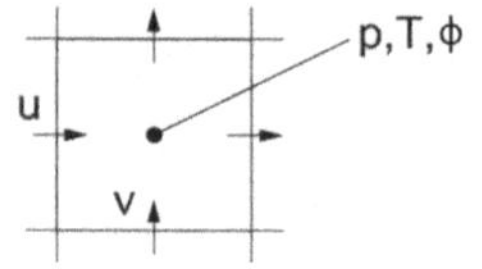

FIGURE 13.2 Staggered grid [1].

of pressure boundary conditions. The momentum eq. (13.31) is decomposed into two fractional steps:

$$\hat{\rho}\frac{\mathbf{u}^* - \mathbf{u}^n}{\Delta t} = -\hat{\rho}\mathbf{u}^n \cdot \nabla \mathbf{u}^n - \nabla p^{n+1} - \sigma\kappa\nabla H + \hat{\rho} g + \nabla \cdot \hat{\mu}\nabla \mathbf{u}^* + \nabla \cdot \hat{\mu}\nabla \mathbf{u}^{nT} \tag{13.33}$$

$$\mathbf{u}^{n+1} = \mathbf{u}^* - \Delta t\left(p^{n+1} - p^n\right)/\hat{\rho} \tag{13.34}$$

First, eq. (13.33) is solved using the pressure evaluated at the previous time step. Thereafter, the resulting velocity, $\boldsymbol{u}^*$, which does not satisfy the continuity equation, is corrected as given by eq. (13.34). When substituting eq. (13.34) into eq. (13.33) and then comparing with eq. (13.31), the projection error can be estimated to be of the order of $\Delta t\left(p^{n+1} - p^n\right)$. Using eq. (13.34) and continuity eq. (13.32), the governing equation for pressure is obtained as:

$$\nabla \cdot \frac{1}{\hat{\rho}}\nabla p^{n+1} = \nabla \cdot \frac{1}{\hat{\rho}}\nabla p^n + \frac{1}{\Delta t}\nabla \cdot \boldsymbol{u}^* - \frac{1}{\Delta t}\frac{\dot{m}^{n+1}}{\hat{\rho}^2} \cdot \nabla \hat{\rho} - \dot{V}_{\text{micro}} \tag{13.35}$$

While discretizing the differential equations spatially, a second-order central difference method is used for the diffusion terms and the interfacial curvature κ. However, to prevent numerical oscillations, a second-order ENO method described by Chang et al. [8] is adopted for the convection terms and the distance function, $|\nabla\phi| = 1$. The discretized equations are solved iteratively by a line-by-line TDMA (Tridiagonal-Matrix Algorithm) supplemented by Gauss Seidel method, which was suggested by Patankar [9]. To enhance the rate of iteration convergence, a relaxation factor is obtained from an orthogonal-residual method. Also, the iterative procedure to solve the pressure equation is combined with a sawtooth type of multigrid method [10].

13.2.3 Macro Region for Nucleate Boiling at High Heat Fluxes

To analyze nucleate boiling at high heat fluxes when bubbles start to merge both vertically and horizontally, Son and Dhir [11] used the sharp-interface LS formulation. The equations governing the conservation of mass, momentum, and energy for each phase are written as

$$\nabla \cdot \boldsymbol{u}_f = 0 \tag{13.36}$$

$$\rho_f\left(\frac{\partial \boldsymbol{u}_f}{\partial t}+\boldsymbol{u}_f\cdot\nabla\boldsymbol{u}_f\right)=-(\nabla p)_f+\rho_f\left[1-\beta_T\left(T_f-T_{\text{sat}}\right)\right]\boldsymbol{g}+\nabla\cdot\mu_f\left(\nabla\boldsymbol{u}+\nabla\boldsymbol{u}^T\right)_f \tag{13.37}$$

$$\rho_f c_f\left(\frac{\partial T_f}{\partial t}+\boldsymbol{u}_f\cdot\nabla T_f\right)=\nabla\cdot k_f(\nabla T)_f \tag{13.38}$$

where the subscript "f" denotes the liquid phase for $\phi > 0$ and the vapor phase for $\phi < 0$. The conservation equations for each phase are coupled through the matching (or jump) conditions at the interface:

$$\boldsymbol{u}_l - \boldsymbol{u}_v = v_{lv}\dot{m}\boldsymbol{n} \tag{13.39}$$

$$\boldsymbol{n}\cdot\left[\mu_l\left(\nabla\boldsymbol{u}+\nabla\boldsymbol{u}^T\right)_l-\mu_v\left(\nabla\boldsymbol{u}+\nabla\boldsymbol{u}^T\right)_v\right]\times\boldsymbol{n}=0 \tag{13.40}$$

$$-p_l+p_v+\boldsymbol{n}\cdot\left[\mu_l\left(\nabla\boldsymbol{u}+\nabla\boldsymbol{u}^T\right)_l-\mu_v\left(\nabla\boldsymbol{u}+\nabla\boldsymbol{u}^T\right)_v\right]\cdot\boldsymbol{n}=\sigma\kappa-v_{lv}\dot{m}^2 \tag{13.41}$$

where $v_{lv} = \rho_v^{-1} - \rho_l^{-1}$. The normal, $\boldsymbol{n}$, to the interface, the interface curvature, κ, and the mass flux, $\dot{m}$, are defined as

$$\boldsymbol{n} = \nabla\phi/|\nabla\phi| \tag{13.42}$$

$$\kappa = \nabla\cdot\boldsymbol{n} \tag{13.43}$$

$$\dot{m} = \rho_f\left(\boldsymbol{U}-\boldsymbol{u}_f\right)\cdot\boldsymbol{n} \tag{13.44}$$

where $\boldsymbol{U}$ is the interface velocity. The interface temperature is specified as a Dirichlet boundary condition, $T_f = T_{\text{sat}}$, and as such, evaporative resistance at the interface is neglected. The mass flux $\dot{m}$ is evaluated from the energy balance at the interface

$$\dot{m} = \mathbf{n}\cdot k_l\nabla T_l/h_{fg} \tag{13.45}$$

For its efficient implementation, $\dot{m}$ is extrapolated into the entire domain (or a narrow band near the interface) by the calculation procedure developed in [12]. Based on the ghost fluid approach [13–17], which is a numerical technique for accurately enforcing the boundary or matching conditions at the interface without being smoothed over several grid spacings, the conservation equations can be rewritten for the liquid–vapor region as

$$\nabla\cdot\boldsymbol{u} = v_{lv}\dot{m}\boldsymbol{n}\cdot\nabla H_\varphi + \dot{V}_{\text{micro}} \tag{13.46}$$

$$\begin{aligned}\hat{\rho}\left(\frac{\partial\mathbf{u}}{\partial t}+\mathbf{u}_f\cdot\nabla\mathbf{u}_f\right)&=-\left[\nabla p+\left(\sigma\kappa-v_{lv}\dot{m}^2\right)\nabla H_\phi\right]+\hat{\rho}\left[1-\beta_T\left(T_f-T_{\text{sat}}\right)\right]\mathbf{g}\\&\quad+\nabla\cdot\hat{\mu}\left[\nabla\mathbf{u}-v_{lv}\dot{m}\mathbf{n}\ \nabla H+\left(\nabla\mathbf{u}-v_{lv}\dot{m}\mathbf{n}\cdot\nabla H_\phi\right)^T\right]\end{aligned} \tag{13.47}$$

$$\rho_f c_{pf}\left(\frac{\partial T_f}{\partial t}+\boldsymbol{u}_f\cdot\nabla T_f\right)=\nabla\cdot\hat{k}_f\nabla T_f \qquad \text{if } \phi\neq 0$$
$$T_f = T_{\text{sat}} \qquad \text{if } \phi = 0 \tag{13.48}$$

where

$$H_\varphi = 1 \quad \text{if } \phi > 0$$
$$= 0 \quad \text{if } \phi \leq 0$$
$$\boldsymbol{u}_l = \boldsymbol{u} + v_{lv}\dot{m}\boldsymbol{n}\left(1 - H_\phi\right)$$
$$\boldsymbol{u}_v = \boldsymbol{u} - v_{lv}\dot{m}\boldsymbol{n}H_\phi$$
$$\hat{\rho} = \rho_v\left(1 - F_\varphi\right) + \rho_l F_\phi$$
$$\hat{\mu}^{-1} = \mu_v^{-1}\left(1 - F_\phi\right) + \mu_l^{-1}F_\phi$$
$$\hat{k}_f = \hat{k}_l = k_l/F_\varphi \quad \text{if } \phi > 0$$
$$\hat{k}_v = k_v/1 - F_\varphi \quad \text{if } \phi < 0$$

Here, H is the discontinuous step function rather than the smoothed step function varying over several grid spacings, and $\boldsymbol{u}_f$ ($\boldsymbol{u}_l$ or $\boldsymbol{u}_v$) is the velocity for each phase, which is extrapolated into the entire domain by using the velocity jump condition given by eq. (13.39). The effective (or interpolated) properties, ($\hat{\rho}$, $\hat{\mu}$, and $\hat{k}_f$), are evaluated from a fraction function, F_φ, which is defined as

$$F_\phi = 1 \quad \text{if } H(\phi_A) = H(\phi_B) = 1$$
$$= 0 \quad \text{if } H(\phi_A) = H(\phi_B) = 0$$
$$= \frac{\max(\phi_A, \phi_B)}{\max(\phi_A, \phi_B) - \min(\phi_A, \phi_B)} \quad \text{otherwise}$$

where ϕ_A and ϕ_B are obtained at the adjacent grid points [19].

In the LS formulation, the LS function ϕ is advanced and reinitialized as

$$\frac{\partial \phi}{\partial t} + \boldsymbol{U}\cdot\nabla\phi = 0 \tag{13.49}$$

$$\frac{\partial \phi}{\partial \tau} = S(\phi_0)\left(1 - |\nabla\phi|\right) \tag{13.50}$$

where

$$U = u_f + \frac{\dot{m}n}{\rho_f} \tag{13.51}$$

$$S(\phi_0) = 0 \qquad \text{if } |\nabla\phi| \le \frac{h}{2}$$
$$= \frac{\phi_0}{\sqrt{\phi_0{}^2 + h^2}} \qquad \text{otherwise} \tag{13.52}$$

The formulation given by eq. (13.52) implies that a near-zero LS rather than $\phi = 0$ is used as the immobile boundary condition during the reinitialization procedure. This formulation of the sign function improves the LS method significantly [19].

To preserve mass conservation free from any errors occurring in numerical implementation of the LS advection and reinitialization procedures, the following volume-correction step is added to the LS formulation:

$$\frac{\partial\phi}{\partial\tau} = (V_v - V_{v0})|\nabla\phi| \tag{13.53}$$

where V_v is a vapor volume computed form ϕ, and V_{v0} is the vapor volume that satisfies mass conservation.

While discretizing the governing equations temporally, the convection and source terms are treated by a first-order explicit scheme and the diffusion terms by a fully implicit scheme as follows:

$$\frac{\phi^{n+1} - \phi^n}{\Delta t} = -U^n \cdot \nabla\phi^n \tag{13.54}$$

$$\frac{T_f{}^{n+1} - T_f{}^n}{\Delta t} = -u_f{}^n \cdot \nabla T_f{}^n + \frac{1}{\rho_f c_{pf}}\left(\nabla \cdot \hat{k}_f \nabla T_f{}^{n+1}\right) \tag{13.55}$$

$$\frac{u^{n+1} - u^*}{\Delta t} = -\frac{1}{\hat{\rho}}\left[\nabla p + \left(\sigma\kappa - v_{lv}\dot{m}^2\right)\nabla H_\phi\right] + \frac{1}{\hat{\rho}}\left(\nabla \cdot \hat{\mu}\nabla u^{n+1} + s_u\right) \tag{13.56}$$

$$\nabla \cdot u^{n+1} = v_{lv}\dot{m}n \cdot \nabla H_\phi \tag{13.57}$$

where

$$u^* = u_f{}^n + \Delta t\left(-u_f{}^n \cdot \Delta u_f{}^n + g\right)$$

$$s_u = \nabla \cdot \hat{\mu}\left(\nabla u^n\right)^T - \nabla \cdot \hat{\mu}\left[\left(v_{lv}\dot{m}n\nabla H_\phi\right)^T + v_{lv}\dot{m}n\nabla H_\phi\right]$$

Here, the effective properties and the step function are evaluated from ϕ^{n+1}, and $\dot{m}$ and from T^{n+1}. The momentum eq. (13.56) and the mass conservation eq. (13.57) are solved by employing the projection method. Equation (13.56) is decomposed into two steps:

$$\frac{\boldsymbol{u}^{**}-\boldsymbol{u}^{*}}{\Delta t}=\frac{1}{\hat{\rho}}\left(\nabla\cdot\hat{\mu}\nabla\boldsymbol{u}^{**}+\boldsymbol{s}_u\right) \tag{13.58}$$

$$\frac{\boldsymbol{u}^{n+1}-\boldsymbol{u}^{**}}{\Delta t}=\frac{1}{\hat{\rho}}\left(\nabla p+\left(\sigma\kappa-v_{lv}\dot{m}^2\right)\nabla H_\phi\right) \tag{13.59}$$

Substituting eq. (13.59) into the mass conservation eq. (13.57), the governing equation for pressure is obtained as

$$\nabla\cdot\frac{1}{\hat{\rho}}\nabla p=\frac{\nabla\cdot\boldsymbol{u}^{**}-\left(v_{lv}\dot{m}\boldsymbol{n}\cdot\nabla H_\phi\right)}{\Delta t}-\nabla\cdot\frac{1}{\hat{\rho}}\left(\sigma\kappa-v_{lv}\dot{m}^2\right)\nabla H_\phi \tag{13.60}$$

13.2.4 Results of Numerical Simulations of Nucleate Boiling and Experimental Validation

In carrying out numerical simulations, characteristic length, l_0, characteristic velocity, u_0, and characteristic time, t_0, are defined as

$$l_0=\sqrt{\sigma/g\left(\rho_l-\rho_v\right)};\qquad u_0=\sqrt{gl_0};\qquad t_0=l_0/u_0 \tag{13.61}$$

The dimensionless temperature is defined as

$$\theta=\frac{T-T_{\text{sat}}}{T_w-T_{\text{sat}}} \tag{13.62}$$

During numerical simulations of partial nucleate boiling, Son et al. [1] used the properties of water at 1 atm. The computation domain for axisymmetric cases was chosen to be $(R/l_0, Y/l_0)=(1, 3)$ so that the bubble growth was not affected by the computational boundary. To start the computations, the initial fluid temperature profile was taken to be linear in the natural convection thermal boundary layer and fluid velocity was set equal to zero. The initial thermal boundary layer thickness, δ_T, was evaluated from the correlation for the turbulent natural convective heat transfer on an upward facing flat plate (refer to Kays and Crawford [20]) as:

$$\delta_T=7.14\left(v_l\alpha_l/g\beta_T\Delta T\right)^{1/3} \tag{13.63}$$

In reality, partial nucleate boiling on a horizontal surface is a cyclic process; hence, the computations should be carried out over several cycles until no cycle-to-cycle change in the bubble growth pattern or in the temperature profile in the thermal boundary layer is observed. The computational domain was divided into 96×288 grid points, and a dimensionless time step of 5×10^{-4} was used.

13.2.4.1 Single-Bubble Dynamics and Associated Heat Transfer

Figure 13.3 shows the bubble growth pattern during one cycle for $\Delta T = 6.17\text{K}$ and $A = -8.5\times10^{-21}\,\text{J}$. As mentioned earlier, the solution for the macro region is coupled with the solution for the micro region. The results for the micro region corresponding at $t/t_0 = 38.38$ are plotted in Figure 13.4. The value of the radius of the bubble base, $(r/l_0,\ y/l_0) = (R_1/l_0,\ h/2l_0) = (0.189,\ 0.005)$ at $t/t_0 = 38.38$ in Figure 13.3, obtained in the macro region, serves as a boundary condition for the governing equations for the micro region shown in Figures 13.4a and 13.4b. From Figure 13.4b, it is seen that due to the contribution of disjoining pressure, the slope of the interface decreases to zero rapidly near the nonevaporating region ($\delta = \delta_0$). As the thickness of the liquid microlayer decreases, the interface temperature rises up to the wall temperature as shown in Figure 13.4c. Also, a significant evaporative heat transfer is observed in the micro region (Figure 13.4d). Nusselt number drops to zero very rapidly at the inner radius of the microlayer where rate of evaporation vanishes. The apparent contact angle (38° from Figure 13.4a) and the total energy removal integrated over the micro region plotted in Figure 13.4d are used in the calculations for the macro region. In the calculations, for a given dispersion constant (A), the apparent contact was kept constant during the bubble growth process. In reality, however, the dynamic contact

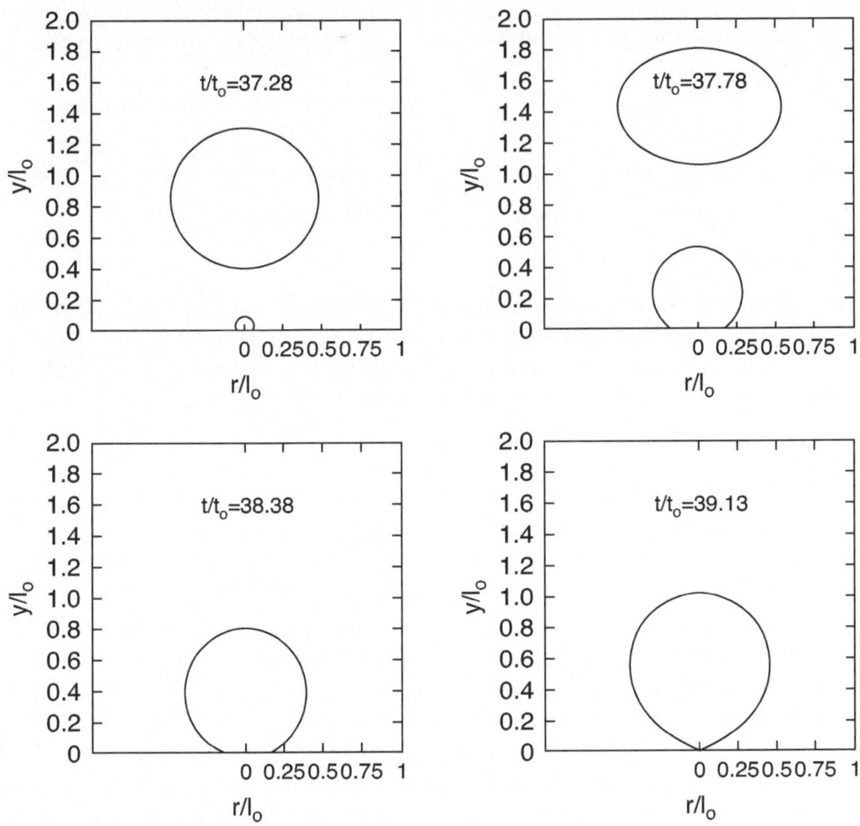

FIGURE 13.3 Bubble growth pattern for $\Delta T = 6.17$ K and $A = -8.5\times10^{-21}$ J [1].

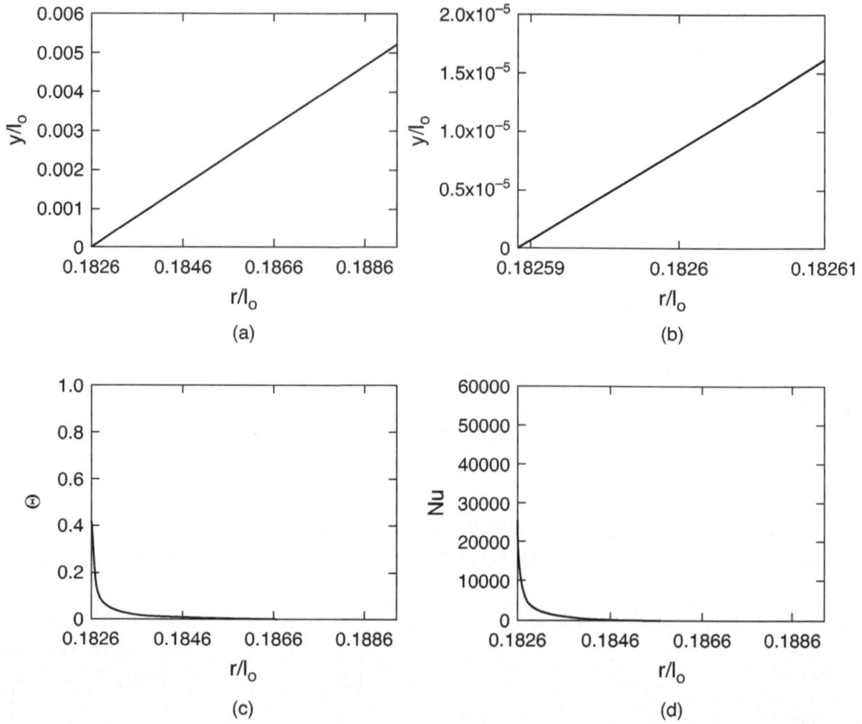

FIGURE 13.4 Results for the micro region at $\frac{t}{t_0} = 38.38$: (a), (b) interface shape, (c) interfacial temperature and (d) Nusselt number for $\Delta T = 6.17$ K and $A = -8.5 \times 10^{-21}$ J [1].

angles (advancing and receding contact angles) are different from the static contact angle. The hysteresis of contact angle on a real solid surface was not modeled.

It is seen from Figure 13.3 that during the early period of bubble growth ($37.28 < t/t_0 < 38.38$), the location where the liquid–vapor interface contacts, the wall moves outward. However, the horizontal movement of the bubble base is retarded by the wall friction and the surface tension forcing the bubble shape to satisfy the geometric condition (contact angle). This in turn tends to push the upper portion of the bubble outward and, as a result, leads to increased buoyancy force. The nonuniform bubble motion induces a clockwise vortex in the liquid layer. When liquid vortex becomes stronger with the increase of the bubble diameter, the bubble base starts to shrink (moves inward) at ($38.38 < t/t_0 < 39.13$). The departure process occurs rapidly because of the reduction of downward force due to surface tension as the bubble base shrinks.

The flow field in and around a bubble growing on the wall and for a detached bubble is shown in Figure 13.5a. During the early period of bubble growth, the liquid around the bubble is seen to be pushed out. A circulatory flow pattern inside the bubble as well as in the liquid outside is clearly seen for the freely rising detached bubble. Figure 13.5b shows the flow pattern in and around the bubble shortly before detachment. On the upper portion of the bubble, the liquid is being pushed outward,

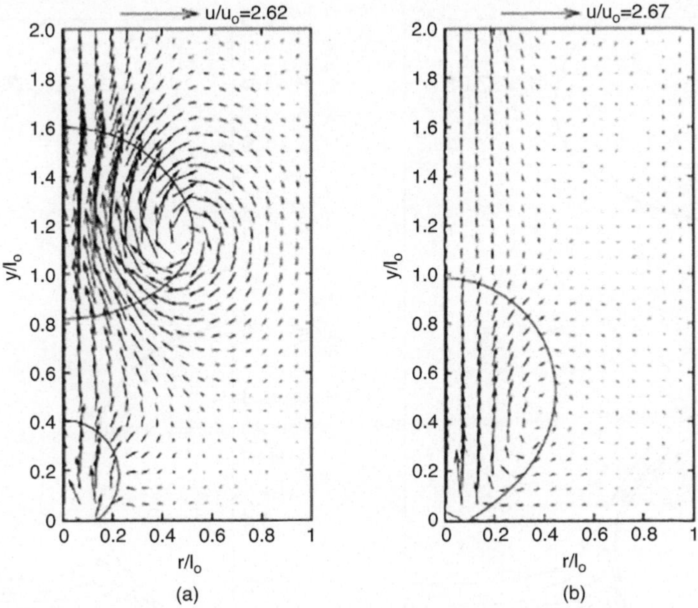

FIGURE 13.5 Flow patterns during growth and detachment of single bubbles for $\Delta T = 6.17$ K and $A = -8.5\times10^{-21}$ J: (a) $\frac{t}{t_0} = 37.58$ and (b) $\frac{t}{t_0} = 39.03$ [1].

whereas the liquid flow is radially inward in the lower portion of the bubble. The temperature field at the heated surface and around the bubble is shown in Figure 13.6. The crowding of the isotherms underneath the bubble is reflective of the very high heat flux in that region. Initially, when a bubble locates inside the thermal boundary layer, the bubble grows as a result of evaporation all around of the bubble interface including the micro region. As the bubble grows out of the thermal layer, the energy required for evaporation is supplied through only the portion around the bubble base. Also it is seen from Figure 13.6b that just after bubble detachment, a thin thermal layer forms on the area vacated by the departing bubble. The thickness of the thermal layer increases as the heat diffuses into the liquid adjacent to the wall.

Figure 13.7 shows temporal variation of predicted Nusselt number $\left(hl_0/k_l\right)$ based on the area averaged heat flux at the wall, Nu_{av}, and the local Nusselt number at the edge of the domain, $\mathrm{Nu}_{r=R}$. Because of the uncertainty in the specification of the initial condition, magnitude of Nusselt number is seen to initially change from cycle to cycle. However, after about 14 cycles, the steady-state condition appears to have been achieved. $\mathrm{Nu}_{\mathrm{micro}}$ is the Nusselt number based on the contribution of the microlayer. It is seen that for this set of calculations, the microlayer contributes about 20% to the total energy input to the bubble. Also, it is noted that local heat transfer at $r = R$ is 80% higher than that obtained from the correlation for the turbulent natural convection. This indicates that the liquid vortex associated with a growing and departing bubble is much stronger than the motion induced by pure natural convection. Nusselt number is seen to increase as the bubble base expands and decreases during the detachment phase as bubble base shrinks.

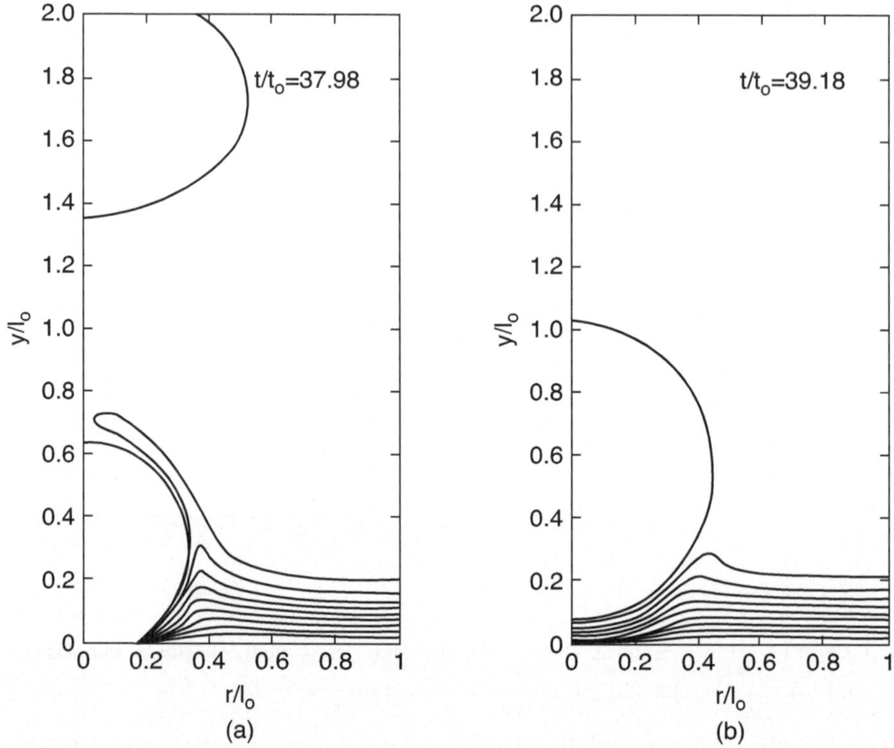

FIGURE 13.6 Temperature field during bubble growth with temperature interval of 0.617 K for $\Delta T = 6.17$ K and $A = -8.5\times10^{-21}$ J [1].

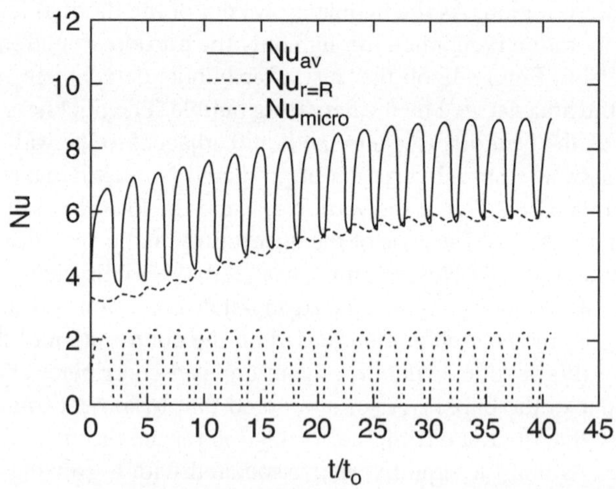

FIGURE 13.7 Variation of Nusselt number with time for various bubble growth cycles for $\Delta T = 6.17$ K and $A = -8.5\times10^{-21}$ J [1].

In Figure 13.8, a comparison of the bubble growth predicted from the numerical calculations is made with the data reported by Siegel and Keshock [21] for $\Delta T = 6.17\text{K}$ and $\varphi = 38°$. The numerical predictions are bounded by the upper and lower limits of the data. The model also correctly predicts the bubble diameter at departure and the time for departure.

A similar comparison of the predictions of the bubble diameter as a function of time is made in Figure 13.9 with the data obtained from the experiments reported in [1]. The data and predictions plotted in Figure 13.9 are for saturated water at

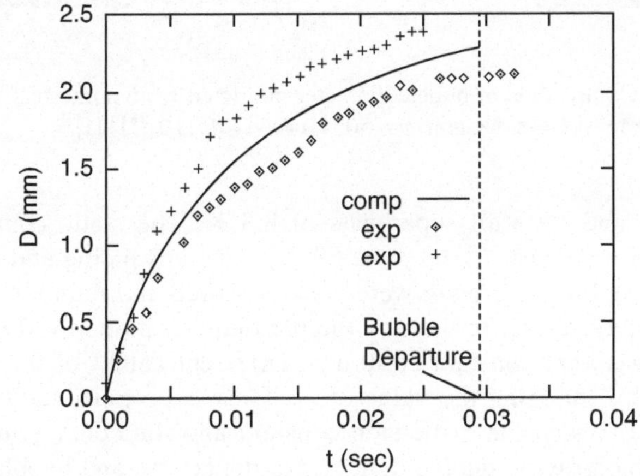

FIGURE 13.8 Comparison of bubble diameter as a function of time predicted from numerical simulation with experimental data obtained by Siegel and Keshock [21] for $\Delta T = 6.17$ K and $\varphi = 38° \left(A = -8.5 \times 10^{-21}\text{J}\right)$ [1].

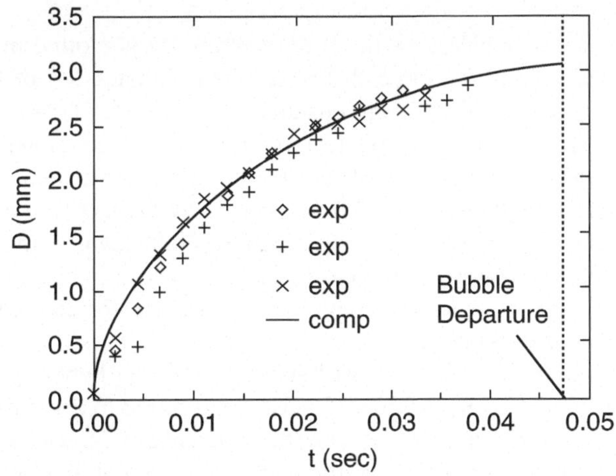

FIGURE 13.9 Comparison of prediction of bubble diameter at departure with data for $\Delta T = 8.5K$ and $\varphi = 50° \left(A = -14.4 \times 10^{-21} J\right)$ [1].

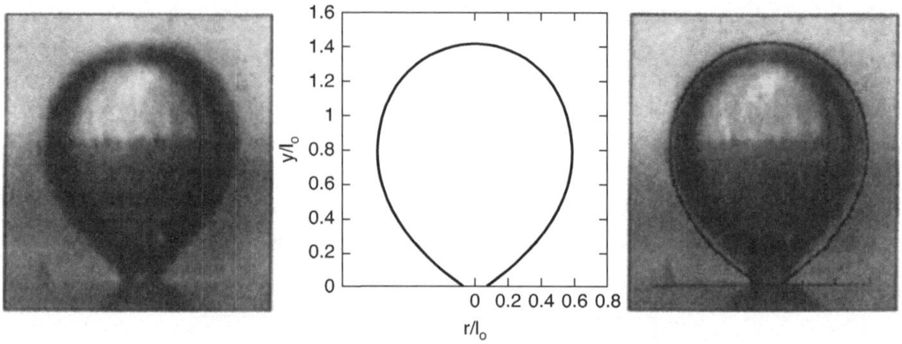

FIGURE 13.10 Comparison of bubble diameter predicted from numerical simulation with experimental data for $\Delta T = 8.5$ K and $\varphi = 50°$ $\left(A = -14.4 \times 10^{-21}\,\text{J}\right)$ [1].

1 atm pressure and for wall superheats of 8.5 K. The static contact angle for water on silicon wafer was found to be 50 deg. The advancing and receding contact angles during bubble growth were also measured and found to vary between ±5 deg of the static value. In making the predictions, plotted in Figure 13.9, the contact angle was kept constant at 50 deg. Different values of the contact angle were obtained by changing the value of A. Thus, a proportionality relationship was developed between the dispersion constant and the contact angle. The predictions of bubble growth, bubble diameter at departure, and bubble growth time are found to be in good agreement with the data except that the model generally tends to overpredict the growth period. In Figure 13.10, the bubble shape just prior to departure, obtained from the visual observations for a superheat of 8.5 K, is compared with that obtained from numerical simulation. Overall, the agreement between the two is quite good. However, some differences are observed in the neck region.

The dependence of bubble growth on the wettability is plotted in Figure 13.11. Computations were made for three different contact angles (φ) of 25°, 30°, and 38°, which correspond to dispersion constants of -3×10^{-21} (J), -5×10^{-21} (J), and -8.5×10^{-21} (J). It is seen from Figure 13.11 that as the contact angle increases, the bubble growth period increases. Also, the bubble diameter at departure becomes larger as the contact angle increases. This is caused by the fact that as the contribution of surface tension increases with increase in the contact angle, the vapor volume required for bubble departure also increases. The increase of bubble departure diameter with contact angle is generally in agreement with the correlations proposed by Fritz [23] and Staniszewski [24].

Figure 13.12 shows the effect of wall superheat on the bubble growth. As ΔT increases from 6.17 to 12.5 K, the growth period was found to decrease from 29.5 to 20.5 ms. Also, the bubble diameter at departure increases with wall superheat. This indicates that for a fixed contact angle, the bubble diameter at departure depends on the growth rate (bubble dynamics). In other words, aside from buoyancy and surface tension, inertia plays a role.

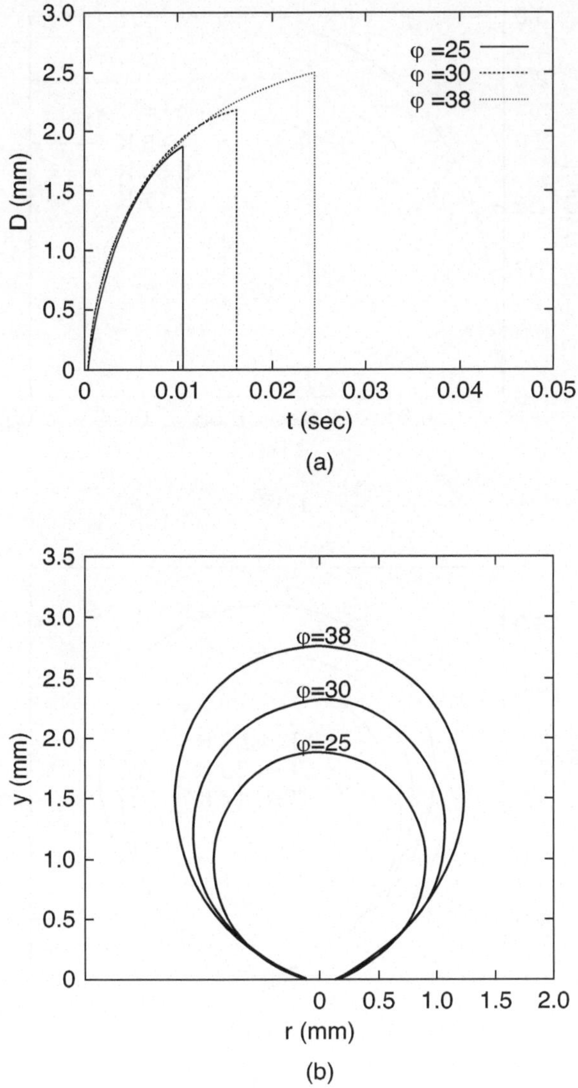

FIGURE 13.11 Effect of contact angle (φ) on bubble growth: (a) variation of bubble diameter with time and (b) bubble shape at departure for $\Delta T = 8.5$ K [1].

Nusselt numbers based on heat transfer coefficient averaged over the cell area are plotted in Figure 13.13. During an early period of bubble growth, the Nusselt number increases because the area of the microlayer increases as the bubble base expands. Thereafter, the Nusselt number decreases during the bubble departure period as the bubble base shrinks. Also, it is seen from Figure 13.13 that as the wall superheat increases, the Nusselt number increases. This is caused by the face that the bubble growth rate increases with ΔT as more energy from the thermal layer is utilized for evaporation.

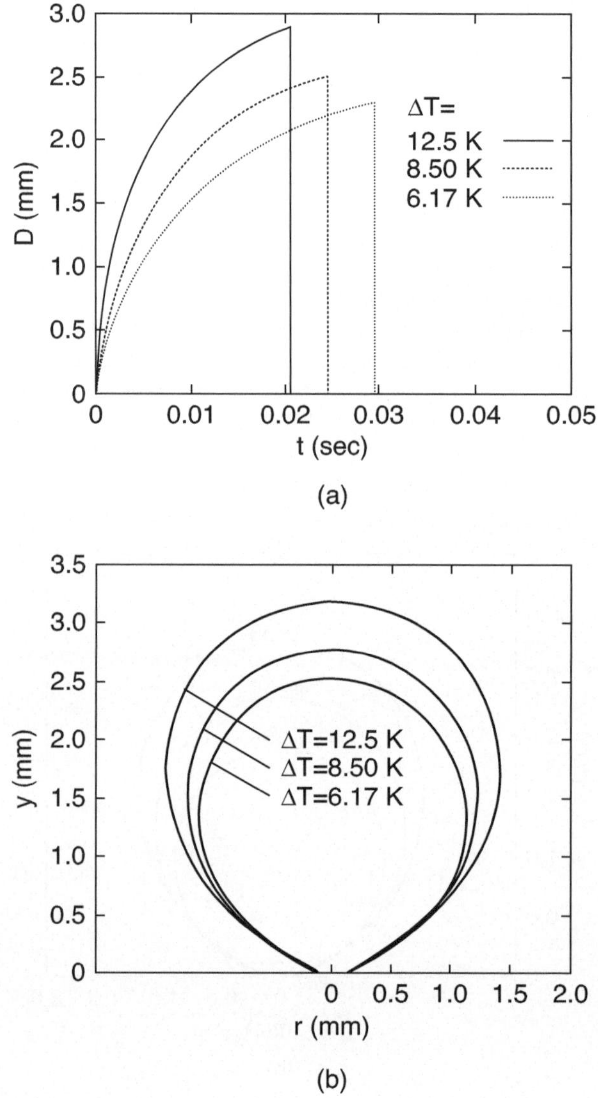

FIGURE 13.12 Effect of wall superheat (ΔT) on bubble growth: (a) variation of bubble diameter with time and (b) bubble shape at departure for $\varphi = 38°$ [1].

13.2.4.2 Bubble Merger at a Nucleation Site and Formation of Vapor Columns

During numerical simulations of bubble growth and merger process, the waiting period is one of the important parameters affecting the bubble merger pattern. The waiting period for a nucleation of a cavity (10 μm) is practically zero under the assumption that the wall temperature remains constant. In most experiments, a constant heat flux condition is applied. As such, the wall temperature decreases momentarily when the

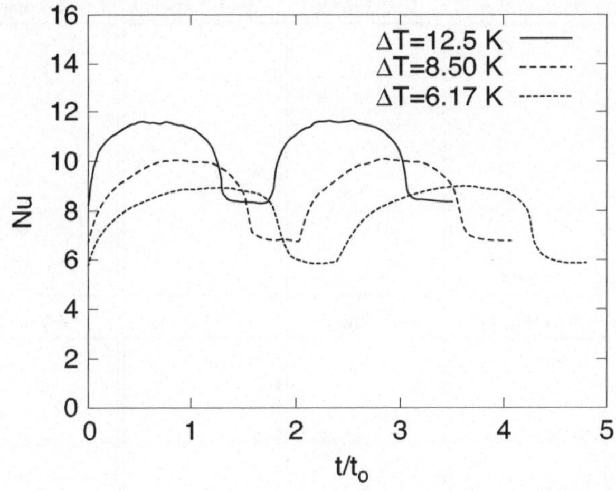

FIGURE 13.13 Effect of wall superheat (ΔT) on Nusselt number as a function of time for $\varphi = 38°$ [1].

cold pool liquid fills the area vacated by the departing bubble. The liquid layer and the heater surface have to be heated to a temperature necessary to initiate the next bubble. The waiting period decreases with wall superheat and can be predicted when a conjugate problem involving conduction in the solid is solved simultaneously with the flow and temperature field in the liquid. In the work of Son et al. [22], the waiting period, τ_w, was varied parametrically because they did not solve the conjugate problem.

Figure 13.14 shows the bubble merger at a particular site when the waiting period is arbitrarily chosen to be 1.28 ms. After the departure of the merged bubble at 5.5 ms, the bubble merger process is interrupted until a single bubble grows on the surface and departs. Thereafter, the merger process begins all over ($t > 32.2$ ms). This indicated that for a given wall superheat, the type of bubble merger pattern depends on the waiting period. For a given size cavity, the waiting period decreases with increase in wall superheat. Thus, because of distribution of different size cavities on a particular surface, bubble merger will be observed to occur at varying wall superheats. The merger of bubbles in the vertical direction gives appearance of the formation of a vapor column at $t = 4.9$ ms, the configuration of which is consistent with the observation of Gaertner [25].

In Figure 13.15, a comparison of the shapes of merged bubbles predicted from numerical simulation is made with those observed in the experiments. It is seen that bubble shapes calculated numerically are in excellent agreement with those observed experimentally.

13.2.4.3 Lateral Bubble Merger and Formation of Mushroom Type Bubbles

Lateral bubble merger process in pool nucleate boiling was investigated both numerically and experimentally by Mukherjee and Dhir [26]. They solved three-dimensional conservation equations for laminar incompressible fluids using Semi-Implicit Method for Pressure-Linked Equations (SIMPLE). The interface shape was captured using Level Set (LS) method as described earlier. Figure 13.16 compares numerically

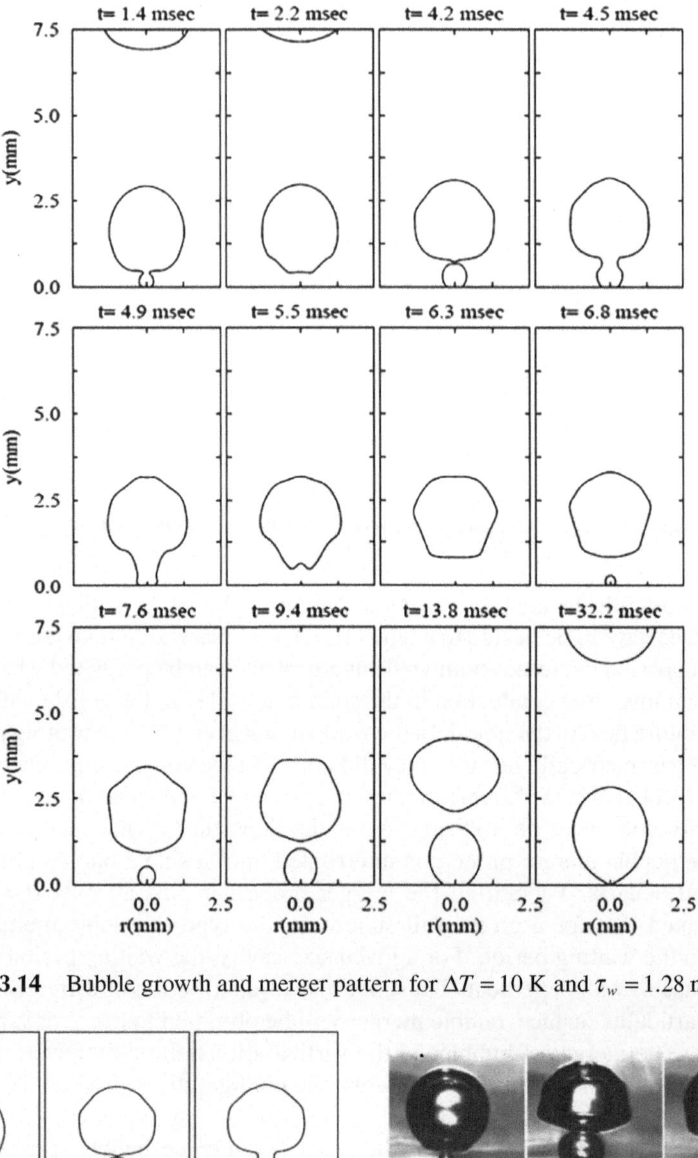

FIGURE 13.14 Bubble growth and merger pattern for $\Delta T = 10$ K and $\tau_w = 1.28$ ms [22].

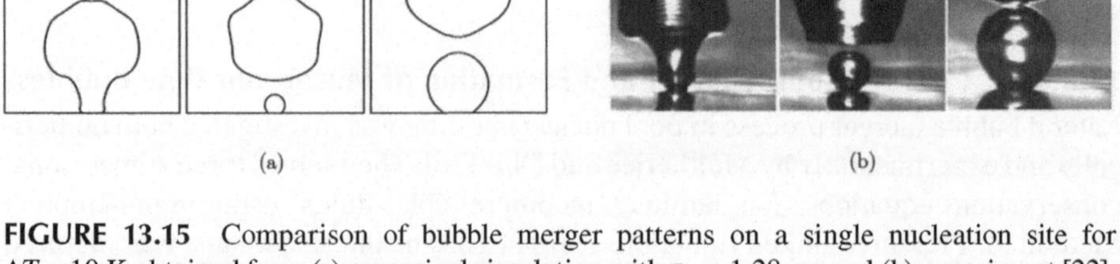

FIGURE 13.15 Comparison of bubble merger patterns on a single nucleation site for $\Delta T = 10$ K obtained from (a) numerical simulation with $\tau_w = 1.28$ ms and (b) experiment [22].

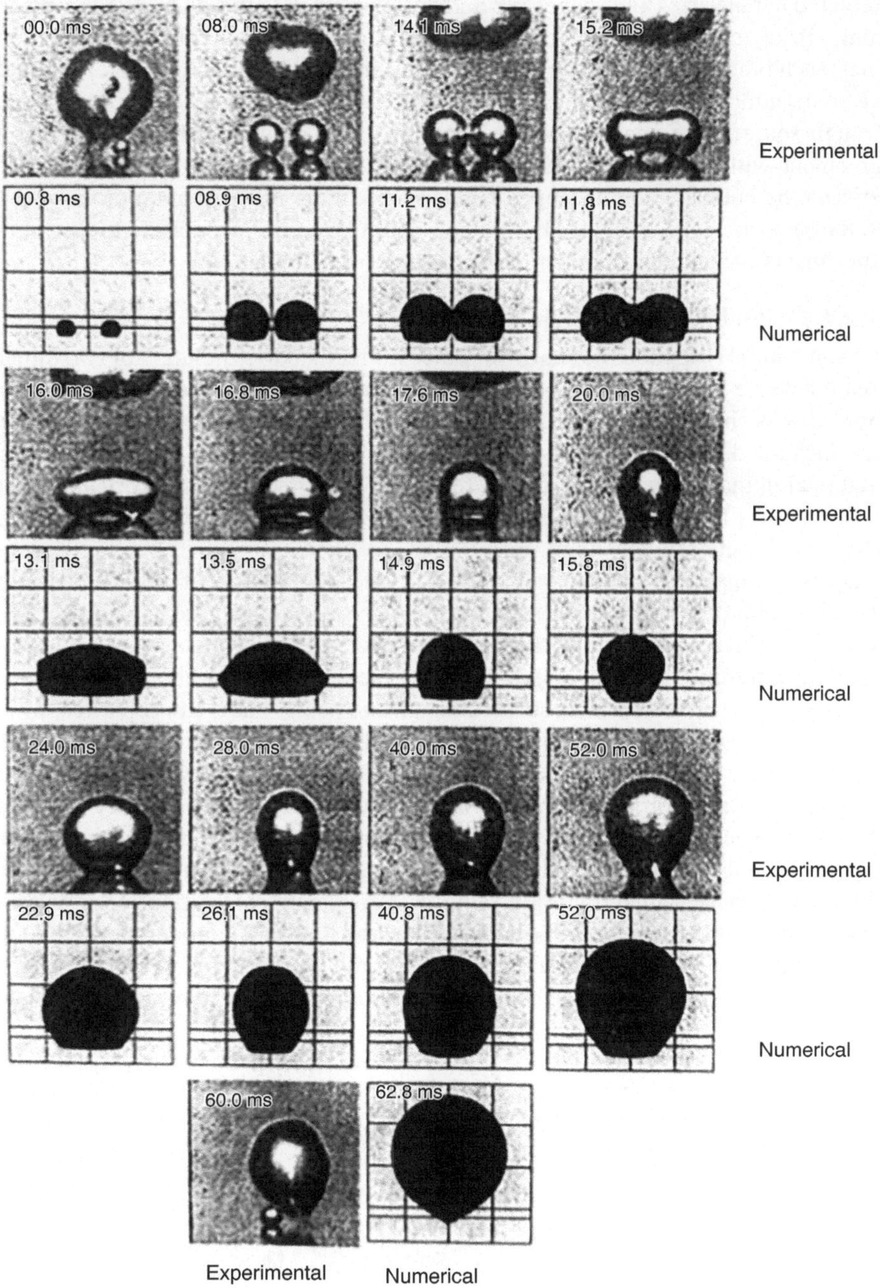

FIGURE 13.16 Numerical results and experimental observations of two-bubble lateral merger at 5°C wall [26].

predicted bubble sizes and shapes pre and post merger with those observed in experiments. Prior to merger, the two bubbles are almost similar in size and shape and touch each other at 14 ms. After merger, a mushroom type of bubble is formed with two stems attached to the heated surface. Subsequently, the merged vapor mass goes through some shape changes before departure. The predicted shapes are in good agreement with those observed in experiments, i.e., formation of a vapor bridge between the bubbles, liquid trapped between the stems and formation of vapor tails (16.8 ms) in the experiments. The trapped liquid layer, however, stays longer in the experiments compared to prediction from numerical simulations.

13.2.4.4 Pool Boiling at High Heat Fluxes

Son and Dhir [11] applied the sharp-interface LS method to simulation of two dimensional nucleate boiling at high heat fluxes with bubble merger at neighboring nucleation sites as shown in Figure 13.17. The contact angle was taken to be 35 degrees, for which the data for the active site density are available in the literature. The computational domain was chosen to be a region of $0 \le x \le 2$ and $0 \le y \le 50$. The width of domain has little effect on the bubble dynamics and heat transfer as long as the active site density remains fixed.

During computations of nucleate boiling, the distribution of active nucleation sites is specified by the number density of active sites, N, and the spacing between sites, s, as depicted in Figure 13.18. The site density depends not only on wall superheat and wettability but also on several other factors, including the procedure used in preparing the heater surface, surface finish, and heater material properties, which cannot be easily quantified. As such, N is treated as an externally provided variable. For a given N, the spacing s is determined so that the primary bubble (or vapor column) forms near $x = 0$, and the bubbles do not touch the side boundary at $x = 2l_0$. They also varied parametrically the waiting period, τ_w, between formation of two consecutive bubbles at a given site because the prediction of τ_w in most experiments with a constant heat flux condition requires a conjugate calculation for simultaneously solving conduction in the solid as well as the flow and temperature field in the liquid. Figure 13.19 shows the bubble growth and merger pattern for $\Delta T = 25°C$, $N = 5/4$, $s = 0.5$, and $\tau_w = 0.1$. The locations of active sites are the same as depicted

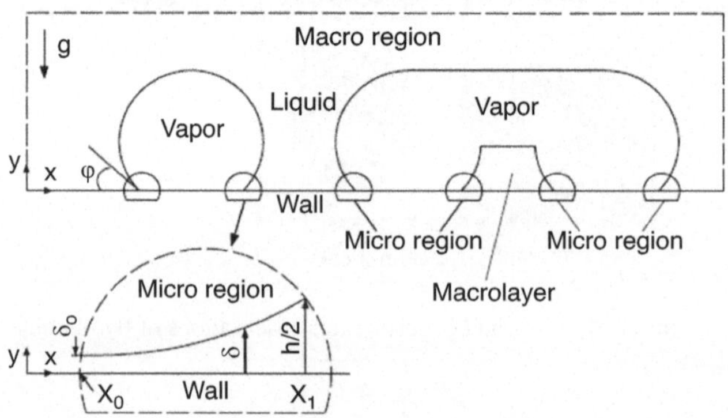

FIGURE 13.17 Macro and micro regions used in numerical simulation [11].

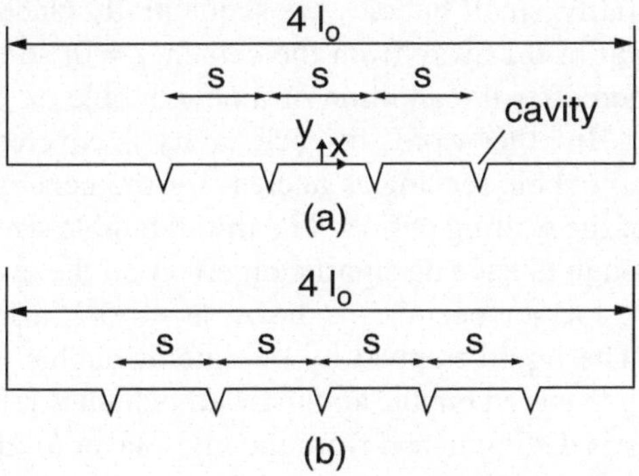

FIGURE 13.18 Configuration of active nucleation sites: (a) $N = 4/4$ and (b) $N = 5/4$ [11].

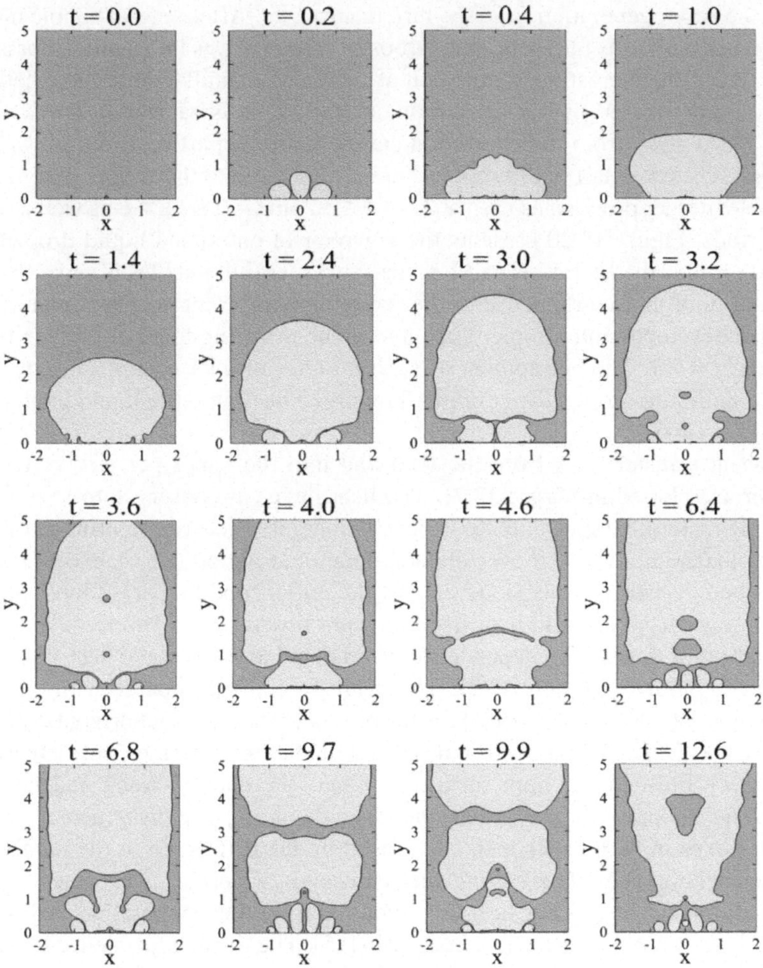

FIGURE 13.19 Bubble growth and merger for $\Delta T = 25°C$, $N = 5/4$, $s = 0.5$, and $\tau_w = 0.1$ [11].

in Figure 13.18b. Initially, small bubbles are sequentially placed on the heater surface with time interval of 0.1 away from the center, $x = 0$, so that a vapor column forms near $x = 0$. Thereafter the addition of a new bubble on each cavity depends on the waiting period. In other words, if each cavity is covered with liquid during a period of τ_w, a small bubble serving as nucleus for the next cycle is placed on the cavity after the end of the waiting period. The initial bubble size chosen in the computations is small enough to have no significant effect on the overall bubble growth. As the lateral bubble merger occurs, as shown at $t = 0.2$, a large merged bubble appears near $x = 0$. The liquid trapped by the lateral bubble merger forms a thin layer between the large merged bubble and the wall, which is referred to as a macrolayer. The macrolayer is distinguished from the microlayer (or micro region), which forms just near the bubble–solid contact, as depicted in Figure 13.17. The periodical formation of the liquid macrolayer is observed during the whole period of bubble evolution. The large bubble is fed from below by smaller bubbles (or vapor stems) as well as the vapor generation from the thin macrolayer. After several bubble mergers, the large bubble forms into vapor mushroom, which evolves into a tall vapor column after $t = 3.2$. It is interesting to note that at $t = 9.7$, the small bubbles attached to the wall are outnumbered by the nucleation cavities. This is caused by the fact that a bubble slides away from the nucleation cavity before departing from the wall. It is also observed that small liquid droplets are entrapped into the vapor column during the bubble merger process, as seen at $3.0 \le t \le 3.2$ and $4.0 \le t \le 4.6$ as well as during later periods. Figure 13.20 presents the behavior of entrapped liquid droplets with time. At $t = 4.4$, the liquid located between two merging bubbles is separated from the liquid pool and entrapped into the merged vapor column. The isolated liquid portion is developed into droplet form due to the restoring force of surface tension, as seen at $t = 5.8$. The entrapment of such a liquid droplet occurs subsequently at $t = 6.2$. The liquid droplets are eventually merged back into the liquid pool as demonstrated at $t = 6.6$.

The heat transfer rates from the wall and into the vapor per unit surface area of heater are plotted in Figure 13.21. The heat fluxes are observed to vary in time, depending on the changing bubble shape with mergers occurring continuously. When the heat fluxes are averaged over the computational period ($t \ge 6$), except the early period when the heat transfer is affected by the initial condition, it is found that about 51% of energy from the wall is utilized in vapor production, whereas 49% goes into superheating of liquid. The dependence of time and area averaged heat flux on wall superheat is plotted in Figure 13.22. The results are also compared with the experimental data obtained by Gaertner [25] for nucleate boiling on a horizontal plate and by Wang and Dhir [27] on a vertical plate. The data of Gaertner were obtained on a 2/0 emery finished platinum surface, whereas the data of Wang and Dhir were obtained on a copper surface with a contact angle of 35°. The disagreement between the heat fluxes in both experiments is caused by the difference in the number density of nucleation sites. The boiling surface used by Gaertner is expected to have a very small contact angle. The heat flux obtained from the computation varies as ΔT, whereas it varies as ΔT^3 in [27] and $\Delta T^{3.75}$ in [25]. The linear dependence of heat flux on wall superheat obtained from numerical simulations is due to the fixed nucleation site density. In reality, the nucleation site density increases with superheat.

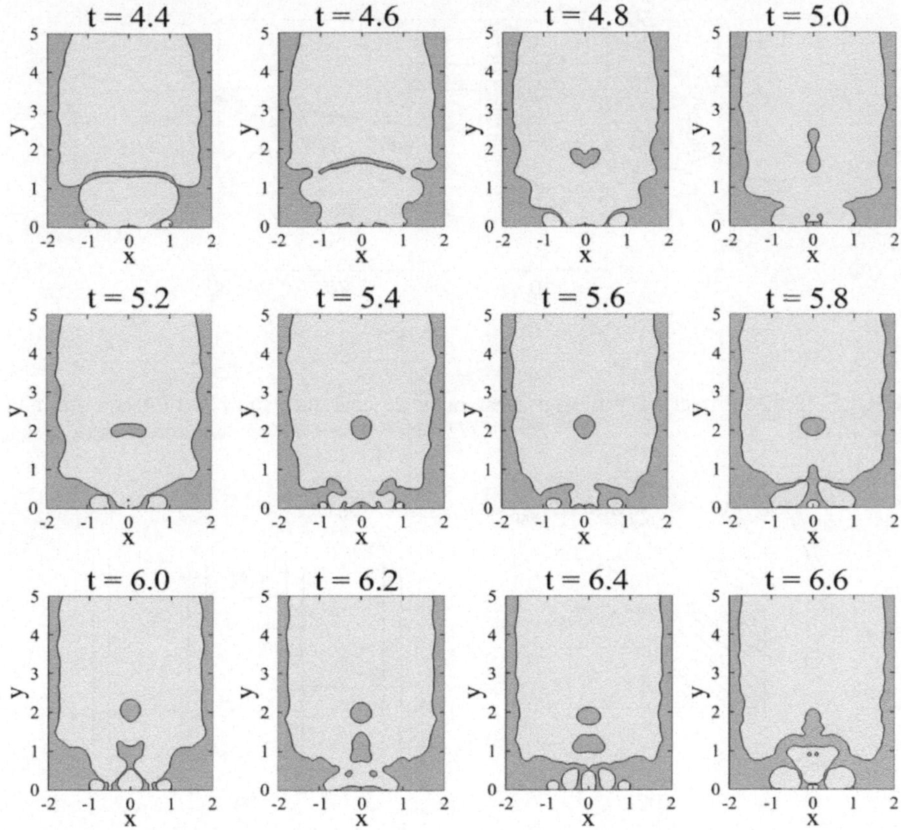

FIGURE 13.20 Behavior of liquid droplets entrapped during bubble growth and merger for $\Delta T = 25°C$, $N = 5/4$, $s = 0.5$, and $\tau_w = 0.1$ [11].

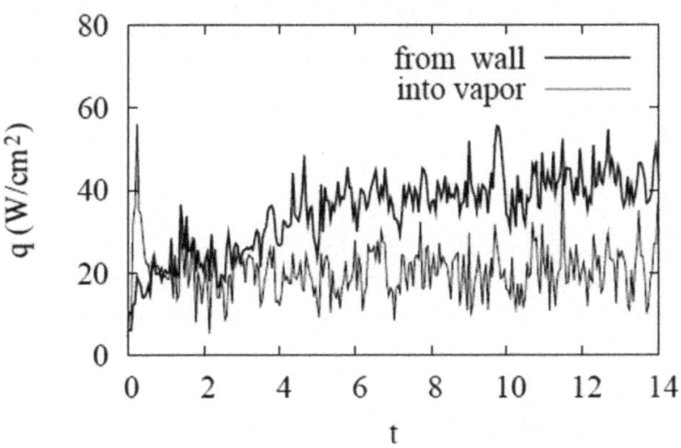

FIGURE 13.21 Temporal variation of heat fluxes from wall and into vapor for $\Delta T = 25°C$, $N = 5/4$, $s = 0.5$, and $\tau_w = 0.1$ [11].

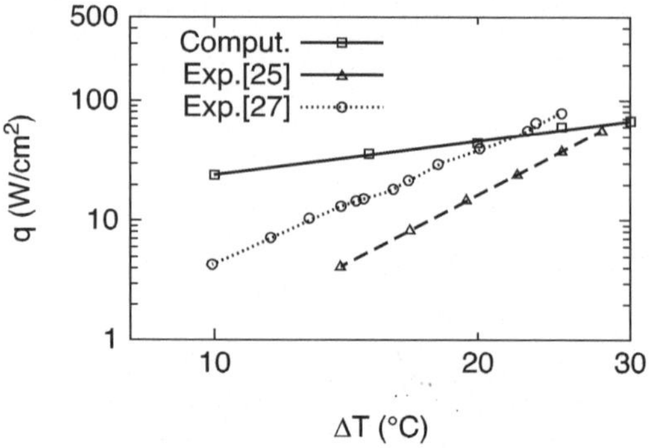

FIGURE 13.22 Effect of wall superheat on wall heat flux for $N = 17/4$, $s = 0.2$, and $\tau_w = 0.1$ [11].

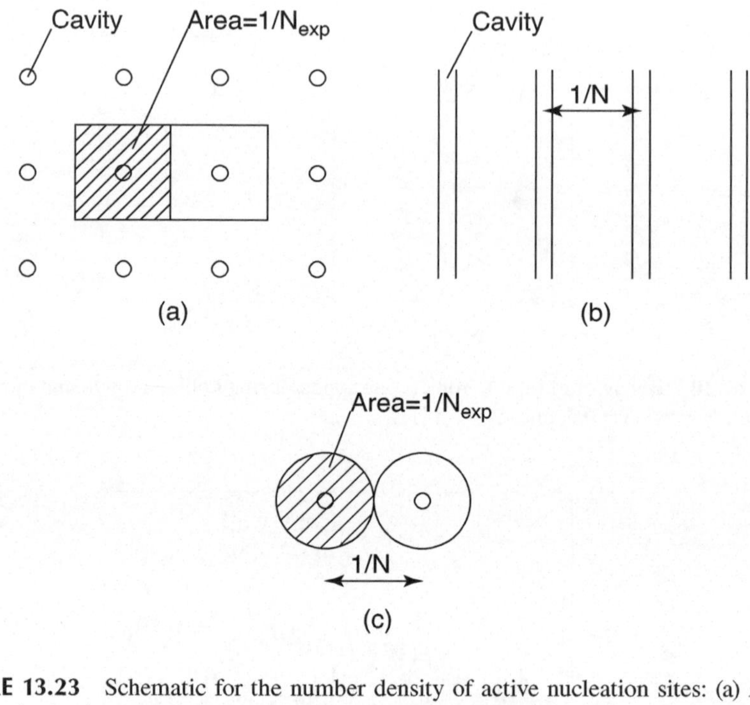

FIGURE 13.23 Schematic for the number density of active nucleation sites: (a) N_{exp}, site density obtained experimentally, (b) N, site density used in the two-dimensional computation, (c) the proposed relation between N and N_{exp} and surrounding area in computations [11].

To simulate nucleate boiling on a real surface, Son and Dhir [11] used the experimental data for the number density of active sites obtained by Wang and Dhir [27]. In their two-dimensional computations, the site density, N, was evaluated from the experimental data, N_{exp}, which was for a two-dimensional grid. Referring to Figure 13.23, they derived a relation

TABLE 13.1
Data for the Distribution of Active Nucleation Sites Used in the Computations of Son and Dhir [11]

ΔT (°C)	$16N_{exp}\left(\#/l_o^2\right)$	$4N\left(\#/l_o\right)$	$s\left(l_o\right)$
10	1	$0.9 \rightarrow 1$	–
13	4	$1.8 \rightarrow 2$	1
15	10	$2.8 \rightarrow 3$	1
20	80	$7.9 \rightarrow 8$	0.4
25	350	$16.6 \rightarrow 17$	0.2

$$\frac{1}{N_{exp}} = \frac{\pi}{4}\left(\frac{1}{N}\right)^2 \quad \text{or} \quad N = \sqrt{\frac{\pi N_{exp}}{4}} \tag{13.64}$$

The values of N and N_{exp} are listed in Table 13.1. Here, s is the spacing between sites used in the computations. Figure 13.24 shows the effect of wall superheat on bubble growth and merger while the site density is also varied with the wall superheat. It is observed from Figure 13.24a that at the low superheat, ΔT = 13°C, discrete bubbles are released from the heater and the bubble merger occurs occasionally. When the wall superheat is increased to ΔT = 20°C, the bubble growth pattern changes as shown in Figure 13.24b. The bubble merger occurs both in the vertical and horizontal directions, which leads to formation of a large vapor column and mushroom type of bubble, respectively. The large bubble is supported by vapor stems connecting the vapor mass to the heater surface. The periodic formation of macrolayer underneath the merger bubble is also observed. This transition in the bubble growth and merger pattern is consistent with the observation of Gaertner [25].

The dependence of bubble growth and merger on the waiting time, τ_w, is presented in Figure 13.25. At $\tau_w = 0.1$, a vapor column is observed to form. It is noted that the small bubbles attached to the wall are outnumbered by the nucleation cavities, as shown at $t = 11.2$ and $t = 11.6$. The bubble growth pattern shift to a discrete bubble mode when τ_w is increased to 1 while keeping ΔT and N constant. This indicates that the waiting period is also one of the important parameters determining the bubble growth pattern. Figure 13.26 shows the variation of heat flux with wall superheat for different waiting periods. The site density, N, listed in Table 13.1 is used in the computations. For a given waiting period, the heat fluxes vary as $\Delta T^{3.1}$ for $\Delta T < 15$°C but its variation decreases to $\Delta T^{1.6}$ for the higher wall superheat. As the waiting period is reduced, the heat flux is found to increase significantly. The effect is pronounced in the high superheat range.

Several expressions for the waiting period are available in the literature. However, their application requires knowledge of the cavity size or the thermal properties of a heater, as such is not so straightforward. As a practical correlation for the waiting period, the following equation was presented by Basu et al. [28] while developing a model for the wall heat flux partitioning during subcooled flow boiling:

$$\tau_w = 139.1\Delta T^{-4.1}/t_0 \tag{13.65}$$

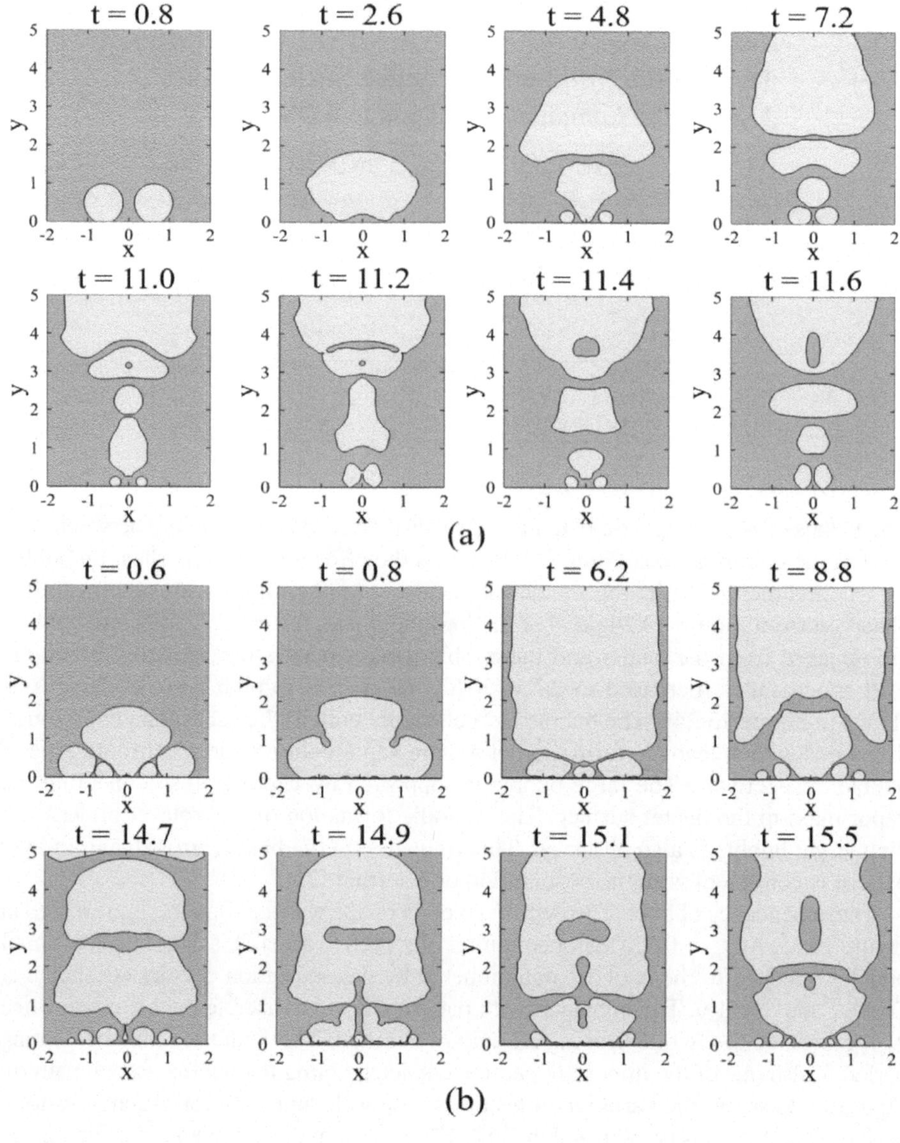

FIGURE 13.24 Effect of wall superheat on bubble growth and merger for variable site densities and $\tau_w = 0.1$: (a) $\Delta T = 13°C$, $N = 2/4$, $s = 1$, and (b) $\Delta T = 20°C$, $N = 8/4$, $s = 0.4$ [11].

When using the above correlation, the heat flux as a function of wall superheat is plotted in Figure 13.26. The experimental data [25,27] and three correlations available in the literature [29–31] are also included in Figure 13.27. The correlation of Rohsenow [29] for pool boiling is expressed as

$$q = \mu_l h_{lv} \sqrt{\frac{g(\rho_l - \rho_v)}{\sigma}} \left(\frac{C\rho_l \Delta T}{C_s h_{lv}} \right)^3 \mathrm{Pr}_l^{-3} \tag{13.66}$$

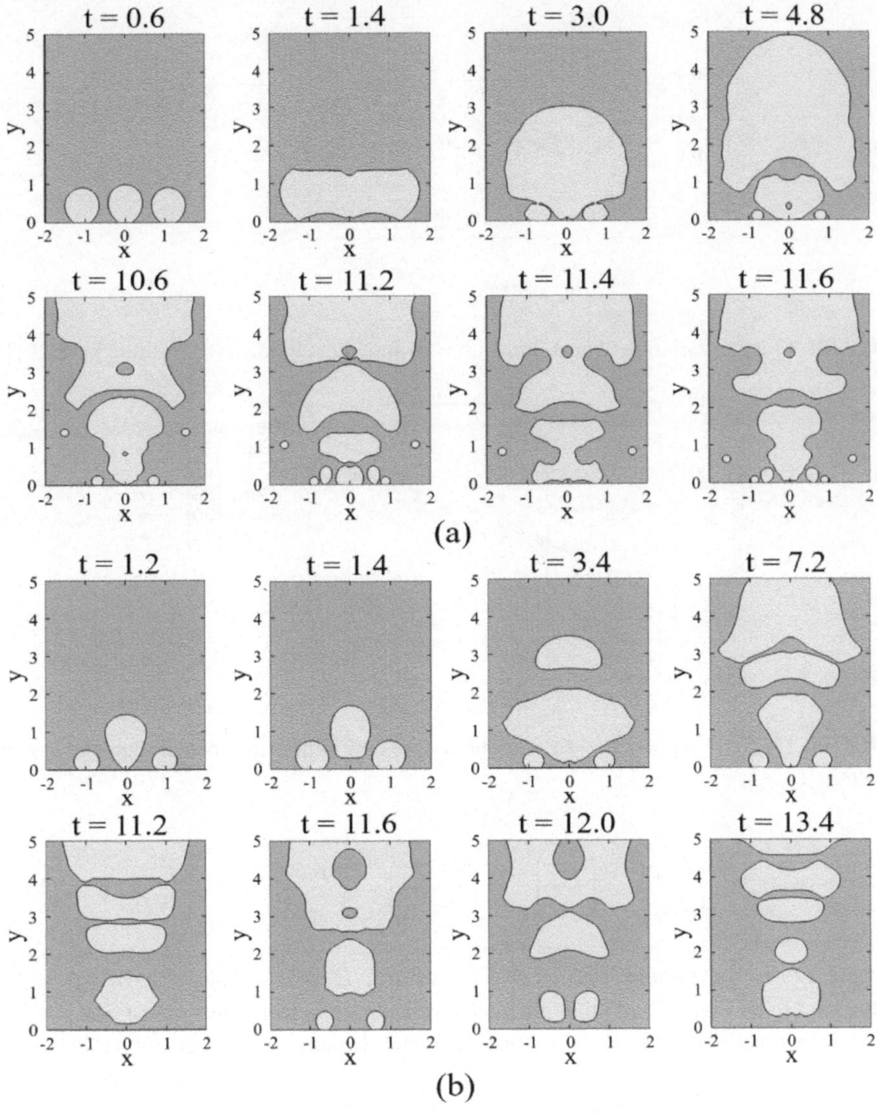

FIGURE 13.25 Effect of waiting period on bubble growth and merger for $\Delta T = 15°C$, $s = 1$, and $N = 3/4$: (a) $\tau_w = 0.1$ and (b) $\tau_w = 1$ [11].

where C_s depends on heater material and fluid combination, and its value is 0.013 for water–copper combination. As shown by Liaw and Dhir [30], C_s can be quantified as a function of contact angle. Its value is 0.02 for a contact angle of 35°. A more comprehensive correlation of Stephan and Abdelsalam [31], which was obtained from nearly 5,000 existing experimental data for nucleate boiling on horizontal surfaces, is expressed as

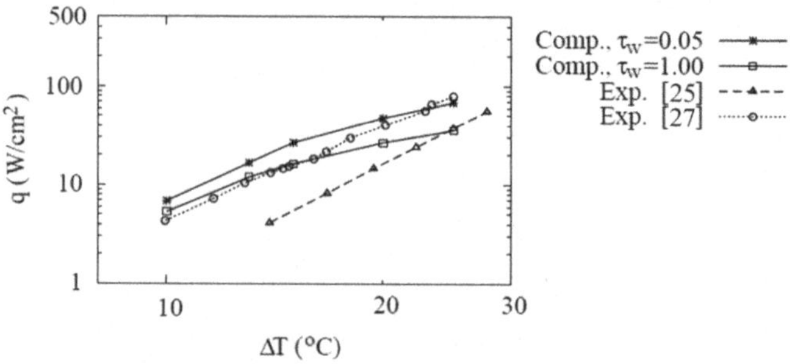

FIGURE 13.26 Variation of heat flux with wall superheat for different waiting periods [11].

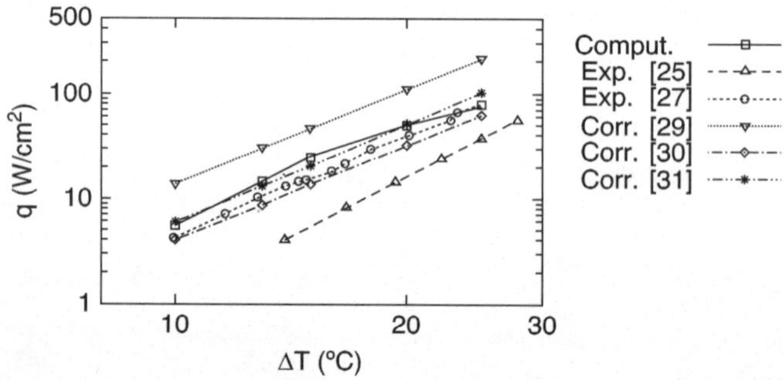

FIGURE 13.27 Comparison of the numerical results with experimental data and correlations [11].

$$\frac{qD_d}{k_l \Delta T} = 0.23\left(\frac{qD_d}{k_l T_{\text{sat}}}\right)^{0.674}\left(\frac{\rho_v}{\rho_l}\right)^{0.297}\left(\frac{h_{lv} D_d^{\,2}}{\alpha_l^{\,2}}\right)^{0.371}\left(\frac{\rho_l-\rho_v}{\rho_l}\right)^{-1.73}\left(\frac{\rho_l \alpha_l^{\,2}}{\sigma D_d}\right)^{0.35} \tag{13.67}$$

where D_d is the bubble diameter at departure, $D_d = 0.0208\varphi\sqrt{\sigma/g\left(\rho_l - \rho_v\right)}$. It is observed from Figure 13.27 that the heat fluxes obtained from the work of Son and Dhir [11] are comparable to the experimental data with the active site density given in [27] and are within at most ±25% deviation with those predicted from Stephan and Abdelsalam's correlation. Also, the dependence of the computed heat flux on wall superheat is consistent with the experimental data and are three correlations except near ΔT = 25°C, where the variation of the computed heat flux on wall superheat becomes weaker.

A three-dimensional (3D) simulation of nucleate boiling was performed on the domain, $0 \le x \le 2$, $0 \le y \le 20$, and $0 \le z \le 2$. The grid spacing was chosen as $h = 1/32$, which is much coarser than $h = 1/96$ used for two-dimensional (2D) computations. The 3D evolution of bubbles is demonstrated in Figure 13.28. The bubble growth and merger pattern, including frequent bubble mergers in the vertical and lateral directions and the formation of a large vapor column or mushroom type bubble, is qualitatively comparable to the 2D results plotted in Figure 13.25b. However, the formation of macrolayer underneath the merger bubble is not captured clearly in the 3D computation using a coarse grid. Figure 13.29 shows the wall heat fluxes obtained from 2D

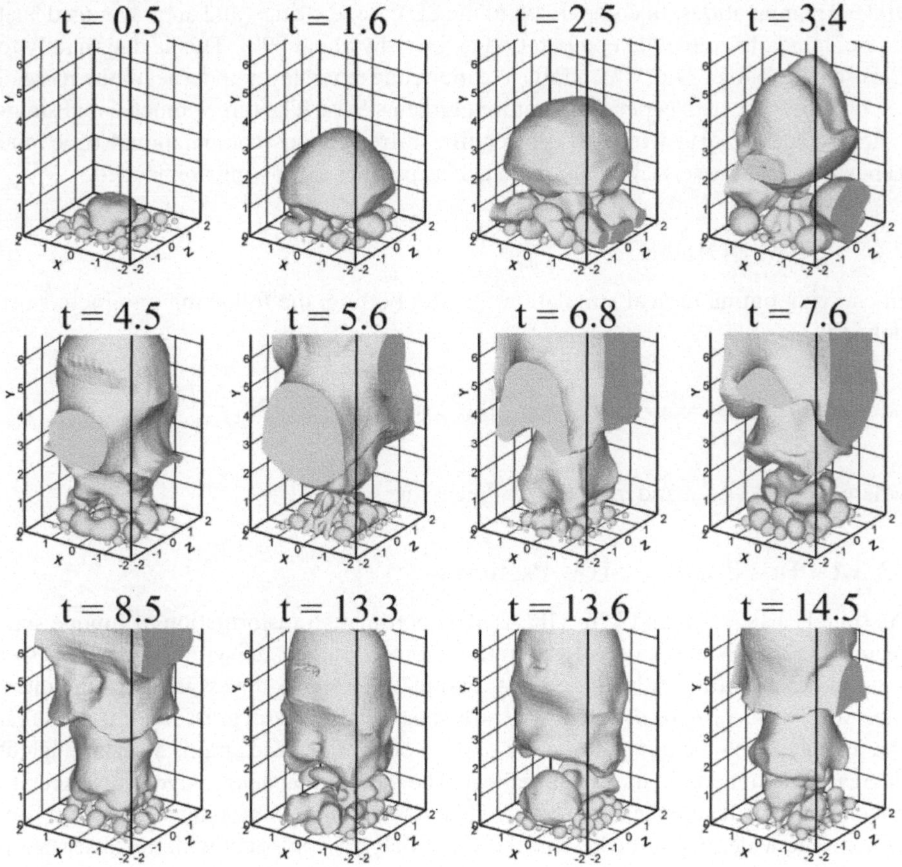

FIGURE 13.28 Bubble growth and merger obtained from three-dimensional computations for $\Delta T = 20°C$, $N_{exp} = 81/16$, $s = 0.35$, and $\tau_w = 0.1$. [11].

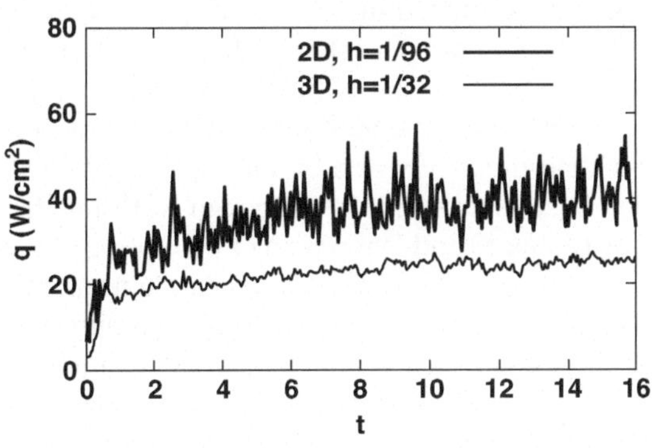

FIGURE 13.29 Wall heat fluxes obtained from two- and three-dimensional computations for $\Delta T = 20°C$ and $\tau_w = 0.1$ [11].

and 3D computations. In comparison to the 2D case, the time- and area-averaged heat flux obtained from the 3D computation is lower by about 39%. This is due mainly to the use of a coarser grid, which fails to capture the macrolayer underneath the merged vapor bubble where steeper temperature gradients exist. The three-dimensional simulation on a finer grid with $h = 1/96$ requires a tremendous computational time, and hence, efficient numerical techniques such as parallel computing are needed.

13.3 FILM BOILING

In carrying out numerical simulation for film boiling, the following parameters are defined:

$$\text{Re} = \frac{\rho_v u_o l_o}{\mu_v}; \quad \text{Pr} = \frac{c_{pv}\mu_v}{k_v}; \quad \text{Ja} = \frac{c_{pv}\Delta T}{h_{fg}}; \quad \text{Nu} = \frac{l_o q_w}{k_v \Delta T}$$

where $\Delta T = T_w - T_{\text{sat}}$, and q_w is the wall heat flux.

13.3.1 Film Boiling at Low Pressures

As stated earlier, Son and Dhir [18] used a coordinate transformation technique supplemented by a numerical grid-generation method in their axisymmetric analysis of saturated film boiling at low pressures. Figure 13.30 shows the evolution of the interface on a flat surface for different wall superheats (Ja) at 1 atm pressure. It is seen that the time at which the upward moving portion of the interface attains a certain height decreases with increase in wall superheat. The base of the vapor bulge is also seen to widen as the wall superheat increases. This in turn reduces the extent of the thin film region. The widening of the base of the vapor bulge is consistent with the observation made by Lienhard and Dhir [32] during film boiling on horizontal cylinders.

The bubble shapes at departure for different wall superheats are compared in Figure 13.31. As the wall superheat increases, the bubble becomes larger, and the shape is closer to a mushroom rather than a sphere. The calculated height and the effective radius of bubbles are listed in Table 13.2. Here the effective radius is estimated by assuming the bubble to be a sphere. The bubble breakoff radius predicted numerically is found to compare well with the observations of Berenson [33] and Lienhard and Dhir [34]. However, the bubble radii reported by Hosler and Westwater [35] are much larger than those reported by others and obtained from the calculations. The bubble height at departure is a function of wall superheat and is comparable to that reported by Lienhard and Dhir during film boiling of acetone on cylinders. Lienhard and Dhir did not report the wall superheat; hence, a direct comparison is not possible. Bubble release period predicted from the numerical analysis is also listed in Table 13.2. It is noted that the predicted values for water are comparable to those reported by Hosler and Westwater. While comparing the bubble radii and heights predicted numerically with those obtained experimentally, the possibility of presence of radiation in the experiments was checked. According to Siviour and Ede [36], the film boiling heat transfer rate including the radiation effect assuming simple superposition of the two modes of heat transfer is expressed as:

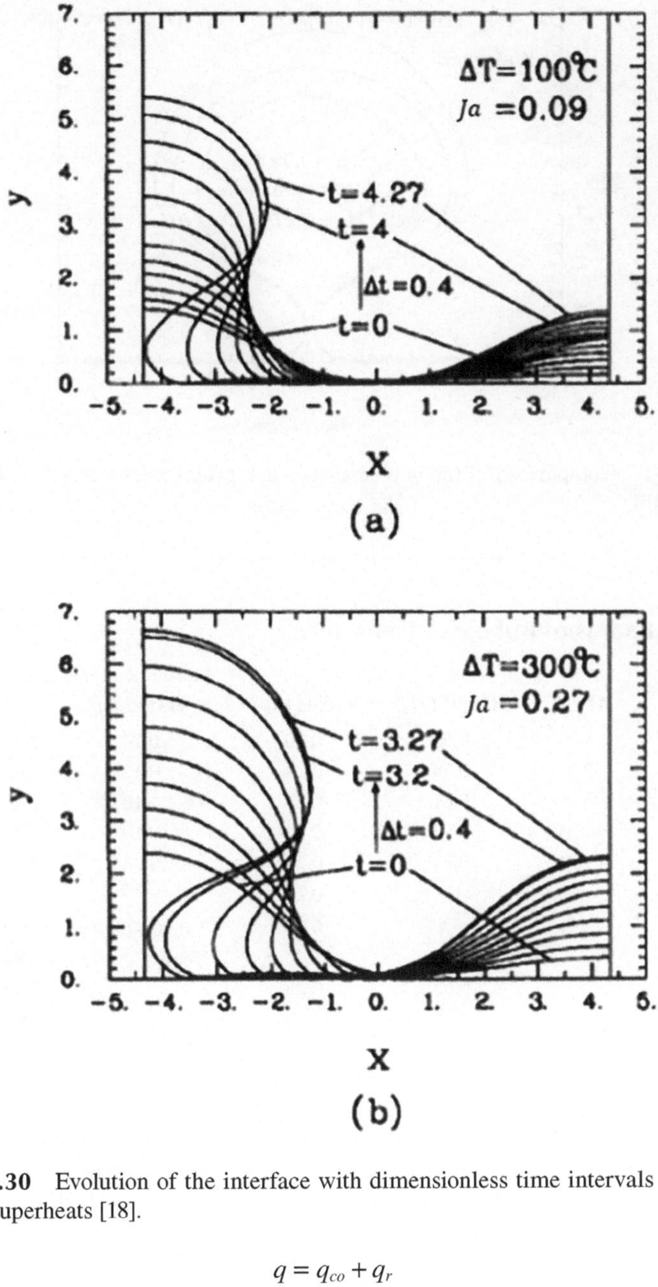

FIGURE 13.30 Evolution of the interface with dimensionless time intervals of 0.4 at different wall superheats [18].

$$q = q_{co} + q_r \tag{13.68}$$

$$q_r = B\epsilon_w \sigma_s \left(T_w{}^4 - T_{\text{sat}}{}^4\right) \tag{13.69}$$

$$B = \frac{3}{4} + \frac{1}{4}\frac{q_r}{q_{co}}\left[2.62 + \frac{q_r}{q_{co}}\right]^{-1} \tag{13.70}$$

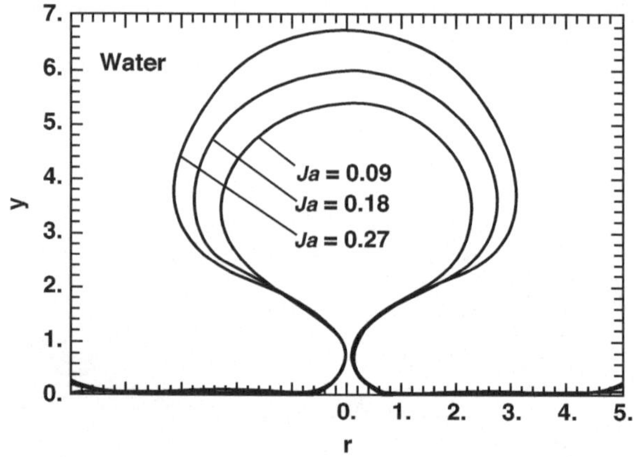

FIGURE 13.31 Comparison of the bubble shapes just prior to departure for different wall superheats [18].

TABLE 13.2
Height and Radius of Bubble at Departure

Fluid	ΔT (°C)	Height (λ_{d2})	Radius (λ_{d2})	Bubble Release Period (s)	References
Water	100	0.50	0.20	0.14	Son and Dhir [18]
Water	200	0.54	0.23	0.12	Son and Dhir [18]
Water	300	0.61	0.26	0.10	Son and Dhir [18]
Benzene	120	0.52	0.22	0.10	Son and Dhir [18]
R-113	100	0.51	0.21	0.08	Son and Dhir [18]
n-pentane			0.22		Berenson
Water	158		0.35	0.20±0.10	Hosler and Westwater [35]
Acetone		0.65	0.25		Lienhard and Dhir [34]
Water (100 atm.)	100	0.47	0.16	0.46	Son and Dhir [18]

where q_{co} is film boiling heat flux without radiation, q_r is radiative heat flux and ϵ_w is the emissivity of the surface. In eq. (13.69), the radiation absorptivity of liquid was assumed to be 1. Usually, the heat transfer surface used in the film boiling experiments such as those referred to in Table 13.2 is well polished. The radiative emissivities of polished metal surfaces such as copper and inconel are less than 0.2 (Siegel and Howell [37]). If $\epsilon_w = 0.2$, $\frac{q_r}{q_{co}}$ is 1.8% for water at $\Delta T = 100°C$. When the wall superheat is increased to 300°C, $\frac{q_r}{q_{co}}$ increases to 4.3%. Even if $\epsilon_w = 0.5$, $\frac{q_r}{q_{co}}$ is less than 11% at the highest superheat investigated in this work. Therefore, as long as the wall superheat is not very high and the surfaces are not dirty, the effect of radiation can be neglected.

In Figure 13.32, the velocity vectors at $t = 2.3$ during film boiling of water for $\Delta T = 200°C$ and 1 atm pressure are plotted. It is seen that upward movement of the interface is much stronger in the primary bubble region (near $x = -R$) than in the secondary bulge (near $x = R$). The vapor injected into the primary bubble through the thin film induces a clockwise vortex motion near the interface. The liquid flow induced by the evolving interface decays with distance from the interface.

The x-direction velocity profiles in the thin film region are plotted in Figure 13.33. The maximum velocity occurs at the location ($x = -1.5$) where the film is thinnest. As x increases, the magnitude of horizontal velocity decreases due to increases in

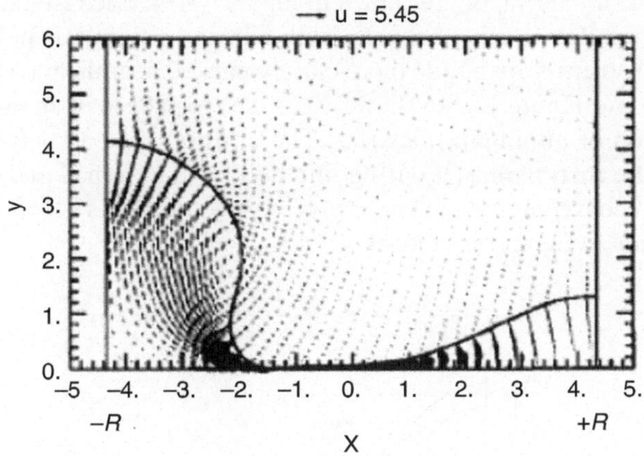

FIGURE 13.32 Velocity vectors at $t = 2.3$ for $\Delta T = 200°C$ in film boiling [18].

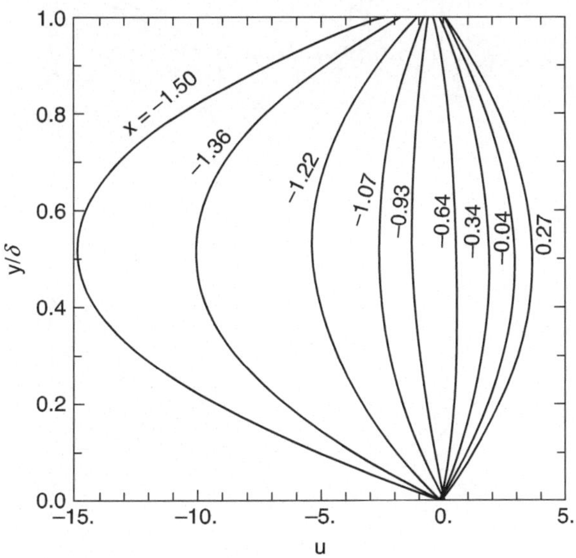

FIGURE 13.33 Radial velocity profiles at $t = 2.3$ for $\Delta T = 200°C$ [18].

the film thickness and is zero near $x = -0.64$ where the direction of x-directional velocity changes. This indicates that at $t = 2.3$ the vapor generated at the interface is fed into the primary and the secondary bubble but is not symmetrical. Also, the vapor velocity profiles are found to be almost parabolic and symmetric with respect to $y = 0.5\delta$. The boundary condition at the liquid–vapor interface is closer to the no-slip condition rather than the slip condition. This is expected from the fact that the liquid viscosity and density at low pressures are much larger than the vapor viscosity and density. It should be noted that in Berenson's model described in Chapter 7, the film thickness was assumed to be constant in the x direction, but vapor velocity did increase in the direction of vapor flow.

The isotherms in the vapor are shown in Figure 13.34. Each isotherm almost follows the interface. The spacing between isotherms is uniform except in the primary bubble where isotherms are a little more closely packed near the interface due to the upward vapor flow. Figure 13.34b shows the isotherm pattern in an expanded region near the location of minimum thickness. The uniform spacing between isotherms indicates that the energy transfer during film boiling is governed mainly by conduction rather than convection as the vapor film is very thin. This is consistent with the assumption made in Berenson's model.

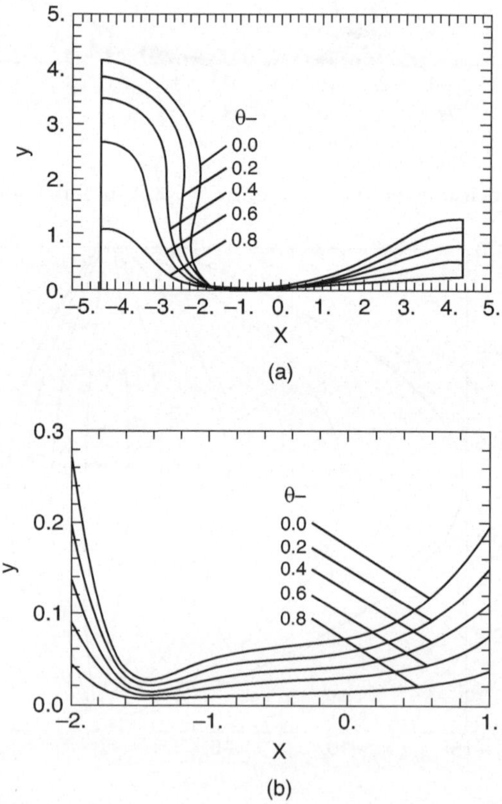

FIGURE 13.34 Isotherms at $t = 2.3$ for $\Delta T = 200°C$: (a) in the vapor over the computational domain and (b) in expanded region near the location of minimum thickness [18].

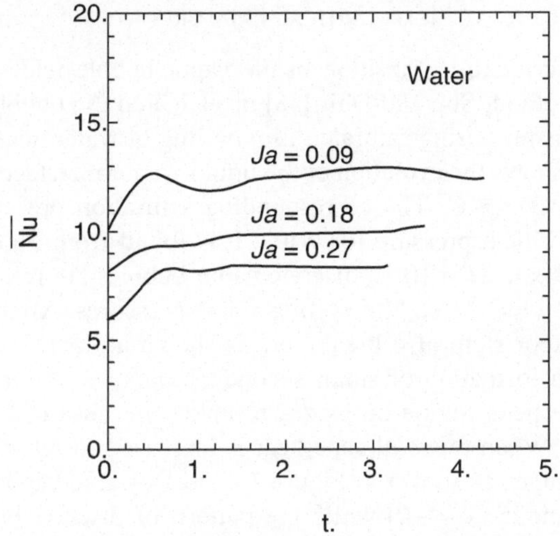

FIGURE 13.35 Nusselt numbers based on heat transfer coefficient averaged over the cell area for different wall superheats [18].

The average Nusselt numbers, $\overline{\text{Nu}}$, based on heat transfer coefficient averaged over the cell area are plotted in Figure 13.35. With an increase in wall superheat, or Ja, the vapor film thickness in the thin-film region increases, and as a result, the heat-transfer coefficient decreases. The average Nusselt number also shows some temporal dependence, but it is much less than that seen in the local heat-transfer coefficient. The Nusselt numbers based on the area- and time-averaged heat-transfer coefficients obtained by integrating the area under the curves are about 30 to 35% lower than those obtained from Berenson's correlation as given in eq. (13.71).

$$\text{Nu}_B = 0.42\left[\frac{\rho_v(\rho_l-\rho_v)gh'_{fg}}{k_v\mu_v\Delta T}\right]^{1/4}\left[\frac{\sigma}{g(\rho_l-\rho_v)}\right]^{3/8} \tag{13.71}$$

where

$$h'_{fg} = h_{fg} + 0.5c_{pv}\Delta T$$

Also, the numerically calculated time- and area-averaged heat-transfer coefficients are closer to the lower bound of Hosler and Westwater's data [35]. One possible reason for the underprediction of the heat-transfer rate in the numerical simulation is the use of axisymmetric analysis instead of three-dimensional analysis, which is more appropriate for low-pressure film boiling on a horizontal surface. It should be noted that numerical simulation results did not include thermal coupling with the solid. Nevertheless, the analysis provides a better physical description of the film-boiling process on a horizontal surface. The numerical analysis, such as that used by Son and Dhir, can be a useful simulation tool for investigation of the effect of various system variables on the film-boiling process, including the thermal properties of the heater material.

13.3.2 Film Boiling at Near-Critical Pressures

At pressures near critical, a transition in the vapor bubble release pattern occurs. By using the LS method, Son and Dhir [38] investigated the bubble-release pattern-associated heat transfer during saturated film boiling of water near the critical pressure. Figure 13.36 shows the evolution of the liquid–vapor interface for different wall superheats at $T_{sat} = 373.3°C$. The corresponding saturation pressure is 21.90 MPa, which is 0.99 the critical pressure for water. It is found from Figure 13.36a that at the low wall superheat, $\Delta T = 10°C$, discrete vapor bubbles are released alternatively at the nodes and antinodes like film boiling at low pressures. After the vapor bubble pinches off, the vapor stem also breaks off as shown at $t = 157$ and $t = 169$. Such a breakoff leads to formation of small secondary bubbles. After bubble breakoff, the interface in the peak region drops down rapidly because of the restoring force of surface tension. When the wall superheat is increased to $\Delta T = 22°C$, the bubble release pattern changes as shown in Figure 13.36b. A stable vapor column starts to be formed on the node ($x = -R$) while the pattern of discrete bubble release still exists on the antinode ($x = R$). At a higher superheat, $\Delta T = 30°C$, stable columns are formed on both the node and the antinode. Figure 13.36c shows that the interface is nearly stationary except the fact that bubbles are released from the upper end of the vapor columns. Also, at the nearly steady-state condition, the interface is found to be almost symmetric with respect to $x=0$. These hydrodynamic transitions in bubble release pattern are consistent with those observed by Reimann and Grigull [39]. They observed that during film boiling of water on 0.1 mm diameter platinum wire at $T_{sat} = 373.3$, vapor left in several different modes as shown in Figure 13.37. With increase in heat flux, the modes of vapor removal shift from (a) to (d). The modes (a), (b), and (c) are similar to the bubble release pattern simulated numerically. It is interesting to note that the heat fluxes for Figure 13.37b and c are comparable to the heat fluxes, $q_w = 21.6\left(\text{W/cm}^2\right)$ and $q_w = 29.9\left(\text{W/cm}^2\right)$ in Table 13.3, obtained numerically for different bubble release modes though the geometry of the boiling surface is not the same.

Nusselt numbers based on heat transfer coefficient averaged over a cell for one cycle are listed in Table 13.3. In this table, the values of Nu are compared with the predictions from Berenson's model as described in Chapter 7 and the correlation of Klimenko [41] eqs. (13.72) and (13.73).

$$\text{Nu}_K = 1.90\times10^{-1}\text{Gr}^{1/3}\,\text{Pr}^{1/3}\,f_1; \quad \text{for} \quad \text{Gr} < 4.03\times10^5 \tag{13.72}$$

$$\text{Nu}_K = 2.16\times10^{-2}\text{Gr}^{1/2}\,\text{Pr}^{1/3}\,f_2; \quad \text{for} \quad \text{Gr} < 4.03\times10^5 \tag{13.73}$$

where

$$f_1 = 1 \quad \text{for Ja} > 0.71$$

$$= 0.89\,\text{Ja}^{-1/3} \quad \text{for Ja} < 0.71$$

$$f_2 = 1 \quad \text{for Ja} > 0.50$$

$$= 0.71\,\text{Ja}^{-1/2} \quad \text{for Ja} < 0.50$$

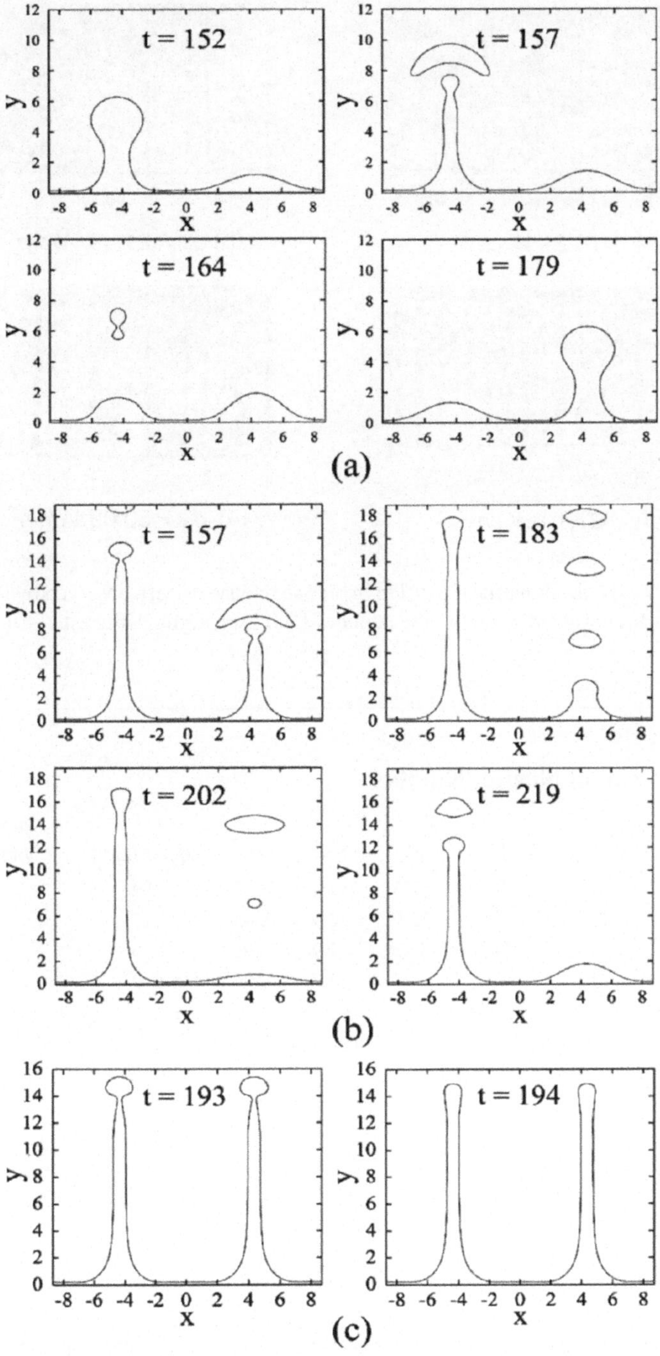

FIGURE 13.36 Evolution of the interface for different wall superheats at $T_{sat} = 373.3°C$: (a) $\Delta T = 10°C$, (b) $\Delta T = 22°C$ and (c) $\Delta T = 30°C$ [38].

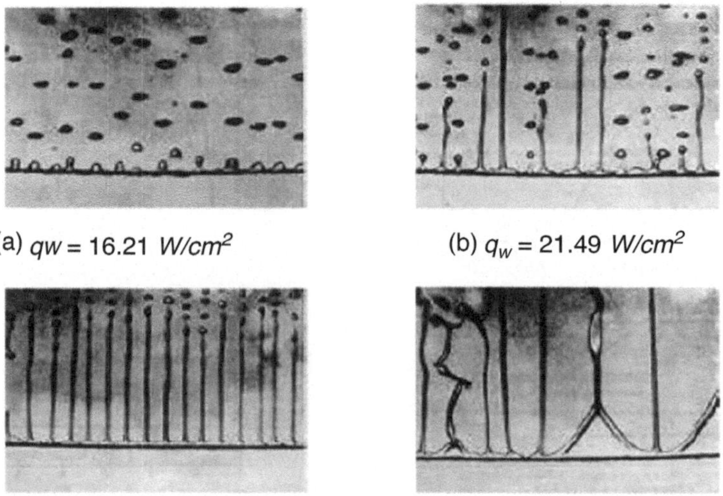

(a) $qw = 16.21\ W/cm^2$ (b) $q_w = 21.49\ W/cm^2$

(c) $qw = 27.10\ W/cm^2$ (d) $q_w = 30.05\ W/cm^2$

FIGURE 13.37 Hydrodynamic transition in bubble release pattern observed by Reimann and Grigull [39], which was presented in the paper of Dhir and Taghavi-Tafreshi [40].

TABLE 13.3
Comparison of Nusselt Number

T_{sat} (°C)	ΔT (°C)	Ja	Nu_B^*	Nu_k	Nu	q_w(W/cm²)	No.of Stable Vapor Jets
373.3	10	40.3	8.53	12.9	8.44	9.28	0
373.3	20	80.7	8.49	12.9	8.85	19.5	0
373.3	22	88.7	8.48	12.9	8.91	21.6	1
373.3	30	121.	8.47	12.9	9.06	29.9	2

* For Nu_B, $h'_{fg} = h_{fg} + 0.5c_{pv}\Delta T = h_{fg}(1 + 0.5\,\text{Ja})$.

where

$$G_r = \left(\frac{P_r^2 g_{lo}^3}{\mu_v}\right)^2 \left(\frac{P_l}{P_v} - 1\right)$$

$$\text{Ja} = c_{pv}\Delta T / h_{fg}$$

Near the critical pressure, the Nusselt numbers predicted from the correlations are observed to remain almost constant at large Ja, while those predicted numerically increase slightly with ΔT. This increase results from the fact that the formation of stable vapor columns at high wall superheat increases the pressure difference between the peak and the valley region of the interface, which provides more efficient flow passages for vapor removal. Such hydrodynamic transition in bubble release pattern was not considered in Berenson's and Klimenko's models, where the ratio of bubble height to the bubble spacing was assumed to be constant regardless of the wall superheat. It is interesting to note that the Nusselt numbers at higher superheats obtained from the numerical simulation are bounded by those predicted from Berenson's and Klimenko's correlations. Table 13.3 shows that the Nusselt numbers obtained in the numerical work are closer to the values obtained from the Berenson's model than those predicted by the Klimenko's correlations, which are about 31%–19% higher.

Table 13.3 lists the number of stable jets as well as the average Nusselt numbers for different wall superheats. It is found that a critical value of $\text{Ja} = c_{pv}\Delta T/h_{fg}$ exists at which the bubble release pattern changes. Approximately, for $\text{Ja} \geq 120$, stable vapor columns are formed on both the nodes and antinodes. For $85 \leq \text{Ja} \leq 120$, vapor release occurs in the forms of stable vapor columns and bubbles. However, for $\text{Ja} \leq 80$, no stable vapor jets exist and bubbles are released cyclically from nodes and antinodes. As discussed in Chapter 7, the latter condition has been found to prevail for film boiling at low pressures.

The numerical simulations of film boiling at high pressures presented in this section as well as at low pressures carried out in the earlier work (Son and Dhir [18]) provide a completely theoretical basis for correlation of the Nusselt number for film boiling on a horizontal surface. The range of parameters varied during the computations were as follows:

$$\text{Gr} = 210 - 2.2 \times 10^6 \qquad \text{Pr} = 0.72 - 420 \qquad \text{Ja} = 0.09 - 140$$

The Nusselt numbers obtained numerically are fitted within at most ±6% deviation by the following equation:

$$\text{Nu} = 0.265(\text{Gr}\,\text{Pr})^{1/4}\left(\text{Ja}^{-1} + 0.5 + 1.3\,\text{Ja}^{0.25}\right)^{1/4} \tag{13.74}$$

In eq. (13.74), radiation contribution was not included. Therefore, the correlation is valid as long as T_w is less than 400°C. It should be noted that for low Ja at low pressures, the Nusselt numbers vary as $\text{Ja}^{-1/4}$, which is consistent with Berenson's prediction even though the Nusselt numbers obtained from the numerical simulations are about 34% lower than those obtained from Berenson's model. However, for high Ja, the numerically predicted Nusselt numbers increase slightly with Ja unlike Berenson's prediction. It must be noted that this correlation of Nusselt number for film boiling was obtained purely from numerical simulation, which has not been done in the past, even though the computations were restricted to two-dimensional cases. Furthermore, near critical pressures, the cell fractional area occupied by an evolving vapor bubble or a vapor column is much smaller than that for an evolving vapor bubble during film boiling at low pressures. This in turn causes the film boiling near critical pressures to be closer to a two-dimensional (axisymetric) situation than is film boiling at low pressures.

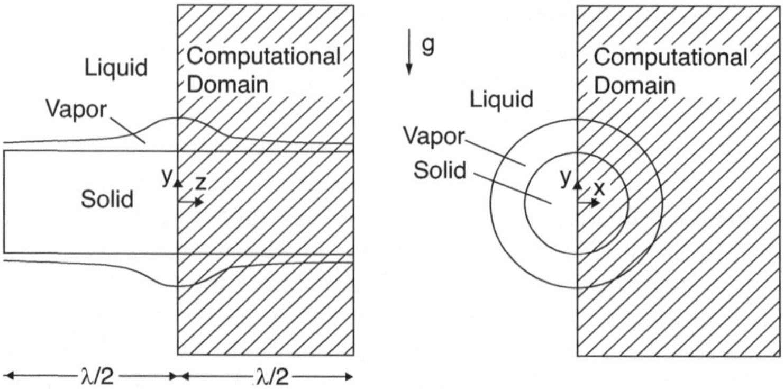

FIGURE 13.38 Computational domain [42].

13.3.3 Film Boiling on a Horizontal Cylinder

Son and Dhir [42] applied the sharp-interface LS method to three-dimensional computations of film boiling in a hexahedral domain including a horizontal cylinder located at $(x, y) = (0,0)$, as depicted in Figure 13.38. The slip boundary condition was imposed at all of the computational boundaries except the top of the domain, which was treated as a free surface. The following parameters, which are evaluated from the properties of saturated water at 1 atm, were used for all of the computations presented here for $g = g_e$:

$$\frac{\rho_v}{\rho_l} = 6.24\times10^{-4} \qquad \frac{\mu_v}{\mu_l} = 4.26\times10^{-2} \qquad l_o\sqrt{\frac{g}{g_e}} = 2.5\text{ mm}$$

$$\frac{\text{Gr}}{D'^3}\sqrt{\frac{g}{g_e}} = 6.01\times10^{-5} \qquad \text{Pr}_v = 0.99 \qquad \text{Ja} = 0.18\ \left(\text{for } T_w - T_{\text{sat}} = 200°\text{C}\right)$$

A theoretical model for describing Taylor waves in film boiling on a horizontal cylinder was developed by Lienhard and Wong [43]. Adding the transverse component of surface tension, which was evaluated from a simple assumed cylindrical interface configuration, they derived an expression for the "most dangerous" wavelength on a horizontal cylinder as

$$\lambda_d = \lambda_{dF}/\sqrt{1+2/D'^2} \tag{13.75}$$

where λ_{dF} is the "most dangerous" wavelength on a plane horizontal surface. Replacing the dimensionless cylinder diameter, $D' = \dfrac{D}{\sqrt{\dfrac{\sigma}{g(\rho_l-\rho_v)}}}$, with the actual diameter of the cylindrical interface while accounting for vapor film thickness, D_I', Lienhard and Sun [44] modified eq. (13.75) as

$$\lambda_d = \lambda_{dF}/\sqrt{1+2/D_I'^2} \tag{13.76}$$

They evaluated D_I' using a simple model equation for the vapor film thickness. The experimentally observed wavelengths were found to lie in the range of +60% and −25% of those predicted from eq. (13.76).

In film boiling on a cylinder, the most dangerous wavelength, which varies with the vapor blanket thickness, cannot be predetermined but should be obtained as part of the solution. As such, the computations were carried out for various wavelengths. Figure 13.39 shows evolution patterns of the liquid–vapor interface for different wavelengths at $D' = 0.5$ and $g = 1g_e$. The computational domain was taken to be a region of $|x| \leq 3.3$, $-3.6 \leq y \leq 11.5$, and $|z| \leq 0.5\lambda$. The fluid and solid regions were represented by the LS function, $\psi = \sqrt{x^2 + y^2} - 0.25$. To save computational time, they used nonuniform meshes with the ratio of two adjacent intervals of 1.1 except near the cylinder surface, $|x| \leq 0.3$ and $|y| \leq 0.3$, where the grid spacing was uniform at 0.01. The grid spacing in the z direction was chosen as 0.057. Initially, the interface was disturbed sinusoidally as $\phi = x\sqrt{x^2 + y^2} - \left[0.4 + 0.015\cos\left(2\pi z/\lambda\right)\right]$. The vapor temperature profile was taken to be linear, and fluid velocity was set to be zero. It is found from Figure 13.39a that for $\lambda = 0.33\lambda_{dF}$, which corresponds the most dangerous wavelength predicted from Lienhard and Wong's equation (13.74), the formation and release of a bubble occur in a nearly two-dimensional pattern. During the early period, the growth of the interface in the peak region at $z = 0$ is much faster than in the valley region at $|z| = 0.5\lambda$. However, for $t \geq 1.75$, the z-direction variation (or disturbance) of the interface is not further developed. This inconsistency with the prediction of Lienhard and Wong can be explained by noting that the diameter of the computed interface, D_I', is much larger than the cylinder diameter, D', in contrast to the assumption, $D_I' \simeq D'$, used in the analysis of Lienhard and Wong. When $D_I' \geq 2D'$, as seen at $t = 1.75$, the critical wavelength, which can be obtained from $\lambda_c = \lambda_d/\sqrt{3}$ and eq. (13.75), is larger than $0.33\lambda_{dF}$. When the wavelength is increased to $\lambda = 0.5\lambda_{dF}$, the evolution of the interface occurs in a three-dimensional pattern, as demonstrated in Figure 13.39b. Discrete vapor bubbles are released alternatively at the node, $z = 0$, and antinodes, $|z| = 0.5\lambda$, of the upper portion of the cylinder. A similar evolution pattern of the interface is observed in Figure 13.39c for $\lambda = 0.67\lambda_{dF}$. The period of bubble release measured at $z = 0$ is $2.22t_o$, which is much smaller than the period of $2.99t_o$ for $\lambda = 0.5\lambda_{dF}$. This means that the most unstable wavelength (or the most probable bubble spacing) is closer to $0.67\lambda_{dF}$ rather than $0.5\lambda_{dF}$. As λ is further increased to λ_{dF}, which is the most unstable wavelength for film boiling on a flat plate, the bubbles are released at more locations, as shown in Figure 13.39d. During the early period, vapor bubbles are released alternatively at $z = 0$ and $|z| = 0.5\lambda$. However, after $t = 5.7$, bubbles release also occurs at $|z| = 0.3\lambda$. This indicates that the most dangerous wavelength for film boiling shrinks on a small-diameter cylinder, as experimentally observed in [43,44].

The interface shape predicted from the numerical simulation for $\lambda = 0.67\lambda_{dF}$ is compared in Figure 13.40 with that observed in the experiments [45] for film boiling of cyclohexanol at $D' = 0.5$ and $\lambda = 0.70\lambda_{dF}$. It is seen that bubble release pattern calculated numerically is in agreement with that obtained experimentally except that the computed bubble size at departure is smaller than that observed in the experiment. This is possibly caused by the fact the vapor volume flux per unit surface area of cylinder, $\dot{V}_v$, obtained under the computational condition is much smaller than the experimental condition: $\dot{V}_v = 4.25\times10^{-2}\,\mathrm{m^3/m^2s}$ in the computation, whereas $\dot{V}_v = 13.8\,\mathrm{m^3/m^2s}$ in the experiment. Increased vapor volume flux in the experiments

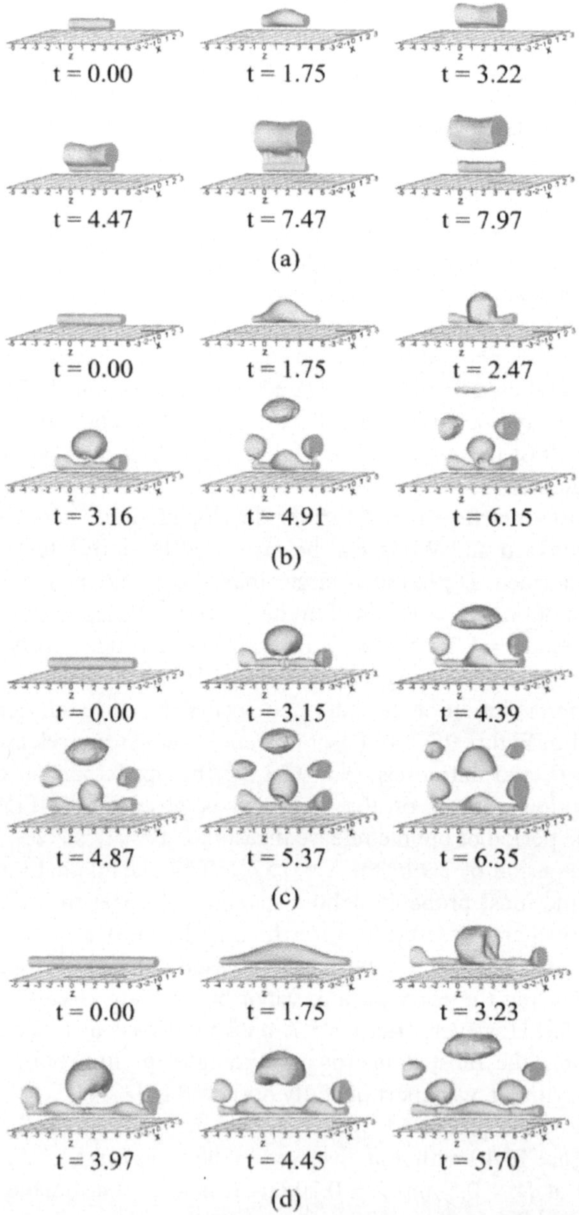

FIGURE 13.39 Evolution of the liquid-vapor interface for different wavelengths at $D' = 0.5$ and $g = 1g_e$: (a) $\lambda = 0.33\lambda_{dF}$, (b) $\lambda = 0.5\lambda_{dF}$, (c) $\lambda = 0.67\lambda_{dF}$, and (d) $\lambda = \lambda_{dF}$ [42].

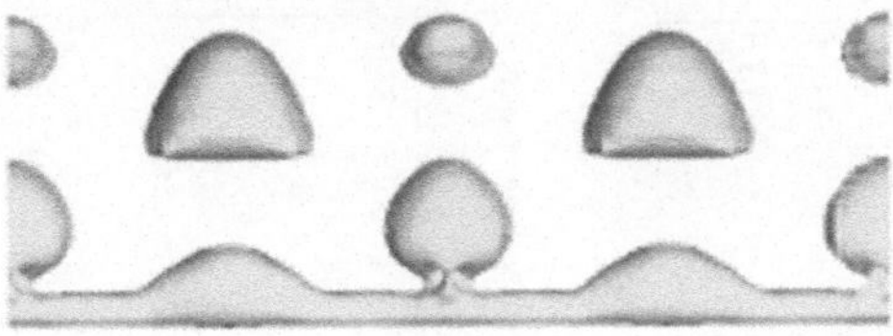

(a)

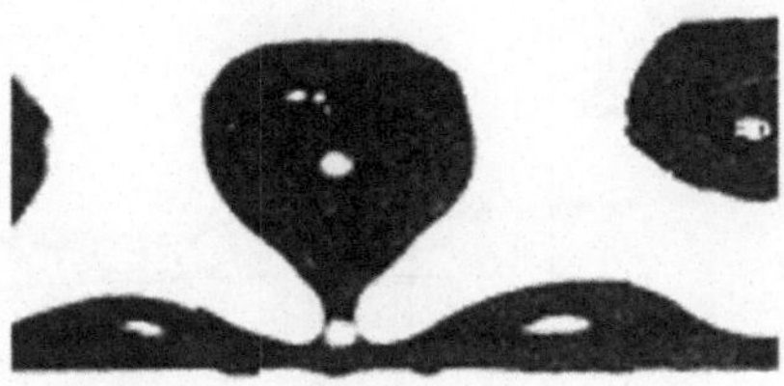

(b)

FIGURE 13.40 Comparison of interface shapes during film boiling: (a) numerical simulation [42] for water at $D' = 0.5$ and $\lambda = 0.67\lambda_{dF}$ and (b) experiment for cyclohexanol at $D' = 0.46$ and $\lambda = 0.70\lambda_{dF}$ [45].

is associated with the much reduced vapor density at the low pressures (about 1 kPa) in comparison with that used in the analysis. The simulation for the condition of the experiment is possible without any change in the present numerical method, but it requires quite a large computational domain to accommodate the departing bubbles, which become larger with the increase in vapor volume flux at very low pressure.

The spatial variation of the interface position and Nusselt number on different positions in the z direction for $D' = 0.5$, $g = g_e$, and $\lambda = 0.67\lambda_{dF}$ is plotted in Figure 13.41 after a quasi-steady state condition has been achieved. The Nusselt number is defined as $Nu = Dq_w/k_v(T_w - T_{sat})$. It is seen that most of heat is transferred via the thin film on the lower portion of the cylinder. The heat flux on the upper portion is very small except near $z = 0.14\lambda$, which corresponds to the valley of interfacial wave formed on the upper portion of the cylinder. The circumferential variations of vapor film thickness and local Nusselt number obtained from numerical simulation are quite different from the predictions from Bromley's [46] two-dimensional model, which does not include the effect of interfacial wave in the z direction. It was observed that the heat transfer coefficient also varied in time, depending on the interfacial motion. After the bubble pinches off, the heat transfer coefficient increases as the vapor bulge on the top portion of the cylinder grows, which results in the increase of the buoyancy force that pulls up and hence thins the vapor film surrounding the cylinder. The heat transfer coefficient drops during the bubble pinch-off process in which the surface tension force dominates over the buoyancy force, thus restoring the disturbed interface. The Nusselt numbers

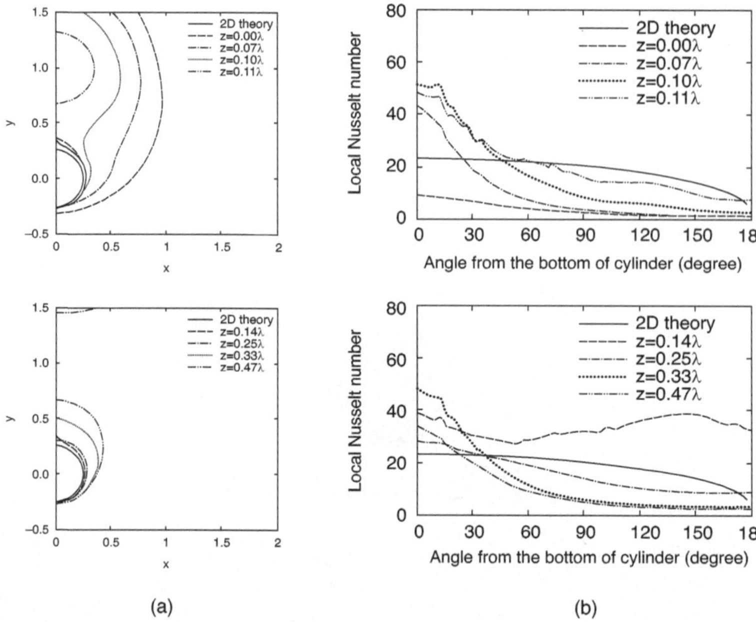

FIGURE 13.41 Results at $D' = 0.5$, $g = 1g_e$ and $\lambda = 0.67\lambda_{dF}$: (a) interface shape and (b) local Nusselt number defined as $\bar{h}D/k_v$ [42].

TABLE 13.4
Predicted Nusselt Numbers

	$g = 1\ g_e$			$g = 0.01\ g_e$	
***D* (mm)**	**0.125**	**1.25**	**12.5**	**1.25**	**12.5**
D/l_o	0.05	0.5	5	0.05	0.5
Bromley [46]	2.84	16.0	90.0	5.05	28.4
Breen and Westwater [47]	12.7	20.0	92.6	22.5	35.5
Sakurai et al. [48]	5.57	16.4	89.0	8.93	27.9
Son and Dhir [42]	5.40	15.4	96.2	9.37	25.3

averaged over the computational period except the first cycle of bubble release are listed in Table 13.4 for $1g_e$ and $0.01g_e$. Except for the smallest diameter cylinder, the values are in good agreement with those predicted from the predictions from Bromley [46], Breen and Westwater [47], and Sakurai et al. [48] correlations. As expected for a fixed g, Nusselt numbers increase as cylinder diameter increases. The effect of reduced gravity is to decrease the Nusselt number.

The effect of dimensionless cylinder diameter, D' or D/l_o, on the most dangerous wavelength and average Nusselt number is shown in Figure 13.42. It is seen that for small value of D', the most dangerous wavelengths obtained from the numerical study as well as Lienhard and Sun's experimental study [44] are larger than those

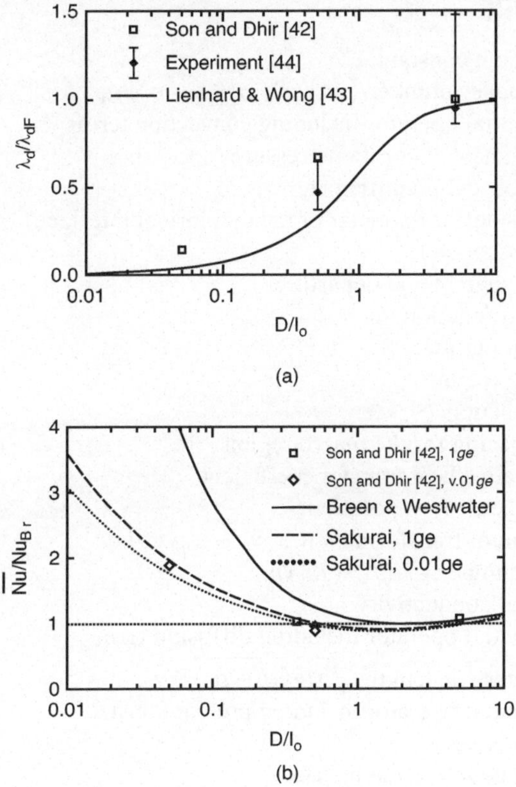

FIGURE 13.42 Effect of dimensionless cylinder diameter, D/l_o, on (a) the most dangerous wavelength and (b) Nusselt number [42].

predicted from Lienhard and Wong's eq. (13.75). The most dangerous wavelength is determined by not only the cylinder diameter but also the vapor layer thickness on the upper portion of the cylinder, as discussed by Dhir and Lienhard [45]. If the vapor film thickness is included in Lienhard and Wong's equation, its prediction for the most dangerous wavelength will be closer to the numerical result. In Figure 13.42b, the Nusselt numbers based on average heat transfer coefficient and normalized with prediction from Bromley's [46] two-dimensional model and correlations obtained from the numerical simulation are compared with those from Breen and Westwater [47] and Sakurai et al. [48]. The numerical results match within at most ±10% deviation with Sakurai et al.'s predictions.

13.4 SUMMARY

In this chapter, we have developed a scheme to numerically simulate pool nucleate and film boiling. Results of numerical simulations are compared with experimental results including bubble sizes and shapes. Film boiling has also been investigated at pressures near critical and at reduced gravity condition.

NOMENCLATURE

A	Dispersion constant
Ar	Archimedes number, $(2\pi)^3 g l_0^3 \rho_v (\rho_l - \rho_v)/\mu_v^2$
C	Differential operator including convection terms
c_p	Specific heat at constant pressure
D	Bubble or cylinder diameter
$\hat{D}_I$	Dimensionless diameter of the cylindrical interface
D_c	Cavity diameter
D_c	Bubble diameter at departure
F	Fraction function
G_r	Grashof number
g	Gravity vector
H	Step function
h	Grid spacing for the macro region
h_{ev}	Evaporative heat transfer coefficient
h_{fg}	Latent heat of evaporation
J	Jacobian of transformation, $x_\xi y_\eta - x_\eta y_\xi$
Ja	Jakob number, $c_{pv}(T_w - T_{\text{sat}})/h_{fg}$
k	Thermal conductivity
L	Differential operator including diffusion terms
l_0	Characteristic length, $\sqrt{\sigma/g(\rho_l - \rho_v)}$
$M1$, $M2$	Differential operator including pressure terms
$\dot{\mathbf{m}}$	Mass flux vector
$\dot{m}$	Mass flux across the interface
n	Unit normal vector
N	Number density of active nucleation sites
Nu	Nusselt number, $l_0 q_w/k_l \Delta T$ for nucleate boiling, $l_0 q_w/k_v \Delta T$ for film boiling
Nu_B	Nusselt number predicted from Bromley's correlation
$\overline{\text{Nu}}$	Nusselt number averaged over a cell
p	Pressure
P	Control function for grid generation
p'	Pressure correction
Pr	Prandtl number, $\mu c_p/k$
q	Heat flux
Q	Control function for grid generation
q_{co}	Film boiling heat flux without radiation
q_r	Radiative heat flux
R	Radius of computational domain
R_v	Gas constant
R_0	Radius of dry region beneath a bubble
R_1	Radial location of the interface at $y = h/2$
r	Radial coordinate
Re	Reynolds number, $\rho u_0 l_0/\mu$
s	Spacing between the active sites
S	Sign function

t	Time
T	Temperature
ΔT	Temperature difference, $T_w - T_{sat}$
t_0	Characteristic time, l_0/u_0
u	r or x-directional velocity
$\boldsymbol{u}$	Velocity vector
$\overline{u}$	Contravariant velocity component
$\boldsymbol{U}$	Interfacial velocity vector
u_0	Characteristic velocity, $\sqrt{gl_0}$
$\dot{V}_{\text{micro}}$	Rate of vapor volume production from the microlayer
$\Delta\dot{V}_{\text{micro}}$	Control volume near the micro region
v	y-directional velocity
$\overline{v}$	Contravariant velocity component
v_{lv}	$v_{lv}\rho_v^{-1} - \rho_l^{-1}$
Y	Height of computational domain
y	Vertical coordinate

GREEK LETTERS

α	Thermal diffusivity
β_T	Coefficient of thermal expansion
δ	Liquid or vapor film thickness
δ_0	Nonevaporating liquid film thickness
δ_T	Thermal layer thickness
η	Transformed coordinate
θ	Dimensionless temperature, $(T - T_{sat})/\Delta T$
κ	Interfacial curvature
λ	Wavelength
λ_{d2}	2-dimensional "most dangerous" wavelength
λ_{dF}	"most dangerous" wavelength for a flat plate
μ	Dynamic viscosity
ν	Kinematic viscosity
ξ	Transformed coordinate
ρ	Density
σ	Surface tension
σ_s	Stefan–Boltzmann constant
τ	Artificial time
τ_w	Waiting time
ϕ	Level-set function or general dependent variable
φ	Contact angle
ψ	Level-set function from the solid surface

SUPERSCRIPTS

*	Intermediate time step
n	Time step of $n\Delta t$

SUBSCRIPTS

e	Earth
f	Liquid or vapor
int	Interface
l, v	Liquid, vapor
r, t, x, y	$\partial/\partial r$, $\partial/\partial t$, $\partial/\partial x$, $\partial/\partial y$
sat, w	Saturation, wall
ξ, η	Partial differentiation with respect to ξ, η

REFERENCES

1. Son, G., Dhir, V.K., and Ramanujapu, N. (1999), "Dynamics and Heat Transfer associated with a Single Bubble during Nucleate Boiling on a Horizontal Surface." *J. Heat Transfer*, Vol. 121, pp. 623–631.
2. Wayner, P.C. Jr. (1992), "Evaporation and Stress in the Contact Line Region." *Proceedings of the Engineering Foundation Conference on Pool and External Flow Boiling*, V. K. Dhir and A. E. Bergles, eds., Santa Barbara, CA, pp. 251–256.
3. Lay, J.H. and Dhir, V.K. (1995), "Shape of a Vapor Stem During Nucleate Boiling of Saturated Liquids." *J. Heat Transfer*, Vol. 117, pp. 394–401.
4. Sussman, M., Smereka, P., and Osher, S. (1994), "A Level Set Approach for Computing Solutions to Incompressible Two-Phase Flow." *J. Comput. Phys.*, Vol. 114, pp. 146–159.
5. Chorin, A.J. (1968), "Numerical Solution of the Navier-Stokes Equations." *Math. Comput.*, Vol. 22, pp. 745–762.
6. Kim, J. and Moin, P. (1985), "Application of a Fractional-Step Method to Incompressible Navier-Stokes Equations." *J. Comput. Phys.*, Vol. 59, pp. 308–323.
7. Bell, J.B. and Colella, P. (1989), "A Second-Order Projection Method for the Incompressible Navier-Stokes Equations." *J. Comput. Phys.*, Vol. 85, pp. 257–283.
8. Chang, Y.C., Hou, T.Y., Merriman, B., and Osher, S. (1996), "A Level Set Formulation of Eulerian Interface Capturing Methods for Incompressible Fluid Flows." *J. Comput. Phys.*, Vol. 124, pp. 449–464.
9. Patankar, S.V. (1980), *Numerical Heat Transfer and Fluid Flow*, pp. 44–47, 61–66, Hemisphere, Washington, DC.
10. Sonneveld, P. and Wesseling, P. (1985), "Multigrid and Conjugate Gradient Methods as Convergence Acceleration Techniques." *Multigrid Methods for Integral and Differential Equations*, D.J. Paddon and H. Holstein, eds., Oxford, Clarendon Press, pp. 117–167.
11. Son, G. and Dhir, V.K. (2008), "Numerical Simulation of Nucleate Boiling on a Horizontal Surface at High Heat Fluxes." *Int. J. Heat Mass Transfer*, 51, pp. 2566–2582.
12. Adalsteinsson, D, & Sethian, J A (1999), "The Fast Construction of Extension Velocities in Level SetMethods". *J. Comput. Phys.*, Vol. 148, pp. 2–22.
13. Fedkiw, R. P., Aslam, T., Merriman, B., and Osher, S. (1999), "A Non-Oscillatory Eulerian Approach to Interfaces in Multimaterial Flows (the Ghost Fluid Method)." *J. Comput. Phys.*, Vol. 152, pp. 457–492.
14. Kang, M., Fedkiw, R. P., and Liu, X.-D. (2000), "A Boundary Condition Capturing Method for Multiphase Incompressible Flow." *J. Sci. Comput.*, Vol. 15, pp. 323–360.
15. Nguyen, D. Q., Fedkiw, R. P., and Kang, M. (2001), "A Boundary Condition Capturing Method for Incompressible Flame Discontinuities." *J. Comput. Phys.*, Vol. 172, pp. 71–98.
16. Gibou, F., Fedkiw, R. P., Cheng, L.-T., and Kang, M. (2002), "A Second-Order-Accurate Symmetric Discretization of the Poisson Equation on Irregular Domains." *J. Comput. Phys.*, Vol. 176, pp. 205–227.

17. Marella, S., Krishnan, S., Liu, H., and Udaykumar, H.S. (2005), "Sharp Interface Cartesian Grid Method I: An Easily Implemented Technique for 3D Moving Boundary Computations." *J. Comput. Phys.*, Vol. 210, pp. 1–31.
18. Son, G. and Dhir, V.K. (1997), "Numerical Simulation of Saturated Film Boiling on a Horizontal Surface." *J. Heat Transfer*, Vol. 119, pp. 525–533.
19. Juric, D. and Tryggvason, G. (1998), "Computations of Film Boiling." *Int. J. Multiphase Flow*, Vol. 24, pp. 387–410.
20. Kays, W.M. and Crawford, M.E. (1980), *Convective Heat and Mass Transfer*, McGraw-Hill Book Company, New York, p. 328.
21. Siegel, R. and Keshock, E.G. (1964), "Effects of Reduced Gravity on Nucleate Boiling Bubble Dynamics in Saturated Water." *AICHE J.*, Vol. 10, pp. 509–517.
22. Son, G., Ramanujapu, N. and Dhir, V.K. (2002), "Numerical Simulation of Bubble Merger Process on a Single Nucleation Site during Pool Nucleate Boiling." *J. Heat Transfer*, Vol. 124, pp. 51–62, presented at ASME-JSME Conference in San Diego, CA.
23. Fritz, W. (1935), "Maximum Volume of Vapor Bubbles." *Physik Zeitschr.*, Vol. 36, pp. 379–384.
24. Staniszewski, B.E. (1959), *Nucleate Boiling Bubble Growth and Departure*, Tech. Rept. 16, Div. Sponsored Res., Massachusetts Institute of Technology, Cambridge, Mass.
25. Gaertner, R.F. (1965), "Photographic Study of Nucleate Pool Boiling on a Horizontal Surface." *ASME J. Heat Transfer*, Vol. 87, pp. 17–29.
26. Mukherjee, A. and Dhir, V.K. (2004) "Study of Lateral Merger of Vapor Bubbles during Nucleate Pool Boiling." *ASME J. Heat Transfer*, Vol. 126, pp. 1023–1039.
27. Wang, C.H. and Dhir, V.K. (1993), "Effect of Surface Wettability on Active Nucleation Site Density during Pool Boiling of Saturated Water." *J. Heat Transfer*, Vol. 115, pp. 659–669.
28. Basu, N., Warrier, G.R., and Dhir, V.K. (2005), "Wall Heat Flux Partitioning during Subcooled Flow Boiling: Part 1. Model Development." *J. Heat Transfer*, Vol. 127, pp. 131–140.
29. Rohsenow, W.M. (1952), "A Method of Correlating Heat Transfer Data for Surface Boiling of Liquids." *J. Heat Transfer*, Vol. 74, pp. 969–976.
30. Liaw, S.P. and V.K. Dhir (1989), "Void Fraction Measurement during Saturated Pool Boiling of Water on Partially Wetted Vertical Surfaces." *J. Heat Transfer*, Vol. 111, pp. 731–783.
31. Stephan, K. and Abdelsalam, M. (1980), "Heat Transfer Correlation for Natural Convection Boiling." *Int. J. Heat Mass Transfer*, Vol. 23, pp. 73–87.
32. Lienhard, J.H. and Dhir, V.K. (1973), "Extended Hydrodynamic Theory of the Peak and Minimum Pool Boiling Heat Fluxes", NASA CR-2270.
33. Berenson, P.J. (1961), "Film Boiling Heat Transfer from a Horizontal Surface." *J. Heat Transfer*, Vol. 83, pp. 351–362.
34. Lienhard, J.H. and Dhir, V.K. (1980), "On the Prediction of the Minimum Pool Boiling Heat Flux", *J. Heat Transfer*, Vol. 102, pp. 457–460.
35. Hosler, E.R. and Westwater, J.W. (1962), "Film Boiling on a Horizontal Plate." *ARS J.*, Vol. 32, pp. 553–560.
36. Siviour, J.B. and Ede, A.J. (1970), "Heat Transfer in Subcooled Pool Film Boiling." *Proceedings of the Fourth International Heat Transfer Conference*, U. Grigull and E. Hahne, eds., Paris-Versailles, France, Vol. 5, B3.12.
37. Siegel, R. and Howell, J.R. (1981), *Thermal Radiation Heat Transfer*, Hemisphere, Washington, DC.
38. Son, G. and Dhir, V.K. (1998), "Numerical Simulation of Film Boiling Near Critical Presures with a Level Set Method." *J. Heat Transfer*, Vol. 120, pp. 183–192.

39. Reimann, M. and Grigull, U. (1975), "Warmeubergang bei freier Konvektion und Filmsieden im kritischen Gebiet von Wasser und Kohlendioxid." *Warmeund Stoffubertragung*, Vol. 8, pp. 229–239.
40. Dhir, V.K. and Taghavi-Tafreshi, K. (1981), "Hydrodynamic Transitions during Dripping of a Liquid from Underside of a Horizontal Tube." ASME Paper, 81-WA/HT-12.
41. Klimenko, V.V. (1981), "Flim Boiling on Horizontal Plate – New Correlation." *Int. J. Heat Mass Transfer*, Vol. 24, pp. 69–79.
42. Son, G. and Dhir, V.K. (2008), "Three-Dimensional Simulation of Saturated Film Boiling on a Horizontal Cylinder." *Int. J. Heat Mass Transfer*, Vol. 51, pp. 1156–1167.
43. Lienhard, J.H. and Wong, P.T.Y. (1964), "The Dominant Unstable Wavelength and Minimum Heat Flux during Film Boiling on a Horizontal Cylinder." *J. Heat Transfer*, Vol. 86, pp. 220–226.
44. Lienhard, J.H. and Sun, K.H. (1970), "Effects of Gravity and Size upon Film Boiling from Horizontal Cylinders." *J. Heat Transfer*, Vol. 92, pp. 292–298.
45. Dhir, V.K. and Lienhard, J.H. (1973), "Taylor Stability of Viscous Fluids with Application to Film Boiling." *Int. J. Heat Mass Transfer*, Vol. 16, pp. 2097–2109.
46. Bromley, L.A. (1950), "Heat Transfer in Stable Film Boiling." *Chem. Eng. Prog.*, Vol. 46, pp. 221–227.
47. Breen, B.P. and Westwater, J.W. (1962), "Effect of Diameter of Horizontal Tubes on Film Boiling Heat Transfer." *Chem. Eng. Prog.*, Vol. 58, pp. 67–72.
48. Sakurai, A., Shiotsu, M., and Hata, K. (1990), "A General Correlation for Pool Film Boiling Heat Transfer from a Horizontal Cylinder to Subcooled Liquid. Part 2: Experimental Data for Various Liquids and Its Correlation." *J. Heat Transfer*, Vol. 112, pp. 441–450.

Author Index

Subject Index